Optimization for Decision Making II

Optimization for Decision Making II

Editors

Víctor Yepes
José M. Moreno-Jiménez

MDPI • Basel • Beijing • Wuhan • Barcelona • Belgrade • Manchester • Tokyo • Cluj • Tianjin

Editors

Víctor Yepes
ICITECH, Universitat Politècnica
de València
Spain

José M. Moreno-Jiménez
Universidad de Zaragoza
Spain

Editorial Office
MDPI
St. Alban-Anlage 66
4052 Basel, Switzerland

This is a reprint of articles from the Special Issue published online in the open access journal *Mathematics* (ISSN 2227-7390) (available at: https://www.mdpi.com/journal/mathematics/special_issues/Optimization_Decision_Making_II).

For citation purposes, cite each article independently as indicated on the article page online and as indicated below:

LastName, A.A.; LastName, B.B.; LastName, C.C. Article Title. *Journal Name* **Year**, *Article Number*, Page Range.

ISBN 978-3-03943-607-1 (Hbk)
ISBN 978-3-03943-608-8 (PDF)

Contents

About the Editors vii

Preface to "Optimization for Decision Making II" ix

Juan Aguarón, María Teresa Escobar, José María Moreno-Jiménez and Alberto Turón
The Triads Geometric Consistency Index in AHP-Pairwise Comparison Matrices
Reprinted from: *Mathematics* **2020**, *8*, 926, doi:10.3390/math8060926 . . . . **1**

José García, José V. Martí and Victor Yepes
The Buttressed Walls Problem: An Application of a Hybrid Clustering Particle Swarm Optimization Algorithm
Reprinted from: *Mathematics* **2020**, *8*, 862, doi:10.3390/math8060862 . . . . **17**

Lars Magnus Hvattum, Gregorio Tirado and Ángel Felipe
The Double Traveling Salesman Problem with Multiple Stacks and a Choice of Container Types
Reprinted from: *Mathematics* **2020**, *8*, 979, doi:10.3390/math8060979 . . . . **39**

Inmaculada Flores, M. Teresa Ortuño, Gregorio Tirado and Begoña Vitoriano
Supported Evacuation for Disaster Relief through Lexicographic Goal Programming
Reprinted from: *Mathematics* **2020**, *8*, 648, doi:10.3390/math8040648 . . . . **51**

Adán Rodríguez-Martínez and Begoña Vitoriano
Probability-Based Wildfire Risk Measure for Decision-Making
Reprinted from: *Mathematics* **2020**, *8*, 557, doi:10.3390/math8040557 . . . . **71**

José García, Victor Yepes and José V. Martí
A Hybrid k-Means Cuckoo Search Algorithm Applied to the Counterfort Retaining Walls Problem
Reprinted from: *Mathematics* **2020**, *8*, 555, doi:10.3390/math8040555 . . . . **89**

Julia Monzón, Federico Liberatore and Begoña Vitoriano
A Mathematical Pre-Disaster Model with Uncertainty and Multiple Criteria for Facility Location and Network Fortification
Reprinted from: *Mathematics* **2020**, *8*, 529, doi:10.3390/math8040529 . . . . **111**

José M. Ferrer, M. Teresa Ortuño and Gregorio Tirado
A New Ant Colony-Based Methodology for Disaster Relief
Reprinted from: *Mathematics* **2020**, *8*, 518, doi:10.3390/math8040518 . . . . **129**

Vicent Penadés-Plà, Tatiana García-Segura and Víctor Yepes
Robust Design Optimization for Low-Cost Concrete Box-Girder Bridge
Reprinted from: *Mathematics* **2020**, *8*, 398, doi:10.3390/math8030398 . . . . **153**

Antonio Jiménez-Martín, Faustino Tello and Alfonso Mateos
A Variation of the ATC Work Shift Scheduling Problem to Deal with Incidents at Airport Control Centers
Reprinted from: *Mathematics* **2020**, *8*, 321, doi:10.3390/math8030321 . . . . **167**

Jinbao Jian, Lin Yang, Xianzhen Jiang, Pengjie Liu and Meixing Liu
A Spectral Conjugate Gradient Method with Descent Property
Reprinted from: *Mathematics* **2020**, *8*, 280, doi:10.3390/math8020280 . . . . **195**

José María Moreno-Jiménez, Cristina Pérez-Espés and Pilar Rivera-Torres
Relevant Aspects for an EF3-Evaluation of E-Cognocracy
Reprinted from: *Mathematics* **2020**, *8*, 277, doi:10.3390/math8020277 . . . . **209**

Shenshen Gu and Xinyi Chen
The Basic Algorithm for the Constrained Zero-One Quadratic Programming Problem with k-diagonal Matrix and Its Application in the Power System
Reprinted from: *Mathematics* **2020**, *8*, 138, doi:10.3390/math8010138 . . . . **229**

Xiaonan Ji, Lixia Yu and Jiapei Fu
Evaluating Personal Default Risk in P2P Lending Platform: Based on Dual Hesitant Pythagorean Fuzzy TODIM Approach
Reprinted from: *Mathematics* **2020**, *8*, 8, doi:10.3390/math8010008 . . . . **245**

Tingting He, Guiwu Wei, Jianping Lu, Cun Wei and Rui Lin
Pythagorean 2-Tuple Linguistic VIKOR Method for Evaluating Human Factors in Construction Project Management
Reprinted from: *Mathematics* **2019**, *7*, 1149, doi:10.3390/math7121149 . . . . **259**

Francisco Salas-Molina, David Pla-Santamaria, Ana Garcia-Bernabeu, Javier Reig-Mullor
A Compact Representation of Preferences in Multiple Criteria Optimization Problems
Reprinted from: *Mathematics* **2019**, *7*, 1092, doi:10.3390/math7111092 . . . . **273**

About the Editors

Víctor Yepes is a full professor of Construction Engineering; he holds a Ph.D. degree in civil engineering. He serves at the Department of Construction Engineering, Universitat Politecnica de Valencia, Valencia, Spain. He has been the Academic Director of the M.S. studies in concrete materials and structures since 2007 and a Member of the Concrete Science and Technology Institute (ICITECH). He is currently involved in several projects related to the optimization and life-cycle assessment of concrete structures as well as optimization models for infrastructure asset management. He is currently teaching courses in construction methods, innovation, and quality management. He authored more than 250 journals and conference papers including more than 100 published in journals quoted in JCR. He acted as an Expert for project proposals evaluation for the Spanish Ministry of Technology and Science, and he is a main researcher in many projects. He currently serves as an Editor-in-Chief of the *International Journal of Construction Engineering and Management* and a member of the editorial board of 12 international journals (*Structure & Infrastructure Engineering, Structural Engineering and Mechanics, Mathematics, Sustainability, Revista de la Construcción, Advances in Civil Engineering, Advances in Concrete Construction,* among others).

José M. Moreno-Jiménez is a full professor of Operations Research and Multicriteria Decision Making, and has received degrees in mathematics and economics as well as a Ph.D. degree in applied mathematics from the University of Zaragoza, Spain; where he has been teaching the course since 1980–1981. He is the Head of the Quantitative Methods Area in the Faculty of Economics and Business of the University of Zaragoza from 1997, the Chair of the Zaragoza Multicriteria Decision Making Group from 1996, a member of the Advisory Board of the Euro Working Group on Decision Support Systems from 2017, and a Honorary Member of the International Society on Applied Economics ASEPELT from 2019. He has also been the President of an international scientific society (2014–2018) and the Coordinator of the Spanish Multicriteria Decision Making Group (2012–2015). His research interests are in the general area of Operations Research theory and practice, with emphasis on multicriteria decision making, electronic democracy/cognocracy, performance analysis, and industrial and technological diversification. He has published more than 250 papers in scientific books and journals in the most prestigious editorials and is a member of the editorial board of several national and international journals.

Preface to "Optimization for Decision Making II"

Decision making is one of the distinctive activities of the human being; it is an indication of the degree of evolution, cognition and freedom of the species. In the past, until the end of the 20th century, scientific decision making was based on the paradigms of substantive rationality (normative approach) and procedural rationality (descriptive approach). Since the beginning of the 21st century and the advent of the Knowledge Society, decision making has been enriched with new constructivist, evolutionary and cognitive paradigms that aim to respond to new challenges and needs; especially the integration into formal models of the intangible, subjective and emotional aspects associated with the human factor, and the participation in decision-making processes of spatially distributed multiple actors that intervene in a synchronous or an asynchronous manner.

To help address and resolve these types of questions, this book comprises 16 chapters that present a series of decision models, methods and techniques and their practical applications in the fields of economics, engineering and social sciences. The chapters collect the papers included in the "Optimization for Decision Making II" Special Issue of the *Mathematics* journal, 2020, 8(6), first decile of the JCR 2019 in the Mathematics category.

We would like to thank both the MDPI publishing and editorial staff for their excellent work, as well as the 51 authors who have collaborated in its preparation. The papers cover a wide spectrum of issues related to the scientific resolution of problems; in particular, related to decision making, optimization, metaheuristics, and multi-criteria decision making.

We hope that the papers, with their undoubted mathematical content, can be of use to academics and professionals from the many branches of knowledge (philosophy, psychology, economics, mathematics, decision science, computer science, artificial intelligence, neuroscience and more) that have, from such diverse perspectives, approached the study of decision making, an essential aspect of human life and development.

Víctor Yepes, José M. Moreno-Jiménez

Editors

Article

The Triads Geometric Consistency Index in AHP-Pairwise Comparison Matrices

Juan Aguarón, María Teresa Escobar, José María Moreno-Jiménez * and Alberto Turón

Grupo Decisión Multicriterio Zaragoza (GDMZ), Facultad de Economía y Empresa, Universidad de Zaragoza, Gran Vía 2, 50005 Zaragoza, Spain; aguaron@unizar.es (J.A.); mescobar@unizar.es (M.T.E.); turon@unizar.es (A.T.)

* Correspondence: moreno@unizar.es

Received: 5 May 2020; Accepted: 3 June 2020; Published: 6 June 2020

Abstract: The paper presents the Triads Geometric Consistency Index (*T-GCI*), a measure for evaluating the inconsistency of the pairwise comparison matrices employed in the Analytic Hierarchy Process (AHP). Based on the Saaty's definition of consistency for AHP, the new measure works directly with triads of the initial judgements, without having to previously calculate the priority vector, and therefore is valid for any prioritisation procedure used in AHP. The *T-GCI* is an intuitive indicator defined as the average of the log quadratic deviations from the unit of the intensities of all the cycles of length three. Its value coincides with that of the Geometric Consistency Index (*GCI*) and this allows the utilisation of the inconsistency thresholds as well as the properties of the *GCI* when using the *T-GCI*. In addition, the decision tools developed for the *GCI* can be used when working with triads (*T-GCI*), especially the procedure for improving the inconsistency and the consistency stability intervals of the judgements used in group decision making. The paper further includes a study of the computational complexity of both measures (*T-GCI* and *GCI*) which allows selecting the most appropriate expression, depending on the size of the matrix. Finally, it is proved that the generalisation of the proposed measure to cycles of any length coincides with the *T-GCI*. It is not therefore necessary to consider cycles of length greater than three, as they are more complex to obtain and the calculation of their associated measure is more difficult.

Keywords: Analytic Hierarchy Process (AHP); consistency; Geometric Consistency Index (GCI); triads; cycles

---

1. Introduction

Since the origin of the species, humans have used pairwise comparisons to choose between tangible elements. This has generally involved taking the lesser element as a reference unit, and then indicating how many times the greater element "includes" or "dominates" the lesser. This intuitive method for comparing elements has been formalised in the course of human history. After the seminal work of the Majorcan Ramon Llull (1232–1315) (see [1]), the comparisons technique was formally introduced in the second half of the 19th century [2] and further developed in the early years of the 20th century [3]. Following the substantive rationality that was prevalent in the traditional scientific method, pairwise comparisons were generally utilised to order tangible elements, based on an objective attribute with a known unit of reference or measurement scale.

In the mid-1970s, Thomas L. Saaty [4] instigated and refined a new school of thought in the field of multicriteria decisions—the Analytic Hierarchy Process (AHP). Using a hierarchical model of the problem and pairwise comparison to incorporate the preferences of the decision maker, AHP allows for the integration of the objective aspects associated with the traditional scientific method and the subjective aspects associated with the human factor.

The AHP methodology comprises four stages [5]: (i) modelling, or the construction of a hierarchical model; (ii) valuation, or the incorporation of the preferences by means of pairwise comparisons; (iii) prioritisation, or the calculation of the local priorities associated with the nodes of the hierarchy, by means of an existing prioritisation procedure, and the calculation of the global priorities by means of the principle of hierarchical composition; and (iv) synthesis, or the determination of the total priorities of the alternatives by means of one of the available aggregation procedures.

In addition to integrating the subjective with the objective, AHP permits the explicit evaluation of the consistency of the decision maker at the time of eliciting their judgements, whilst admitting small levels of inconsistency. An earlier analysis of inconsistency can be found in [6–8].

Given a pairwise comparison matrix $A_{(n\times n)} = (a_{ij})$ with $a_{ij}a_{ji} = 1$ and $a_{ij} > 0$, Saaty [4] established that the matrix A is consistent when $a_{ij}a_{jk} = a_{ik}\ \forall i,j,k = 1,\ldots,n$. In the same way [9], A is consistent if all the cycles of length three fulfil $a_{ij}a_{jk}a_{ki} = 1\ \forall i,j,k = 1,\ldots,n$.

Although consistency in AHP is defined in terms of triads of the elements of the matrix $A = (a_{ij})$, its evaluation, following Saaty's proposal (and the majority of methods that are traditionally followed for the evaluation of inconsistency), depends on the prioritisation procedure that is employed, that is to say, the measurement of inconsistency is linked to the prioritisation procedure.

Nevertheless, a group of measures explicitly based on the Saaty's definition of consistency that are not linked to any prioritisation method are being considered. The present work puts forward an inconsistency measure—the Triads Geometric Consistency Index (T-GCI)—which belongs to this last group but that coincides with the Geometric Consistency Index (GCI), a measure of the former group. This fact provides a link between both groups of inconsistency measures.

The most relevant contributions of this work are: (i) the definition of the T-GCI; (ii) the demonstration of its relationship with one of the inconsistency measures most used in AHP, the GCI [10], making it possible to use the properties of the GCI as well as its thresholds to make the T-GCI operative; (iii) the study of the computational complexity of the T-GCI, comparing with that of the GCI; and (iv) the generalisation of the T-GCI to cycles of any length.

After this brief Introduction, the remainder of the article is structured as follows. Section 2 offers a summary of the concept of consistency in AHP and details some measurements for its evaluation that have been used in the scientific literature. Section 3 defines the T-GCI, proves its relationship with the GCI, explains its operational behaviour, gives thresholds for evaluation of inconsistency in AHP and analyses its computational complexity. Section 4 places the T-GCI in a more general context of inconsistency measurements based on cycles of any length and demonstrates that they are coincident with the T-GCI. Section 5 highlights the most important conclusions of the work and suggests some future lines of research.

2. Consistency in the Analytic Hierarchy Process

Defined as the cardinal transitivity in judgements ($a_{ij}a_{jk} = a_{ik}\ \forall i,j,k$), consistency in AHP is a desirable property that reflects a certain rationality, logic or formal coherence in the actions of individuals, especially when they compare intangible aspects for which appropriate measurement scales are not currently available.

Some of the factors that may cause inconsistency are: (i) the ambiguity and complexity of the problem; (ii) the knowledge of the actors in the matter under consideration; (iii) the affective aspects (mood, emotions, personality features, attitudes and motivations) that condition the behaviour of the actors; (iv) the level of attention (errors in the response) during the assessment process [11,12]; and (v) the rationality of the procedure followed when incorporating preferences, especially when working with subjective aspects. Some examples are: the assessment scale [13], the use of extreme values [14], the number of comparisons [15] and the incorrect calculation of priorities [16]. To limit the impact of the aspects that determine the preferences of the decision makers, it is necessary to measure their inconsistency when eliciting the judgements and to set appropriate thresholds for making the considered measure operational.

In the AHP literature, two large groups of inconsistency measures can be distinguished: (i) those linked to the prioritisation procedure followed; and (ii) those based on triads, which explicitly incorporate the definition of consistency given by Saaty. A description of some of the measures (the most relevant and widely utilised) from the two groups is set out below.

Given a positive reciprocal pairwise comparison matrix $A_{(n\times n)} = (a_{ij})$, the Eigenvector (EV) method obtains the priority vector by solving the following system:

$$Aw = \lambda_{\max}(A)w, \text{ with } w_j \geq 0, j = 1, \ldots, n \text{ and } \sum_{j=1}^{n} w_j = 1 \quad (1)$$

where $\lambda_{\max}(A)$ it the maximum eigenvalue of the pairwise comparison matrix A. For this prioritisation method, Saaty [5] proposes the measure of inconsistency known as the Consistency Ratio (CR):

$$CR(A) = \frac{CI(A)}{RI(n)} \quad (2)$$

The numerator of this expression is the Consistency Index, $CI(A)$, which measures the inconsistency of matrix A as the normalised difference between the principal eigenvalue and the size of the matrix:

$$CI(A) = \frac{\lambda_{\max}(A) - n}{n-1} = \frac{1}{n(n-1)} \sum_{i\neq j}^{n} (e_{ij} - 1) \quad (3)$$

where $e_{ij} = a_{ij}w_j/w_i$ is the error obtained when estimating the ratio w_i/w_j through judgement a_{ij}. The value of the CI also corresponds to the average of the deviations of the errors with respect to the unit.

The denominator of Equation (2), $RI(n)$, is the Random Consistency Index. This value removes the influence of the order of the matrix A by obtaining the expected value of the $CI(A)$ considering random matrices of order n with values in the set $\{1/9, 1/7, \ldots, 1/3, 1, 3, \ldots, 9\}$. The values of $RI(n)$, obtained for 100,000 simulations, can be seen in [10].

If a matrix A is consistent, $\lambda_{\max}(A) = n$ and then $CI(A) = CR(A) = 0$. Saaty [5] argued that a consistency ratio of less than 10% ($CR \leq 0.10$) would be permissible. Several authors have criticised this rule and its empirical justification [17,18]. In a later work, Saaty [19] proposed the thresholds 5% for $n = 3$, 8% for $n = 4$ and 10% for $n > 4$.

Since the appearance of the Conventional-AHP [4,5], several prioritisation procedures and inconsistency measures have been proposed in the literature. The most extended prioritisation procedure is the Row Geometric Mean (RGM) or Logarithmic Least Square (LLS) method [5,20]. This method is widely employed due to simplicity of use and its psychological and mathematical properties [10,21–23]. For the RGM, Aguarón and Moreno-Jiménez [10] proposed to use the inconsistency measure named the Geometric Consistency Index (GCI), given by:

$$GCI = \frac{1}{(n-1)(n-2)} \sum_{i,j=1}^{n} \log^2 e_{ij} \quad (4)$$

where $e_{ij} = a_{ij}\omega_j/\omega_i$ and ω is the priority vector obtained using the RGM method, $\omega = (\omega_i) = \left(\prod_{k=1}^{n} a_{ik}^{1/n}\right)$.

Other inconsistency measures related to the prioritisation procedures were proposed by: (i) Golden and Wang [24], who suggested an index to measure the deviations between the pairwise comparison matrix entries and the priorities obtained either by the EV or the RGM; (ii) Stein and Mizzi [25], who proposed an index for the Additive Normalisation prioritisation method [26]; and (iii) Ramík and Korviny [27], who put forward another index for the RGM. Kou and Lin [28] proposed a prioritisation procedure based on similarity measures, the Cosine Maximisation (CM) method, and defined the associated inconsistency measure, the Cosine Consistency Index (CCI).

Koczkodaj [18,29,30] designed one of the first inconsistency measures that does not depend on prioritisation methods and is based on triads:

$$KI(A) = \max_{i<j<k} \left\{ \min \left\{ \left| 1 - \frac{a_{ik}}{a_{ij}a_{jk}} \right|, \left| 1 - \frac{a_{ij}a_{jk}}{a_{ik}} \right| \right\} \right\} \quad (5)$$

Several other inconsistency measures have appeared in this group. Especially noteworthy is that of Pelaez and Lamata [31], who defined an index using the average of all the determinants of the 3×3 submatrices of the pairwise comparison matrix (based on the fact that, if $a_{ij}a_{jk} = a_{ik}$, the corresponding determinant is equal to 0):

$$PLI(A) = \sum_{i=1}^{n-2} \sum_{j=i+1}^{n-1} \sum_{k=j+1}^{n} \left(\frac{a_{ik}}{a_{ij}a_{jk}} + \frac{a_{ij}a_{jk}}{a_{ik}} - 2 \right) / \binom{n}{3} \quad (6)$$

Other indices in the literature have not been located in either of the two previous groups (see [32–38]). Brunelli et al. [39] analysed the proportionality between four of the inconsistency indices whilst Szybowski [40] defined triad and cycle inconsistency indexes that are induced by metrics, and proved that the Koczkodaj's measure is an index of this kind. A survey on the inconsistency measures, their properties and relations can be seen in [41]. The current paper presents two new inconsistency measures based, respectively, on triads and cycles, and proves that, under certain conditions, both coincide with the GCI [10].

3. The Triads Geometric Consistency Measure (T-GCI)

This section presents and analyses a triads-based indicator that explicitly considers Saaty's definition of consistency expressed as $a_{ij}a_{jk}a_{ki} = 1 \ \forall i, j, k = 1, \ldots, n$, that is to say, that for each list of judgements that form a cycle of length 3, the product of their intensities is the unit. The suitability of this definition of consistency is intuitively justified in the following example.

Let the matrices A and B be

$$A = \begin{pmatrix} 1 & 2 & 7 \\ 1/2 & 1 & 3 \\ 1/7 & 1/3 & 1 \end{pmatrix} \qquad B = \begin{pmatrix} 1 & 2 & 9 \\ 1/2 & 1 & 3 \\ 1/9 & 1/3 & 1 \end{pmatrix}$$

It can be seen that in both cases the condition of consistency is violated when considering the only three judgements elicited: $a_{12}a_{23} \neq a_{13}$ $(2 \times 3 \neq 7)$ and $b_{12}b_{23} \neq b_{13}$ $(2 \times 3 \neq 9)$. The two matrices are inconsistent, but it can be intuitively appreciated that matrix A is closer to consistency than matrix B. It can be verified that $CR(A) = 0.002$ and $GCI(A) = 0.008$, as well as that $CR(B) = 0.017$ and $GCI(B) = 0.055$.

To determine which of these two matrices is more inconsistent, it is not necessary to apply a prioritisation method and calculate its associated inconsistency measure. Inconsistency measures based on triads employ the evaluation of the intensities of cycles of length 3. If these intensities are close to the unit, the matrix will be close to consistency, while distant values will indicate a lack of consistency. It is therefore necessary to start by measuring the deviations from the unit of the intensities of cycles of length 3, $d_{ijk} = f\left(a_{ij}a_{jk}a_{ki}\right)$, and then aggregate the deviations of all existing cycles of length 3. In general, we can express this as:

$$M = g\left(d_{123}, \ldots, d_{ijk} \ldots d_{n(n-1)(n-2)}\right), i \neq j \neq k \quad (7)$$

A first possibility is to consider the difference between the intensity of the cycle and the unit ($d_{ijk} = a_{ij}a_{jk}a_{ki} - 1$), and then aggregate the differences additively (M would simply be the sum function).

$$M_1 = \sum_{i \neq j \neq k} \left(a_{ij}a_{jk}a_{ki} - 1\right) \quad (8)$$

This measure contemplates intensity when going through each cycle in both directions. In the case of inconsistency, one of these intensities will always be greater than 1 and the other less than 1. When considering a cycle and its inverse, addends of type $\left(a_{ij}a_{jk}a_{ki} + (a_{ij}a_{jk}a_{ki})^{-1} - 2\right)$ are taken. These addends are always positive ($x + 1/x \geq 2$ if $x > 0$), thus the overall measure will always be non-negative (and null when the matrix is consistent):

$$M_1 = 3 \sum_{i<j<k} \left(a_{ij}a_{jk}a_{ki} + \frac{1}{a_{ij}a_{jk}a_{ki}} - 2 \right) \tag{9}$$

Except for the factor, this last expression was the one followed by the PLI measure proposed by Pelaez and Lamata (Equation (6)).

There are two methods for avoiding the different contribution to the inconsistency measures of cycles with intensity greater than the unit and that of its inverses (intensity less than the unit): (i) consider cycles of only one type, either greater than one or less than one; or (ii) consider the definition itself, as is proposed in the current work.

With the first method, reconsidering $d_{ijk} = a_{ij}a_{jk}a_{ki} - 1$ as the divergence function, the maximum as the aggregation function and only cycles with intensity greater than one, a new family of inconsistency measures is derived:

$$M_2 = \max_{i<j<k} \left\{ a_{ij}a_{jk}a_{ki} - 1 \text{ with } a_{ij}a_{jk}a_{ki} > 1 \right\} \tag{10}$$

Measure M_2 does not aggregate the deviations of the intensity of all the cycles of length 3 with respect to 1; it only considers the extreme discrepancy. Further, the use of the maximum function presents problems from an analytical point of view. It can easily be shown that the Koczkodaj's measure KI (5) corresponds to the M_2 function.

To define inconsistency measures that verify the reversibility in the intensity of the cycles of length 3, that is to say, that the cycles in both directions contribute to the indicator in the same value, this work suggests the use of the logarithms of these intensities. Obviously, the values corresponding to a cycle and its inverse would be compensated, thus it is necessary to consider them in positive terms. The first option, taking absolute values, results in the same analytical problems as the Koczkodaj index. The use of the squares of the logarithms of the intensities is therefore proposed.

With this method, the divergence functions are $d_{ijk} = \left(\log a_{ij}a_{jk}a_{ki} - \log 1\right)^2$, which can be understood as the log quadratic distance between the intensity of the cycle and the unit. Finally, the arithmetic mean function corrected by the number of involved judgements is used as the aggregation function. In the particular case of Equation (11), the denominator indicates the number of terms, $n(n-1)(n-2)$, in the summation multiplied by the number of judgments, 3, that intervene in each summand.

Definition 1. *Given a pairwise comparison matrix, $A_{(nxn)} = (a_{ij})$ with $a_{ij}a_{ji} = 1$ and $a_{ij} > 0$, the Triads Geometric Consistency Index is defined as*

$$T\text{-}GCI(A) = \frac{\sum_{i \neq j \neq k} \log^2 \left(a_{ij}a_{jk}a_{ki}\right)}{3n(n-1)(n-2)} \tag{11}$$

Remark 1. *Since there are $3! = 6$ cycles $(ijk, ikj, jik, jki, kij, kji)$, including the indices i, j, k, that contribute to Equation (11) in the same way: $\left(\log a_{ij} + \log a_{jk} + \log a_{ki}\right)^2$. This expression can be rewritten as:*

$$T\text{-}GCI(A) = \frac{2 \sum_{i<j<k} \log^2 \left(a_{ij}a_{jk}a_{ki}\right)}{n(n-1)(n-2)} \tag{12}$$

3.1. A Link between the Two Groups of Inconsistency Measures

The following result establishes the relationship between the proposed measure ($T\text{-}GCI$), based on triads, and the GCI, based on a prioritisation method (RGM).

Theorem 1. *Given a pairwise comparison matrix, $A_{(nxn)} = (a_{ij})$ with $a_{ij}a_{ji} = 1$ and $a_{ij} > 0$, it holds that*

$$T\text{-}GCI(A) = GCI(A) \tag{13}$$

Proof. See Appendix A. □

This result allows the utilisation of the properties and relationships with other inconsistency measures of the GCI [41,42], as well as its inconsistency thresholds and the decision tools developed for it, when working with the $T\text{-}GCI$. In particular, the procedure proposed for improving the inconsistency [43,44] and the judgements' consistency stability intervals used in group decision making [45–49].

The authors of [42,50–54] posed different properties that inconsistency measures should meet. Brunelli and Fedrizzi [50] give five axiomatic properties for inconsistency indices: (i) the existence of a unique element representing consistency; (ii) invariance under permutation of alternatives; (iii) monotonicity under reciprocity-preserving mapping; (iv) monotonicity on single comparisons; and (v) continuity. Brunelli [52] added a new property: (vi) invariance under inversion of preferences. The authors of [50,52] also proved that the indexes CR, GCI, KI and PLI satisfy the six axiomatic properties; therefore, in line with Theorem 1, these properties are also verified by the $T\text{-}GCI$, as they are verified by the GCI.

In the last decade, many authors [18,41,50–52,54–62] have made efforts to formalise the definition of inconsistency measure; they have demanded a series of properties that guarantee the construction of an axiomatic system. The introduction and justification of reasonable properties may help to narrow the general definition of inconsistency index and identify problematic indices that do not satisfy these requirements [54].

Among the many properties proposed for inconsistency measures are those of Koczkodaj and Szybowski [63] (the normalisation of any inconsistency index) and Mazurek [64] (an upper limit on the value of an inconsistency index). Without discussing the relevance of the reversibility of the intensity of the cycles of length 3 (mentioned above) or these last two properties, it can be said that all of them are fulfilled by the $T\text{-}GCI$; the first by construction, and the other two by slightly adapting their expression. It would be enough to define the Normalised $T\text{-}GCI$ as $1 - e^{-T\text{-}GCI}$. It should be noted that the problem is still unresolved, and this is evidenced by the numerous articles that are regularly published.

In addition to the properties of the indicator, another fundamental aspect when applying it in practice is the existence of thresholds that allow the $T\text{-}GCI$ to be operational. In this case, the thresholds obtained in [10] for the GCI by analogy with the Saaty CR are available as a reference. These values are: $T\text{-}GCI = 0.31$ for $n = 3$, $T\text{-}GCI = 0.35$ for $n = 4$ and $T\text{-}GCI = 0.37$ for $n > 4$.

3.2. Computational Complexity

This section describes a study of the computational complexity of both measures: the GCI and the $T\text{-}GCI$. It allows the selection of the most appropriate expression with regards to the size of the matrix.

The first step in the calculation of the GCI is to determine the priorities. Next, the errors must be identified. Then, the squares of the logarithms of the errors must be calculated, added and divided by the denominator. For the calculation of each priority w_i (there are n in total), $n - 1$ products are needed to obtain $\prod_j a_{ij}$. To calculate the nth root, a logarithm, a division and an exponential are required. In total, $n(n-1)$ products, n logarithms, n divisions and n exponentials. As the cost in computer cycles of the product and the division are usually the same, the divisions are counted as products. Similarly, the exponentials are counted as logarithms. Therefore, to obtain the priorities, $n(n-1) + n = n^2$ products and $n + n = 2n$ logarithms are required. When the priorities have been calculated, the errors (one product and one division) must be obtained for each of the judgments that are above the diagonal, a total of $n(n-1)/2$, then their logarithms must be squared (one product). This equals three products and one logarithm for each error; $3n(n-1)/2$ products and $n(n-1)/2$ logarithms are therefore required. Next, all the squared logarithms of the errors are added, which represents $n(n-1)/2 - 1$ sums. Finally, the result is divided by $n(n-1)$ and multiplied by 2, which gives three products.

The $T\text{-}GCI$ is obtained by calculating the term $\log^2 a_{ij}a_{jk}a_{ki}$ for each of the different triads $(n(n-1)(n-2)/6)$. For this calculation, two products, one logarithm and one additional product (square) are needed, that is, three products and one logarithm. This makes a total of $n(n-1)(n-2)/2$ products and $n(n-1)(n-2)/6$ logarithms. Next, all the squared logarithms are added, which represent $n(n-1)(n-2)/6 - 1$ sums. Finally, the result is divided by $n(n-1)(n-2)$ and multiplied by 6, which means making four products. The total number of operations of each type necessary to obtain both measures are summarised in Table 1 .

Table 1. Computational Complexity of GCI and $T\text{-}GCI$.

Operation	GCI	$T\text{-}GCI$
Sums	$\frac{n(n-1)}{2} - 1$	$\frac{n(n-1)(n-2)}{6} - 1$
Products	$n^2 + \frac{3n(n-1)}{2} + 3$	$\frac{n(n-1)(n-2)}{2} + 4$
Logarithms	$2n + \frac{n(n-1)}{2}$	$\frac{n(n-1)(n-2)}{6}$

It is easy to notice that the calculations needed to obtain the GCI and the $T\text{-}GCI$ are, respectively, of orders $o(n^2)$ and $o(n^3)$. However, for small values of n (which is common in AHP), the constants can be very important. Table 2 shows the number of operations of each type for the values of n commonly employed in AHP.

Table 2. Computational complexity of GCI and $T\text{-}GCI$ for different values of n.

	GCI			$T\text{-}GCI$		
n	**Sums**	**Products**	**Logarithms**	**Sums**	**Products**	**Logarithms**
3	2	21	9	0	7	1
4	5	37	14	3	16	4
5	9	58	20	9	34	10
6	14	84	27	19	64	20
7	20	115	35	34	109	35
8	27	151	44	55	172	56
9	35	192	54	83	256	84

It can be noted that for values of n less than 8 the $T\text{-}GCI$ requires fewer operations (logarithms are more complex, followed by products and sums). Obviously, this is not so important if the calculation of the inconsistency of only one matrix is the objective, but in a simulation study with a large number of matrices, e.g., 10^6, the use of one or other measure can have a significant impact on simulation time.

3.3. Example

Consider the following matrix:

$$A = \begin{pmatrix} 1 & 2 & 3 & 4 \\ 1/2 & 1 & 5 & 6 \\ 1/3 & 1/5 & 1 & 7 \\ 1/4 & 1/6 & 1/7 & 1 \end{pmatrix}$$

The steps for the calculation of both measures are followed in Tables 3 and 4. It can be seen that for this size of matrix the calculation of the $T\text{-}GCI$ is less complicated.

The steps for the calculation of the GCI are (Table 3): (i) the calculation of $\prod_j a_{ij}$ for each row; (ii) the calculation of priorities w_i; (iii) the calculation of the errors e_{ij}; (iv) the calculation of the squared logarithm of the errors; and (v) the value of the GCI.

Table 3. Calculation of the GCI.

i	$\prod_j a_{ij}$	$w_i = \left(\prod_j a_{ij}\right)^{1/4}$	$E = (e_{ij}) = (a_{ij}w_j/w_i)$	$\left(\log^2 e_{ij}\right)$
1	$1 \times 2 \times 3 \times 4 = 24$	2.213	1 1.778 1.120 0.502	0 0.331 0.013 0.475
2	$\frac{1}{2} \times 1 \times 5 \times 6 = 15$	1.968	1 2.100 0.847	0 0.550 0.028
3	$\frac{1}{3} \times \frac{1}{5} \times 1 \times 7 = \frac{7}{15}$	0.827	1 2.352	0 0.732
4	$\frac{1}{4} \times \frac{1}{6} \times \frac{1}{7} \times 1 = \frac{1}{168}$	0.278	1	0
			$\sum_{i<j} \log^2 e_{ij} =$	2.129
			$GCI = \frac{2 \times 2.129}{(4-1) \times (4-2)} =$	0.710

Table 4 shows the calculation of the product and squared logarithms for each of the four triads or cycles of length 3. The $T\text{-}GCI$ is obtained by adding these values and dividing by $n(n-1)(n-2)/6$.

Table 4. Calculation of the $T\text{-}GCI$.

ijk	$a_{ij}a_{jk}a_{ki}$	$\log a_{ij}a_{jk}a_{ki}$	$\log^2 a_{ij}a_{jk}a_{ki}$
123	$2 \times 5 \times \frac{1}{3} = \frac{10}{3}$	1.204	1.450
124	$2 \times 6 \times \frac{1}{4} = 3$	1.099	1.207
134	$3 \times 7 \times \frac{1}{4} = \frac{21}{4}$	1.658	2.750
234	$5 \times 7 \times \frac{1}{6} = \frac{35}{6}$	1.764	3.110
		$\sum_{i<j<k} \log^2 a_{ij}a_{jk}a_{ki} =$	8.516
		$T\text{-}GCI = \frac{2 \times 8.516}{4 \times (4-1) \times (4-2)} =$	0.710

4. Inconsistency Measures Based on Cycles

In general, if A is consistent, it holds [9] that

$$a_{i_1 i_2} a_{i_2 i_3} \cdots a_{i_l i_1} = 1 \qquad \forall i_1, i_2, \ldots, i_l \tag{14}$$

An inconsistency measure can therefore be defined through the intensities of cycles of any length $l\,(>2)$.

Definition 2. *Given a pairwise comparison matrix, $A_{(nxn)} = (a_{ij})$ with $a_{ij}a_{ji} = 1$ and $a_{ij} > 0$, the $l-$Cycles Consistency Index is defined as*

$$I_l(A) = \frac{\sum_{i_1 \neq i_2 \neq \cdots \neq i_l} \log^2 \left(a_{i_1 i_2} a_{i_2 i_3} \cdots a_{i_l i_1}\right)}{l V_{n,l}} \tag{15}$$

where $V_{n,l}$ are the number of l-variations of n elements.

Remark 2. *Clearly, the $T\text{-}GCI(A) = I_3(A)$.*

A cycle of length greater than 3 can be easily broken down into product of cycles of length 3. For example (Figures 1 and 2), when $l = 4$:

$$a_{ij}a_{jk}a_{kl}a_{li} = a_{ij}a_{jk}\left(a_{ki}a_{ik}\right)a_{kl}a_{li} = \left(a_{ij}a_{jk}a_{ki}\right)\left(a_{ik}\ a_{kl}a_{li}\right)$$

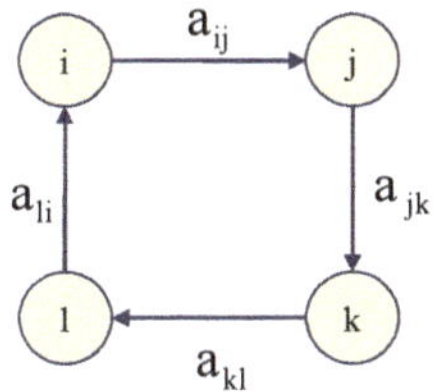

Figure 1. Cycle of length 4.

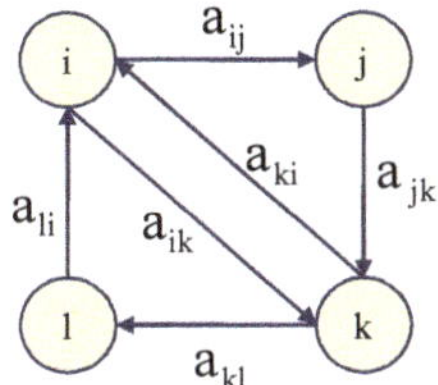

Figure 2. Cycles of length 3.

Therefore, a measure of divergence $d_{ijkl} = f(a_{ij}a_{jk}a_{kl}a_{li})$ can be obtained as $d_{ijkl} = f'(d_{ijk}, d_{ikl})$ and a global measure can be considered as a function of the intensities of the cycles of length 3.

The following result not only allows the expression of the general index $I_l(A)$ to be obtained in terms of cycles of length 3, but also proves that all indexes $I_l(A)$ provide the same value regardless of the value of l.

Theorem 2. *Given a pairwise comparison matrix, $A_{(nxn)} = (a_{ij})$ with $a_{ij}a_{ji} = 1$ and $a_{ij} > 0$, it holds that*

$$I_l(A) = GCI(A) \tag{16}$$

Proof. See Appendix A. □

These cycle-based indices are equivalent, regardless of the cycle length that is considered. Given that the complexity of calculation increases as l increases, it is sufficient to consider the $T\text{-}GCI$ when the objective is to measure the inconsistency from a judgements cycles-based approach.

It can be easily verified that the calculation of the index $I_l(A)$ has complexity $o(n^l)$, with n being the order of the matrix A. From a computational point of view, its calculation for $n > 3$ will not be of interest. Since all these measures are equal, it is faster to obtain the value of $I_3(A) = T\text{-}GCI(A)$.

5. Conclusions

The evaluation of the consistency of decision makers when incorporating their preferences in the AHP, in other words, when eliciting their judgements in pairwise comparison matrices, is one of the most outstanding characteristics of this multi-criteria technique. The initial measures of inconsistency, including the two most widely employed in the scientific literature (CR and GCI), were associated with indexes derived for the prioritisation method used (EV and RGM, respectively). Along with this approach, another method, based on triads, is developed. It evaluates the inconsistency using the definition of a consistent pairwise comparison matrix, or equivalently, in terms of the cycles of length 3, comprising the elements of the matrix.

The current paper introduces a new indicator based on triads ($T\text{-}GCI$) that links the two approaches followed in the evaluation of the inconsistency of AHP. As demonstrated in the paper, this indicator coincides with the GCI, a measure of inconsistency proposed for the RGM. The relationship between the two indicators can take advantage of the properties and characteristics of both to exploit their potential with regards to calculation.

The work does not analyse the relationship between consistency and representativeness of the priority vector derived from the pairwise comparison matrix. The main objective is to propose a new indicator based on triads, the *T-GCI*, which links the two approaches considered when defining inconsistency measures and is more efficient than the *GCI* for matrices with fewer than eight alternatives. This fact is of great significance for simulation studies that require reiterating on numerous occasions the calculation of the inconsistency measure. It is also necessary to emphasise the fact that our work does not deal with obtaining the *T-GCI* for incomplete matrices. Studies analysing the estimation of inconsistency for incomplete matrices can be seen in [65,66].

The relationship between the *T-GCI* and the *GCI*, and the similarities with different inconsistency measures already proved for the *GCI*, can be used to transfer the results obtained for it to other inconsistency measures, based on triads or the priority vectors; in particular, the setting of thresholds to make those inconsistency indices without thresholds operational. Finally, it has been shown that the generalisation of the *T-GCI* to cycles of any length can be expressed in terms of cycles of length 3, which facilitates calculation. Future extensions that are under consideration include how to obtain thresholds for the *T-GCI* that would be calculated directly, or obtained in a similar way to the methods for other related indicators.

Author Contributions: The paper was elaborated jointly by the four authors. Conceptualization, J.M.M.-J.; Data curation, M.T.E. and A.T.; Formal analysis, J.A., M.T.E. and J.M.M.-J.; Investigation, J.A., M.T.E. and A.T.; Methodology, J.M.M.-J.; Software, J.A.; Visualization, A.T.; Writing—review & editing, J.A., M.T.E. and J.M.M.-J. All authors have read and agreed to the published version of the manuscript.

Funding: This research was supported by the Grupo Decisión Multicriterio Zaragoza research group (S35-17R, Government of Aragon) and FEDER funds.

Acknowledgments: The authors would like to acknowledge the work of English translation professional David Jones in preparing the final text, and also the excellent review process carried out by one of the three reviewers.

Conflicts of Interest: The authors declare no conflict of interest.

Appendix A. Proofs

Proof of Theorem 1. Consider the definition of the index $T\text{-}GCI(A)$, where the denominator of that measure is denoted as $D' = 3n(n-1)(n-2)$. It can be seen that the sum extends, without problems, to all subscripts i, j, k.

$$\begin{aligned} T\text{-}GCI(A) &= \frac{1}{D'}\sum_{i,j,k}\log^2\left(a_{ij}a_{jk}a_{ki}\right) = \frac{1}{D'}\sum_{i,j,k}\left(\log a_{ij} + \log a_{jk} + \log a_{ki}\right)^2 = \\ &= \sum_{i,j,k}\left(\log a_{ij}^2 + \log a_{jk}^2 + \log a_{ki}^2 + 2\log a_{ij}\log a_{jk} + 2\log a_{ij}\log a_{ki} + 2\log a_{jk}\log a_{ki}\right) \end{aligned}$$

By symmetry, the first three terms are identical and can be written as

$$\sum_{i,j,k}\log a_{ij}^2 = n\sum_{i,j}\log a_{ij}^2$$

On the other hand, the last three addends are also equal to each other. Thus, the index can be expressed as:

$$\begin{aligned} T\text{-}GCI(A) &= \frac{1}{D'}\left(3\sum_{i,j}\log a_{ij}^2 + 6\sum_{i,j,k}\log a_{ij}\log a_{jk}\right) = \\ &= \frac{1}{3n(n-1)(n-2)}3\left(n\sum_{i,j}\log a_{ij}^2 + 2\sum_{i,j,k}\log a_{ij}\log a_{jk}\right) = \\ &= \frac{1}{n(n-1)(n-2)}\left(n\sum_{i,j}\log a_{ij}^2 + 2\sum_{i,j,k}\log a_{ij}\log a_{jk}\right) \end{aligned} \tag{A1}$$

Now, consider the expression of GCI:

$$GCI = \frac{\sum_{i,j} \log^2 e_{ij}}{(n-1)(n-2)} = \frac{\sum_{i,j} \log^2 a_{ij} w_j / w_i}{(n-1)(n-2)}$$

where $e_{ij} = a_{ij}\omega_j/\omega_i$ is the error obtained when the ratio of priorities ω_i/ω_j is approximated by a_{ij} and $w = (w_i)$ is the priority vector obtained with the RGM method: $\omega_i = \prod_{k=1}^{n} a_{ik}^{1/n}$.

Combining it, and naming $D = (n-1)(n-2)$:

$$\begin{aligned}
GCI &= \frac{1}{D}\sum_{i,j}\log^2 a_{ij}\frac{\omega_j}{\omega_i} = \frac{1}{D}\sum_{i,j}\log^2 a_{ij}\frac{\prod_{k=1}^{n} a_{jk}^{1/n}}{\prod_{k=1}^{n} a_{ik}^{1/n}} = \\
&= \frac{1}{D}\sum_{i,j}\log^2\left(a_{ij}\frac{\prod_{k=1}^{n} a_{jk}^{1/n}}{\prod_{k=1}^{n} a_{ik}^{1/n}}\right) = \frac{1}{D}\sum_{i,j}\left(\log a_{ij} + \frac{1}{n}\sum_{k=1}^{n}\log a_{jk} - \frac{1}{n}\sum_{k=1}^{n}\log a_{ik}^{1/n}\right)^2 = \\
&= \frac{1}{n^2 D}\sum_{i,j}\left(n\log a_{ij} + \sum_{k=1}^{n}\log a_{jk} + \sum_{k=1}^{n}\log a_{ki}\right)^2 = \\
&= \frac{1}{n^2 D}\sum_{i,j}\left(n\log a_{ij} + \sum_{k=1}^{n}\log a_{jk} + \sum_{k=1}^{n}\log a_{ki}\right)\left(n\log a_{ij} + \sum_{l=1}^{n}\log a_{jl} + \sum_{l=1}^{n}\log a_{li}\right) = \\
&= \frac{1}{Dn^2}\left(n^2\sum_{i,j}\log^2 a_{ij} + n\sum_{i,j,l}\log a_{ij}\log a_{jl} + n\sum_{i,j,l}\log a_{ij}\log a_{li} + \right. \\
&+ n\sum_{i,j,k}\log a_{ij}\log a_{jk} + \sum_{i,j,k,l}\log a_{jk}\log a_{jl} + \sum_{i,j,k,l}\log a_{jk}\log a_{li} + \\
&+ \left. n\sum_{i,j,k}\log a_{ij}\log a_{ki} + \sum_{i,j,k,l}\log a_{ki}\log a_{jl} + \sum_{i,j,k,l}\log a_{ki}\log a_{li}\right)
\end{aligned}$$

It is clear that the second, third, fourth, and seventh terms are the same, represented as $n\sum_{i,j,k}\log a_{ij}\log a_{jk}$. Note that the fifth term:

$$\sum_{i,j,k,l}\log a_{jk}\log a_{jl} = \sum_{i}\sum_{j,k,l}\log a_{jk}\log a_{jl} = -n\sum_{j,k,l}\log a_{lj}\log a_{jk}$$

The same with the ninth addend:

$$\sum_{i,j,k,l}\log a_{ki}\log a_{li} = \sum_{j}\sum_{i,k,l}\log a_{ki}\log a_{li} = -n\sum_{i,k,l}\log a_{li}\log a_{ik}$$

The two addends are identical and represented by changing subindices to $-n\sum_{i,j,k}\log a_{ij}\log a_{jk}$ The sixth and eighth addends can also be simplified:

$$\sum_{i,j,k,l}\log a_{jk}\log a_{li} = \sum_{j,k}\log a_{jk}\sum_{i,l}\log a_{li} = 0\times 0 = 0$$

Therefore,

$$\begin{aligned}
GCI &= \frac{1}{Dn^2}\left(n^2\sum_{i,j}\log^2 a_{ij} + n\sum_{i,j,k}\log a_{ij}\log a_{jk} + n\sum_{i,j,k}\log a_{ij}\log a_{jk} + \right. \\
&+ n\sum_{i,j,k}\log a_{ij}\log a_{jk} - n\sum_{i,j,k}\log a_{ij}\log a_{jk} + 0 + \\
&+ \left. n\sum_{i,j,k}\log a_{ij}\log a_{jk} + 0 - n\sum_{i,j,k}\log a_{ij}\log a_{jk}\right) = \\
&= \frac{1}{Dn^2}\left(n^2\sum_{i,j}\log^2 a_{ij} + 2n\sum_{i,j,k}\log a_{ij}\log a_{jk}\right)
\end{aligned}$$

Finally, replacing the denominator D and simplifying:

$$\begin{aligned} GCI &= \frac{1}{n^2(n-1)(n-2)}\left(n^2\sum_{i,j}\log^2 a_{ij} + 2n\sum_{i,j,k}\log a_{ij}\log a_{jk}\right) = \\ &= \frac{1}{n(n-1)(n-2)}\left(n\sum_{i,j}\log^2 a_{ij} + 2\sum_{i,j,k}\log a_{ij}\log a_{jk}\right) \end{aligned} \tag{A2}$$

Equations (A1) and (A2) coincide, thus it is proved that $T\text{-}GCI(A) = GCI(A)$. □

Proof of Theorem 2. This is similar to the proof of Theorem 1. Commencing with the definition of the index $I_l(A)$ to prove that it matches Equation (A2).

$$\begin{aligned} I_l(A) &= \frac{1}{lV_{n,l}}\sum_{i_1\neq i_2\neq\cdots\neq i_l}\log^2\left(a_{i_1i_2}a_{i_2i_3}\ldots a_{i_li_1}\right) = \\ &= \frac{1}{lV_{n,l}}\sum_{i_1\neq\cdots\neq i_l}\left(\log a_{i_1i_2} + \log a_{i_2i_3}\ldots + \log a_{i_li_1}\right)^2 \end{aligned}$$

By developing the summation:

$$\begin{aligned} &\sum_{i_1\neq\cdots\neq i_l}\left(\log a_{i_1i_2} + \log a_{i_2i_3}\ldots + \log a_{i_li_1}\right)\left(\log a_{i_1i_2} + \log a_{i_2i_3}\ldots + \log a_{i_li_1}\right) \\ = &\sum_{i_1\neq\cdots\neq i_l}\left(\log^2 a_{i_1i_2} + \log^2 a_{i_2i_3}\ldots + \log^2 a_{i_li_1}\right) + \\ + \ &2\sum_{i_1\neq\cdots\neq i_l}\left(\log a_{i_1i_2}\log a_{i_2i_3}\ldots + \log a_{i_{l-1}i_l}\log a_{i_li_1}\right) + \\ + &\sum_{i_1\neq\cdots\neq i_l}\left(\ldots\log a_{i_ri_{r+1}}\log a_{i_si_{s+1}}\ldots\right) \end{aligned}$$

In the first summation, squared logarithms of the judgements are grouped. It should be noted that, by symmetry, each of the terms must produce the same sum:

$$\begin{aligned} &\sum_{i_1\neq\cdots\neq i_l}\left(\log^2 a_{i_1i_2} + \log^2 a_{i_2i_3}\ldots + \log^2 a_{i_li_1}\right) = \\ = \ &l\sum_{i_1\neq\cdots\neq i_l}\log^2 a_{i_1i_2} = l\sum_{i_3\neq\cdots\neq i_l}\sum_{i_1\neq i_2}\log^2 a_{i_1i_2} \\ = \ &l(n-2)(n-3)\cdots(n-l+1)\sum_{i_1\neq i_2}\log^2 a_{i_1i_2} \end{aligned}$$

The second summation groups the products of the logarithms of judgements that are adjacent in the cycle, i.e. those that have a common subscript. Again, by symmetry, they are all the same and can be expressed as:

$$\begin{aligned} &2\sum_{i_1\neq\cdots\neq i_l}\left(\log a_{i_1i_2}\log a_{i_2i_3}\ldots + \log a_{i_{l-1}i_l}\log a_{i_li_1}\right) = \\ &2l\sum_{i_1\neq\cdots\neq i_l}\log a_{i_1i_2}\log a_{i_2i_3} = 2l\sum_{i_4\neq\cdots\neq i_l}\sum_{i_1\neq i_2\neq i_3}\log a_{i_1i_2}\log a_{i_2i_3} = \\ = \ &2l(n-3)\cdots(n-l+1)\sum_{i_1\neq i_2\neq i_3}\log a_{i_1i_2}\log a_{i_2i_3} = \end{aligned}$$

Finally, the third sum contains products of logarithms of judgements that have no subscript in common. That each of them is equal to zero can be verified:

$$\sum_{i_1 \neq \cdots \neq i_l} (\log a_{i_r i_{r+1}} \log a_{i_s i_{s+1}}) = \sum_{i_1 \neq \cdots \neq i_l} \sum_{i_r \neq i_{r+1}} \sum_{i_s \neq i_{s+1}} \log a_{i_r i_{r+1}} \log a_{i_s i_{s+1}} =$$

$$= \sum_{i_1 \neq \cdots \neq i_l} \left(\sum_{i_r \neq i_{r+1}} \log a_{i_r i_{r+1}} \right) \left(\sum_{i_s \neq i_{s+1}} \log a_{i_s i_{s+1}} \right) = \sum_{i_1 \neq \cdots \neq i_l} 0 \times 0 = 0$$

Taking everything to the definition of the measure of inconsistency:

$$\begin{aligned} I_l(A) &= \frac{l(n-2)(n-3)\cdots(n-l+1)\sum_{i_1 \neq i_2} \log^2 a_{i_1 i_2}}{ln(n-1)\cdots(n-l+1)} + \\ &\quad + \frac{2l(n-3)\cdots(n-l+1)\sum_{i_1 \neq i_2 \neq i_3} \log a_{i_1 i_2} \log a_{i_2 i_3}}{ln(n-1)\cdots(n-l+1)} = \\ &= \frac{(n-2)\sum_{i_1 \neq i_2} \log^2 a_{i_1 i_2} + 2\sum_{i_1 \neq i_2 \neq i_3} \log a_{i_1 i_2} \log a_{i_2 i_3}}{n(n-1)(n-2)} \end{aligned}$$

Renaming the subscripts:

$$I_l(A) = \frac{(n-2)\sum_{i \neq j} \log^2 a_{ij} + 2\sum_{i \neq j \neq k} \log a_{ij} \log a_{jk}}{n(n-1)(n-2)} \tag{A3}$$

This expression is not exactly equivalent to Equation (A2), but if the latter is slightly modified:

$$GCI = \frac{1}{n(n-1)(n-2)} \left(n \sum_{i,j} \log^2 a_{ij} + 2 \sum_{i,j,k} \log a_{ij} \log a_{jk} \right)$$

Summations are made that allow repetitions of subscripts. Repetitions are removed to compare this expression with the one that was previously obtained.

It is clear that:

$$\sum_{i,j} \log^2 a_{ij} = \sum_{i \neq j} \log^2 a_{ij} \text{ since } a_{ii} = 1$$

Regarding the second term:

$$\begin{aligned} \sum_{i,j,k} \log a_{ij} \log a_{jk} &= \sum_{i \neq j,k} \log a_{ij} \log a_{jk} + \sum_{i,k} \log a_{ii} \log a_{ik} = \\ &= \sum_{i \neq j,k} \log a_{ij} \log a_{jk} + 0 = \\ &= \sum_{i \neq j \neq k} \log a_{ij} \log a_{jk} + \sum_{i \neq j, k=i} \log a_{ij} \log a_{jk} + \sum_{i \neq j, k=i} \log a_{ij} \log a_{jk} = \\ &= \sum_{i \neq j \neq k} \log a_{ij} \log a_{jk} + \sum_{i \neq j} \log a_{ij} \log a_{ji} + \sum_{i \neq j} \log a_{ij} \log a_{jj} = \\ &= \sum_{i \neq j \neq k} \log a_{ij} \log a_{jk} - \sum_{i \neq j} \log a_{ij}^2 + 0 \end{aligned}$$

Grouping everything:

$$\begin{aligned} GCI &= \frac{n\sum_{i \neq j} \log^2 a_{ij} + 2\left(\sum_{i \neq j \neq k} \log a_{ij} \log a_{jk} - \sum_{i \neq j} \log a_{ij}^2\right)}{n(n-1)(n-2)} \\ &= \frac{(n-2)\sum_{i \neq j} \log^2 a_{ij} + 2\sum_{i \neq j \neq k} \log a_{ij} \log a_{jk}}{n(n-1)(n-2)} \end{aligned}$$

This now matches Equation (A3) developed from $I_l(A)$. □

References

1. Colomer, J.M. Ramon Llull: From 'Ars electionis' to social choice theory. *Soc. Choice Welf.* **2013**, *40*, 317–328. [CrossRef]
2. Fechner, G.; Boring, E.; Howes, D. *Elements of Psychophysics*; Number v. 1 in Henry Holt editions in psychology; Holt, Rinehart and Winston: New York, NY, USA, 1966.
3. Thurstone, L.L. A law of comparative judgment. *Psychol. Rev.* **1927**, *34*, 273–286. [CrossRef]
4. Saaty, T.L. A scaling method for priorities in hierarchical structures. *J. Math. Psychol.* **1977**, *15*, 234–281. [CrossRef]
5. Saaty, T.L. *Multicriteria Decision Making: The Analytic Hierarchy Process*; McGraw-Hill: New York, NY, USA, 1980.
6. Gerschgorin, S. Über die Abgrenzung der Eigenwerte einer Matrix. *Izv. Akad. Nauk. SSSR Ser. Mat.* **1931**, *7*, 749–754.
7. Kendall, M.G.; Smith, B.B. On the Method of Paired Comparisons. *Biometrika* **1940**, *31*, 324–345. [CrossRef]
8. Gerard, H.B.; Shapiro, H.N. Determining the degree of inconsistency in a set of paired comparisons. *Psychometrika* **1958**, *23*, 33–46. [CrossRef]
9. Vargas, L.G. A note on the eigenvalue consistency index. *Appl. Math. Comput.* **1980**, *7*, 195–203. [CrossRef]
10. Aguarón, J.; Moreno-Jiménez, J.M. The Geometric Consistency Index: Approximated Thresholds. *Eur. J. Oper. Res.* **2003**, *147*, 137–145. [CrossRef]
11. Schmidt, F.L.; Hunter, J.E. Theory Testing and Measurement Error. *Intelligence* **1999**, *27*, 183–198. [CrossRef]
12. Baby, S. AHP modeling for multicriteria decision-making and to optimise strategies for protecting coastal landscape resources. *Int. J. Innov. Manag. Technol.* **2013**, *4*, 218–227. [CrossRef]
13. Kwiesielewicz, M.; van Uden, E. Inconsistent and contradictory judgements in pairwise comparison method in the AHP. *Comput. Oper. Res.* **2004**, *31*, 713–719. [CrossRef]
14. Danner, M.; Vennedey, V.; Hiligsmann, M.; Fauser, S.; Gross, C.; Stock, S. How Well Can Analytic Hierarchy Process be Used to Elicit Individual Preferences? Insights from a Survey in Patients Suffering from Age-Related Macular Degeneration. *Patient Centered Outcomes Res.* **2016**, *9*. [CrossRef] [PubMed]
15. Bodin, L.; Gass, S.I. On teaching the analytic hierarchy process. *Comput. Oper. Res.* **2003**, *30*, 1487–1497. [CrossRef]
16. Lipovetsky, S.; Conklin, W.M. Robust estimation of priorities in the AHP. *Eur. J. Oper. Res.* **2002**, *137*, 110–122. [CrossRef]
17. Dyer, J.S. Remarks on the Analytic Hierarchy Process. *Manag. Sci.* **1990**, *36*, 249–258. [CrossRef]
18. Koczkodaj, W.; Szwarc, R. On Axiomatization of Inconsistency Indicators for Pairwise Comparisons. *Fundam. Inf.* **2014**, *132*, 485–500. [CrossRef]
19. Saaty, T.L. How to Make a Decision: The Analytic Hierarchy Process. *Interfaces* **1994**, *24*, 19–43. [CrossRef]
20. Crawford, G.; Williams, C. A note on the analysis of subjective judgment matrices. *J. Math. Psychol.* **1985**, *29*, 387–405. [CrossRef]
21. Altuzarra, A.; Moreno-Jiménez, J.M.; Salvador, M. Consensus Building in AHP-Group Decision Making: A Bayesian Approach. *Oper. Res.* **2010**, *58*, 1755–1773. [CrossRef]
22. Csató, L. Characterization of the Row Geometric Mean Ranking with a Group Consensus Axiom. *Group Decis. Negot.* **2018**, *27*, 1011–1027. [CrossRef]
23. Csató, L. A characterization of the Logarithmic Least Squares Method. *Eur. J. Oper. Res.* **2019**, *276*, 212–216. [CrossRef]
24. Golden, B.L.; Wang, Q., An Alternate Measure of Consistency. In *The Analytic Hierarchy Process: Applications and Studies*; Golden, B.L., Wasil, E.A., Harker, P.T., Eds.; Springer: Berlin/Heidelberg, Germany, 1989; pp. 68–81. [CrossRef]
25. Stein, W.; Mizzi, P. The Harmonic Consistency Index for the Analytic Hierarchy Process. *Eur. J. Oper. Res.* **2007**, *177*, 488–497. [CrossRef]
26. Srdjevic, B. Combining different prioritization methods in the analytic hierarchy process synthesis. *Comput. Oper. Res.* **2005**, *32*, 1897–1919. [CrossRef]
27. Ramík, J.; Korviny, P. Inconsistency of pair-wise comparison matrix with fuzzy elements based on geometric mean. *Fuzzy Sets Syst.* **2010**, *161*, 1604–1613. [CrossRef]

28. Kou, G.; Lin, C. A cosine maximization method for the priority vector derivation in AHP. *Eur. J. Oper. Res.* **2014**, *235*, 225–232. [CrossRef]
29. Koczkodaj, W. A new definition of consistency of pairwise comparisons. *Math. Comput. Model.* **1993**, *18*, 79–84. [CrossRef]
30. Duszak, Z.; Koczkodaj, W.W. Generalization of a new definition of consistency for pairwise comparisons. *Inf. Process. Lett.* **1994**, *52*, 273–276. [CrossRef]
31. Peláez, J.; Lamata, M. A new measure of consistency for positive reciprocal matrices. *Comput. Math. Appl.* **2003**, *46*, 1839–1845. [CrossRef]
32. Shiraishi, S.; Obata, T.; Daigo, M. Properties of a Positive Reciprocal Matrix and their Application to AHP. *J. Oper. Res. Soc. Jpn.* **1998**, *41*, 404–414. [CrossRef]
33. Dodd, F.; Donegan, H.; McMaster, T. A statistical approach to consistency in AHP. *Math. Comput. Model.* **1993**, *18*, 19–22. [CrossRef]
34. Jensen, R. *Comparison of Eigenvector, Least Squares, Chi Square and Logarithmic Least Squares Methods of Scaling a Reciprocal Matrix*; Working Paper 127; Trinity University: San Antonio, TX, USA, 1983.
35. Gass, S.; Rapcsák, T. Singular value decomposition in AHP. *Eur. J. Oper. Res.* **2004**, *154*, 573–584. [CrossRef]
36. Monsuur, H. An intrinsic consistency threshold for reciprocal matrices. *Eur. J. Oper. Res.* **1997**, *96*, 387–391. [CrossRef]
37. Chu, A.T.W.; Kalaba, R.E.; Spingarn, K. A comparison of two methods for determining the weights of belonging to fuzzy sets. *J. Optim. Theory Appl.* **1979**, *27*, 531–538. [CrossRef]
38. Wu, Z.; Xu, J. A consistency and consensus based decision support model for group decision making with multiplicative preference relations. *Decis. Support Syst.* **2012**, *52*, 757–767. [CrossRef]
39. Brunelli, M.; Critch, A.; Fedrizzi, M. A note on the proportionality between some consistency indices in the AHP. *Appl. Math. Comput.* **2013**, *219*, 7901–7906. [CrossRef]
40. Szybowski, J. The Cycle Inconsistency Index in Pairwise Comparisons Matrices. *Procedia Comput. Sci.* **2016**, *96*, 879–886. [CrossRef]
41. Brunelli, M. A survey of inconsistency indices for pairwise comparisons. *Int. J. Gen. Syst.* **2018**, *47*, 751–771. [CrossRef]
42. Cavallo, B. Functional relations and Spearman correlation between consistency indices. *J. Oper. Res. Soc.* **2020**, *71*, 301–311. [CrossRef]
43. Aguarón, J.; Escobar, M.T.; Moreno-Jiménez, J.M. Reducing Inconsistency measured by the Geometric Consistency Index in the Analytic Hierarchy Process. *Eur. J. Oper. Res.* **2020**, Forthcoming.
44. Bozóki, S.; Fülöp, J.; Poesz, A. On reducing inconsistency of pairwise comparison matrices below an acceptance threshold. *Cent. Eur. J. Oper. Res.* **2015**, *23*, 849–866. [CrossRef]
45. Aguarón, J.; Escobar, M.T.; Moreno-Jiménez, J.M. Consistency Stability Intervals for a judgement in AHP Decision Support Systems. *Eur. J. Oper. Res.* **2003**, *145*, 382–393. [CrossRef]
46. Moreno-Jiménez, J.M.; Aguarón, J.; Raluy, A.; Turón, A. A Spreadsheet Module for Consistent Consensus Building in AHP-Group Decision Making. *Group Decis. Negot.* **2005**, *14*, 89–108. [CrossRef]
47. Moreno-Jiménez, J.M.; Aguarón, J.; Escobar, M.T. The Core of Consistency in AHP-Group Decision Making. *Group Decis. Negot.* **2008**, *17*, 249–265. [CrossRef]
48. Escobar, M.T.; Aguarón, J.; Moreno-Jiménez, J.M. Some extensions of the precise consistency consensus matrix. *Decis. Support Syst.* **2015**, *74*, 67–77. [CrossRef]
49. Aguarón, J.; Escobar, M.T.; Moreno-Jiménez, J.M. The precise consistency consensus matrix in a local AHP-group decision making context. *Ann. Oper. Res.* **2016**, *245*, 245–259. [CrossRef]
50. Brunelli, M.; Fedrizzi, M. Axiomatic properties of inconsistency indices for pairwise comparisons. *J. Oper. Res. Soc.* **2015**, *66*, 1–15. [CrossRef]
51. Brunelli, M.; Fedrizzi, M. Boundary properties of the inconsistency of pairwise comparisons in group decisions. *Eur. J. Oper. Res.* **2015**, *240*, 765–773. [CrossRef]
52. Brunelli, M. Studying a set of properties of inconsistency indices for pairwise comparisons. *Ann. Oper. Res.* **2017**, *248*, 143–161. [CrossRef]
53. Bozóki, S.; Rapcsák, T. On Saaty's and Koczkodaj's inconsistencies of pairwise comparison matrices. *J. Glob. Optim.* **2008**, *42*, 157–175. [CrossRef]
54. Csató, L. Axiomatizations of inconsistency indices for triads. *Ann. Oper. Res.* **2019**, *280*, 99–110. [CrossRef]

55. Cavallo, B.; D'Apuzzo, L. Investigating Properties of the $\odot$−Consistency Index. In *Advances in Computational Intelligence*; Greco, S., Bouchon-Meunier, B., Coletti, G., Fedrizzi, M., Matarazzo, B.; Yager, R.R., Eds.; Springer: Berlin/Heidelberg, Germany, 2012; pp. 315–327. [CrossRef]
56. Brunelli, M.; Canal, L.; Fedrizzi, M. Inconsistency indices for pairwise comparison matrices: A numerical study. *Ann. Oper. Res.* **2013**, *211*, 493–509. [CrossRef]
57. Kułakowski, K.; Szybowski, J. The New Triad based Inconsistency Indices for Pairwise Comparisons. *Procedia Comput. Sci.* **2014**, *35*, 1132–1137. [CrossRef]
58. Koczkodaj, W.; Szybowski, J. Pairwise comparisons simplified. *Appl. Math. Comput.* **2015**, *253*, 387–394. [CrossRef]
59. Kazibudzki, P. Redefinition of Triad's Inconsistency and its Impact on the Consistency Measurement of Pairwise Comparison Matrix. *J. Appl. Math. Comput. Mech.* **2016**, *2016*, 71–78. [CrossRef]
60. Brunelli, M. Recent advances on inconsistency indices for pairwise comparisons—A commentary. *Fundam. Inf.* **2016**, *144*, 321–332. [CrossRef]
61. Koczkodaj, W.; Urban, R. Axiomatization of inconsistency indicators for pairwise comparisons. *Int. J. Approx. Reason.* **2018**, *94*, 18–29. [CrossRef]
62. Csató, L. Characterization of an inconsistency ranking for pairwise comparison matrices. *Ann. Oper. Res.* **2018**, *261*. [CrossRef]
63. Koczkodaj, W.W.; Szybowski, J. The key properties of inconsistency indicators for a triad in pairwise comparison matrices. *arXiv* **2017**, arXiv:1505.05220v2.
64. Mazurek, J. Some notes on the properties of inconsistency indices in pairwise comparisons. *Oper. Res. Decis.* **2018**, *1*, 27–42.
65. Bozóki, S.; Fülöp, J.; Rónyai, L. On optimal completion of incomplete pairwise comparison matrices. *Math. Comput. Model.* **2010**, *52*, 318–333. [CrossRef]
66. Kułakowski, K.; Talaga, D. Inconsistency indices for incomplete pairwise comparisons matrices. *Int. J. Gen. Syst.* **2020**, *49*, 1–27. [CrossRef]

Article

The Buttressed Walls Problem: An Application of a Hybrid Clustering Particle Swarm Optimization Algorithm

José García [1,†], José V. Martí [2,†] and Víctor Yepes [2,*,†]

1 Escuela de Ingeniería en Construcción, Pontificia Universidad Católica de Valparaíso, Valparaíso 2362807, Chile; jose.garcia@pucv.cl
2 Institute of Concrete Science and Technology (ICITECH), Universitat Politècnica de València, 46022 València, Spain; jvmartia@cst.upv.es
* Correspondence: vyepesp@cst.upv.es
† These authors contributed equally to this work.

Received: 12 May 2020; Accepted: 22 May 2020; Published: 26 May 2020

Abstract: The design of reinforced earth retaining walls is a combinatorial optimization problem of interest due to practical applications regarding the cost savings involved in the design and the optimization in the amount of CO_2 emissions generated in its construction. On the other hand, this problem presents important challenges in computational complexity since it involves 32 design variables; therefore we have in the order of 10^{20} possible combinations. In this article, we propose a hybrid algorithm in which the particle swarm optimization method is integrated that solves optimization problems in continuous spaces with the db-scan clustering technique, with the aim of addressing the combinatorial problem of the design of reinforced earth retaining walls. This algorithm optimizes two objective functions: the carbon emissions embedded and the economic cost of reinforced concrete walls. To assess the contribution of the db-scan operator in the optimization process, a random operator was designed. The best solutions, the averages, and the interquartile ranges of the obtained distributions are compared. The db-scan algorithm was then compared with a hybrid version that uses k-means as the discretization method and with a discrete implementation of the harmony search algorithm. The results indicate that the db-scan operator significantly improves the quality of the solutions and that the proposed metaheuristic shows competitive results with respect to the harmony search algorithm.

Keywords: CO_2 emission; earth-retaining walls; optimization; db-scan; particle swarm optimization

1. Introduction

Retaining walls are structures widely used in engineering for supporting soil laterally. The design of these walls is a problem of interaction between the soil and the structure to retain a material safely and economically. When the height of a cantilever wall becomes important, the volume of concrete required begins to be considerable. From a height of 8–10 m, buttressed walls economize its design. The design of these structures is mainly carried out following rules very much linked to the experience of structural engineers [1]. If the initial design dimensions or material qualities are inadequate, the structure is redefined. With this procedure of trial and error, the different designs obtained do not go beyond a few tests. This process leads to a safe, but not necessarily economic, design [2]. Structural optimization methods have clear advantages over experience-based design.

Presently, the optimum design of reinforced concrete (RC) structures constitutes a relevant line of research. In practical structural optimization problems, the variables used must be discrete, so they are combinatorial optimization problems. However, combinatorial problems are found in a large number of real problems such as allocation resources [3,4], logistics [5], transport [6], routing problems [7,8], scheduling problems [6,9], and engineering design projects [10,11], among others. These problems present a space of solutions that grows exponentially with the variables used, so the metaheuristics, which were inspired by natural phenomena for continuous spaces, are a good approach to obtain optimum solutions to engineering problems. However, two important characteristics of metaheuristics, intensification and diversification, must be preserved to design discrete versions of these algorithms.

While structural optimization began by minimizing the weight or cost of structures [12,13], other objective functions related to the social, Reference [14] and environmental sustainability of structures [15] throughout their entire life cycle have subsequently been incorporated. Reducing the carbon footprint of RC structures is currently investigated as an optimization target. In particular, a hybrid multistart optimization strategic method based on a variable neighborhood search threshold acceptance strategy [16] was used to reduce the cost and carbon-emissions in cantilever retaining walls. A hybrid harmony search together with a threshold acceptance strategy [17,18], the black hole algorithm [17] and a hybrid k-means cuckoo search algorithms [10] were applied to minimize both the economic cost and the CO_2 emissions in counterfort retaining walls. A CO_2 and cost analysis in precast–prestressed concrete road bridges was developed in [15]. In [14] , the importance of the criteria that define social sustainability was analyzed. These criteria considered the complete life of infrastructure. The social sustainability of infrastructure projects was tackled in [14] using Bayesian methods. In recent works [19,20], different meta-heuristics algorithms were used for optimal design of RC retaining walls. The life cycle assessment of earth-retaining walls was analyzed in [21,22].

A strategy that reinforces the results obtained by the metaheuristics has been the hybridization with techniques that deeply modify their way of working. Hybridization is carried out in different ways, the most important of which are: (i) mathematics, integrating mathematical programming and metaheuristics [23], (ii) hybrid heuristics, combining different metaheuristics [24], (iii) symmetrical heuristics, where simulation and metaheuristics are combined [25], and (iv) hybridization between metaheuristics and machine learning.

In this article, we used an emerging line of research that integrates the areas of machine learning and metaheuristic algorithms with the goal of tackle the design of reinforced earth retaining walls problem. This problem presents important challenges in computational complexity since it involves 32 design variables, therefore we have on the order of 10^{20} possible combinations. Therefore, it is interesting to understand how these types of hybrid techniques perform in this problem, in addition to comparing these ones with the state-of-the-art solutions that addressed the design of this type of walls.

The proposed hybrid algorithm uses a machine learning algorithm in a discrete operator that allows continuous metaheuristics to tackle combinatorial optimization problems. In this way, a combinatorial optimization problem, such as the buttressed wall design, can be addressed. The contributions of this work are as follows:

- A hybrid Particle Swarm Optimization (PSO) based on a db-scan clustering technique is proposed. The db-scan is very effective in binary combinatorial problems [6,10]. PSO is often used to solve continuous optimization problems and its tuning is very simple.
- The contribution of the db-scan in the discretization process was studied through a random operator. In addition, the discretization performed by db-scan was compared with methods using k-means [5,26].
- The proposed algorithm is applied to obtain low-carbon, low-cost counterfort wall designs. This hybrid algorithm is compared with an efficient algorithm adapted from the harmony search (HS) proposed in [18].

The rest of this paper is structured as follows: in Section 2 we develop a state-of-the-art of hybridizing metaheuristics with machine learning; in Section 3 we define the optimization problem, the variables involved, and the restrictions; then in Section 4 we detail the discrete db-scan algorithm;

we move on with the experiments and results obtained in Section 5; and conclude with Section 6 in which we summarize the conclusions and new lines of research.

2. Hybridizing Metaheuristics with Machine Learning

Metaheuristics is a broad collection of incomplete optimization techniques inspired by some real-world phenomenon in nature or in the behavior of living beings [27,28]. The objective is to solve problems of high computational complexity in a reasonable execution time so that its optimization mechanism is not significantly affected when the problem to be solved is altered. Then again, the set of techniques capable of learning from a database are the so-called machine learning algorithms [29]. Depending on the learning mode, these techniques are divided into learning by reinforcement, supervised learning, and unsupervised learning. It is common for these algorithms to be used in a wide range of problems such as regression, dimensionality reduction, transformation, classification, time series or anomaly detection and computer vision problems.

The integration of the machine learning techniques with the metaheuristic algorithms can be done basically with two approaches [30]. Either machine learning techniques are used to increase the quality of the solutions and the convergence rates obtained by metaheuristics, or metaheuristic are used to enhance the performance of machine learning techniques [30]. However, metaheuristics often improves the efficiency of an optimization problem concerning machine learning. Based on the work of [30], we propose an extension of the scheme of techniques in which metaheuristics and machine learning are combined (Figure 1).

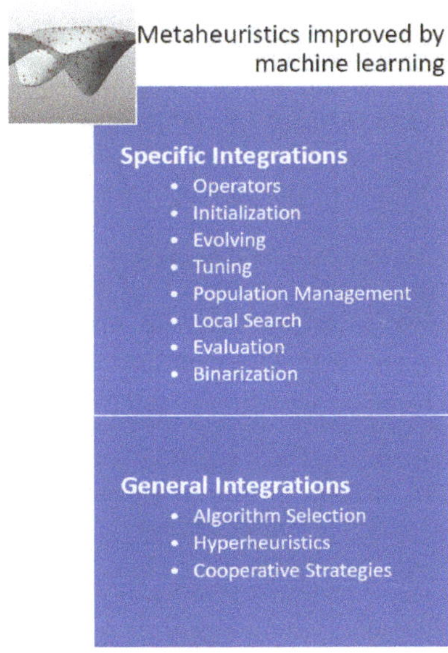

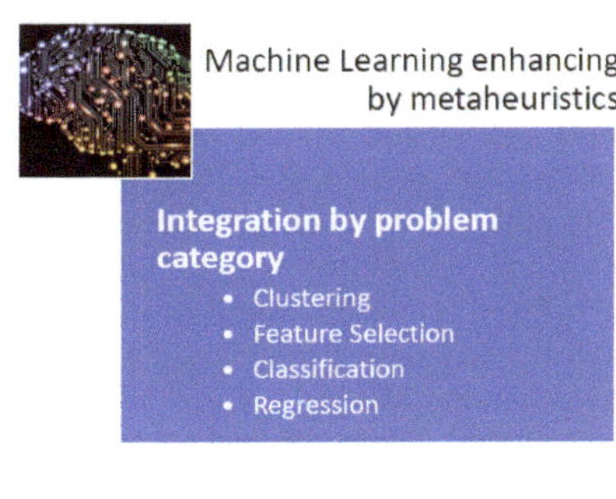

Figure 1. General scheme: Combining Machine Learning and Metaheuristics [10].

Machine learning can be used as a metamodel to determine from a set of metaheuristics the best one for each instance. In addition, specific machine learning operators can also be embedded into a metaheuristic, resulting in three different groups of techniques: hyper-heuristics, algorithm selection, and cooperative strategies [30].

If we automate the design and tuning of metaheuristics to solve a large number of problems, we obtain the so-called hyper-heuristics. The aim of the cooperation strategies is to obtain methods that are more robust by combining the algorithms in a parallel or sequential way. The cooperation mechanism can share the whole solution, or only a part of it. In [31], a multi-objective optimization of an aerogel glazing system through a surrogate model driven by the cross-entropy function was developed with the implementation of the supervised machine-learning method. In [32] the multilevel thresholding image segmentation-based hyperheuristic method was addressed. Finally, in [33] an agent-based distributed framework was proposed to solve the problem of permutation flow stores, and in which each agent is implementing a different metaheuristic.

On the other hand, there are operators that allow enhancing the performance of a metaheuristic integrating machine learning operators. Initialization, population management, solution disruption, binaryization, local search operators and parameter setting and are examples of such operators [30]. Binary operators using unsupervised learning techniques can be integrated into metaheuristics that operate in continuous spaces to perform the binarization process [6]. In [34], a percentile transition-ranking algorithm was proposed as a mechanism to binarize continuous metaheuristics. In [9], the application of Big Data techniques was applied to improve a cuckoo search binary algorithm. The tuning of the parameters of metaheuristics is another line of research of interest. In [35], a tuning method was applied on different sized problem sets for the real-world integration and test order problem. In [36] a semi-automatic approach designs the fuzzy logic rule base to obtain instance-specific parameter values using decision trees. The use of machine learning techniques improves the initiation of solutions, without the need to start it randomly. A cluster-based population initialization framework was proposed in [37] and applied to differential evolution algorithms. In [38] a case-based reasoning technique was applied to initiate a genetic algorithm in a weighted circle design problem. To solve an economic dispatch problem, Hopfield neural networks were used to start solutions of a genetic algorithm [39].

Metaheuristics improve the machine learning algorithms in problems such as feature selection, grouping, classification, feature extraction, among others. Image analysis to identify breast cancer can be enhanced by a genetic algorithm [40] that improves the performance of the Support Vector Machine (SVM). The medical diagnoses and prognoses were tackled in [41] combining swarm intelligence metaheuristics with the probabilistic search models of estimation of distribution algorithms. In [42], the authors used swarm intelligence metaheuristics for the convolution neural network hyper-parameters tuning. In [32], a multiverse optimizer algorithm was used for text documents clustering. An improved normalized mutual information variable selection algorithm for soft sensors in [43] was used to perform the variable selection and validate error information of artificial neural networks. A dropout regularization method for convolutional neural networks was tackled in [44] through the use of metaheuristic algorithms. In [45], a firefly algorithm was combined with the least-squares support vector machine technique to address geotechnical engineering problems. Metaheuristics contributed to the problems of regression, as is the case with the prediction of the strength of high strength concretes in [46]. Another example is the integration of artificial neural networks and metaheuristics for improved stock price prediction [47]. In [48], proposes a sliding-window metaheuristic optimization for predicting the share price of construction companies in Taiwan. In [49] , the least squares support vector machine hybridizing a fruit fly algorithms is applied to simulate the nonlinear system of a MEL time series. Metaheuristics also apply to unsupervised learning techniques, such as clustering techniques. For example, in [50] a metaheuristic optimization was used for a clustering system for dynamic data streams. Metaheuristics have also been integrated into clustering techniques in the search for the centroids that best group the data under a certain metric. A bee colony metaheuristic was used for energy efficient clustering in wireless sensor networks [51]. In [52], a clustering search metaheuristic was applied for the capacitated helicopter routing problem. In [53], a hybrid-encoding scheme was used to find the optimal number of hidden neurons and connection weights in neutral networks. Four metaheuristic-driven techniques were used in [44] to determine the dropout probability in convolutional neural networks. In [54] the bat algorithm and cuckoo search were used to adjust the weights of neural networks. An algorithm was proposed using simulated annealing, differential evolution, and harmony search to optimize convolutional neural networks was proposed in [55]. In [56] long-term short memory trained with metaheuristics were applied in healthcare analysis.

In this paper, the study proposes a hybrid algorithm in which the unsupervised db-scan learning technique to obtain binary versions of the PSO optimization algorithm. This hybrid algorithm was used to obtain a sustainable design buttressed walls. Recently, the db-scan binarization algorithm obtained versions of continuous metaheuristics that have been used to solve the set covering problem [6] and the multidimensional knapsack problem [10] which are NP-hard problems.

3. Problem Definition

This Section describes the optimization problem. First, the equations to be optimized are defined. The variables that define the structure and the parameters applied to solve the problem are described below. In addition, finally, the restrictions and the calculation method applied to verify the structure are summarized.

3.1. Optimization Problem

The goal is to minimize the objective functions F_i for a width of 1 m of a buttress wall. The economic cost in euros will be valued for F_1, and for F_2 the CO_2 equivalent emissions in kg produced in the execution of all parts of the structure. The evaluation of both functions is carried out with precision, and depends on the r construction units used, such as formwork, concrete, steel, excavation and fillings. The values of the units applied to this problem were obtained from [2,18], and were reflected in Table 1. The prices were provided by local Valencian road construction contractors and CO_2 emissions from the BEDEC PR / PCT ITeC (Institute of Construction Technology of Catalonia) database [57]. The objective functions are represented in the following Equation (1). The formula represented values both the cost and the emissions produced during the construction of the wall. It is calculated as the sum of the cost or emissions of each construction unit, being for each unit the product of the unit cost or the unit's emission by its quantity.

$$F_i(x) = \sum_{j=1}^{r} a_{ij} x_j \tag{1}$$

Being a_{ij} the unit value taken from Table 1, for $i = 1$ in cost and $i = 2$ in emissions, corresponding to the measurement of the construction units x_j. The evaluation of Equation (1), in addition to the group of variables, depends on a set of parameters that remain constant throughout the optimization, and as long as the restrictions of the ultimate and service limit states (ULS and SLS) are met.

Table 1. Values per unit of cost and emissions [2,18].

Construction Unit	Cost (€) a_{1j}	(CO_2-eq) a_{2j}
Steel (kg)		
B400	0.56	2.82
B500	0.58	3.02
Concrete in stem (m^3)		
C25/30	56.66	224.34
C30/37	60.80	224.94
C35/45	65.32	265.28
C40/50	70.41	265.28
C45/55	75.22	265.91
C50/60	80.03	265.95
Stem formwork (m^2)	21.61	1.92
Backfill (m^3)	5.56	28.79
Concrete in foundation (m^3)		
C25/30	50.65	224.34
C30/37	54.79	224.94
C35/45	59.31	265.28
C40/50	64.40	265.28
C45/55	69.21	265.91
C50/60	74.02	265.95

3.2. Problem Design Variables

Design variables allow defining the structure. These variables are discrete and modified by the solution search algorithm during the optimization process. There are 3 groups of variables: geometric, material qualities and reinforcing steels. In total there are 32 variables. The definition, arrangement and characteristics of the variables are defined in [58].

Of the set of variables, 24 are shown in Figures 2–4. Figures 2 and 3 represent the configuration of the reinforcement (A1–12), with the diameter and number of steel bars. Three flexural reinforcing bars defined as A1, A2 and A3 contribute to the main bending of the stem. A4 represents the vertical reinforcement of the base at the rear of the stem. The secondary longitudinal reinforcement is provided by A5 for shrinkage and thermal effects on the stem. A6 represents the longitudinal reinforcement of the buttress. The area of the reinforcing bracket from the bottom of the buttress is provided by A7 and A8. A9 and A11 define the upper and lower heel reinforcements and A12, the shear reinforcement in the footing. Finally, the longitudinal effects on the toe are defined by A10. Figure 4 represents most of the geometric variables.These variables are the thickness of the stem (b), the thickness of the footing (cz), the thickness of the buttresses (ec), the length of the heel (lt), the length of the toe (lp), the distance between the buttresses (d), two classes of steel B500S and B400S, and six classes of concrete between C25/30 and C50/60 by discrete intervals of 5 MPa. Table 2 details the set of discrete variables with the ranges of the values they can take. The possible combinations constitute the space solutions of the problem. For the case of a 12 m high wall, the solution space is of the order of 10^{20}.

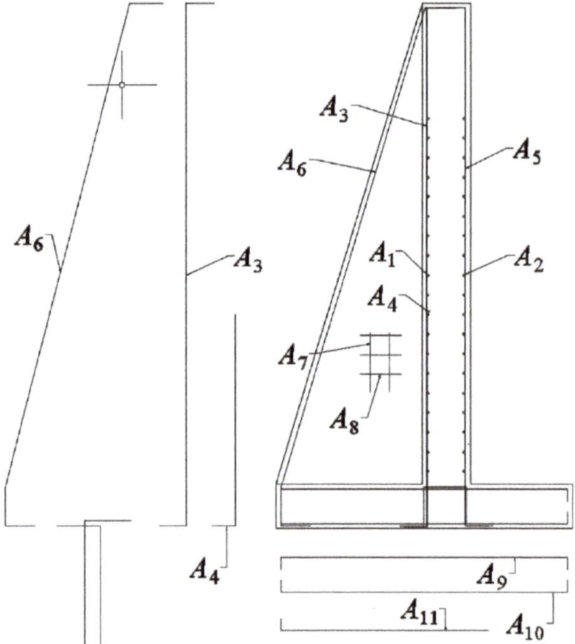

Figure 2. Reinforcement variables for the design of earth-retaining walls [58].

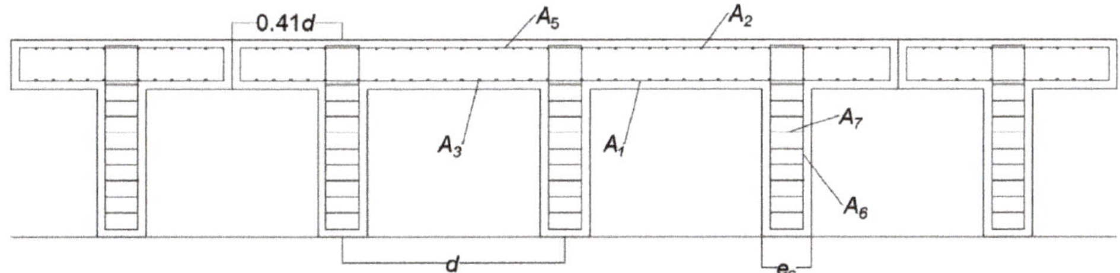

Figure 3. Earth-retaining buttressed wall. Floor cross-section [58].

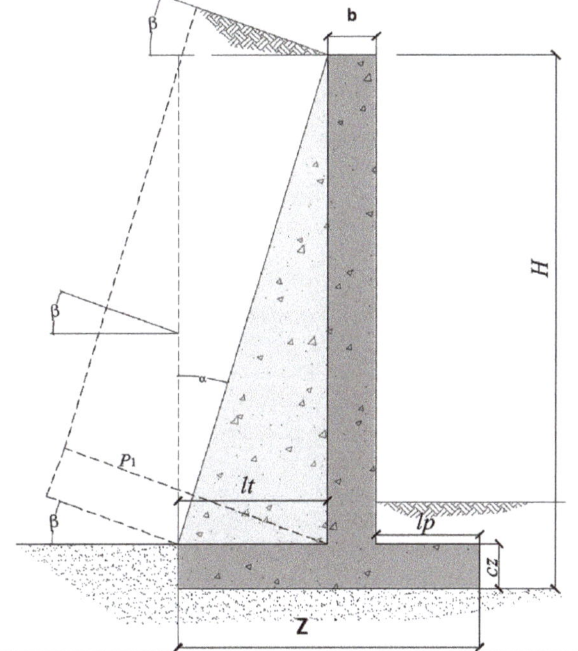

Figure 4. Problem design parameters [58].

Table 2. Design variables.

Variables	Lower Bound	Upper Bound	Increment
c	H/20	H/6	1 cm
d	H/4	H/2	1 cm
b	20 cm	219	1 cm
p	20 cm	610	1 cm
t	20 cm	619	1 cm
e_c	20 cm	219	1 cm
f_{ck}			25, 30, 35, 40, 45, 50 Mpa
f_{yk}			400, 500 Mpa
A_1 to A_{10} Φ			6, 8, 10, 12, 16, 20, 25, 32 mm
n	1	12	1 rebar
A_{11} to A_{12} Φ			6, 8, 10, 12, 16, 20, 25, 32 mm
n	4	10	1 rebar

3.3. Problem Design Parameters

All the cases analyzed in this study are completely defined by constant contour values called problem parameters. These parameters are described in [58] and are represented in Table 3.

Table 3. Problem design parameters values.

Parameter Type	Parameter Considered	Value
Geometric:		
	Foundation depth, H2	2 m
Geotechnical and relative to the load:		
	Bearing capacity	0.3 MPa
	Fill slope	0
	Base-friction coefficient, μ	tg 30°
	Wall-fill friction angle, δ	0°
	Uniform load on top of the fill, γ	10 kN/m^2
Safety coefficient:		
	Against sliding, γ_{fs}	1.5
	Against overturning, γ_{fo}	1.8
	For loading (EHE)	Normal
	Of concrete (ULS)	1.5
	Of steel (ULS)	1.15
Ambient exposure (EHE)		IIa

3.4. Problem Constrains

The viability of the structure is verified as described in [58], in accordance with the Spanish technical standards defined in [59] and the detailed recommendations for foundations in road works [60]. The bending and shear limit states, and the cracking limit state are verified. The structure is checked according to the stem stress distribution [61] for non-cohesive granular materials. The rectangular distribution of soil stresses in the foundations is considered according to the criteria in [62]. The structural hyperstatic model assumes that the top of the stem functions as a cantilever, while the bottom of the stem is embedded between the footing and the lower part of the buttress.

The bending stress verified in the horizontal T-shaped cross section is performed taking into account the effective width, in accordance with [63]. The checking of the mechanical shear and the flexural capacity is carried out considering the equations expressed in [62] and with the verifications in [59]. To assess the controls against overturning and sliding, and the limit of soil stresses, the effect of buttresses is included [58]. It is taken into account that the favorable overturning moments are sufficiently greater than the unfavorable overturning moments with a safety coefficient for frequent events. A slip safety coefficient and a coefficient of base-friction foundation against sliding are also considered.

4. The db-Scan Discrete Algorithm

As a first step, the algorithm generates a set of valid solutions. These solutions are randomly generated. In this procedure, first, it is validated if the solution variables are started. In the case that they are not all started, the variables are started randomly. Once all the variables are generated, the next step is to verify if the solution obtained is valid. In the event that it is not valid, all variables are removed and regenerated. The detail of the initiation procedure is shown in Figure 5. Subsequently, PSO is used to produce a new solution in the continuous space. The PSO algorithm will be described in Section 4.1. Subsequently, the db-scan operator is applied to the continuous solution in order to transform continuous movements into groups associates to transition probabilities. The db-scan operator will be detailed in Section 4.2. After the db-scan operator generates the groups, the discretization operator applies the corresponding transition probability to each group, generating a new discrete solution. The discretization operator is detailed in Section 4.3. Finally, the new solution is validated and, in case it meets the restrictions, it is compared with the best solution obtained. If the new value is higher, it replaces the current one. The detailed flow chart of the hybrid algorithm is shown in Figure 6.

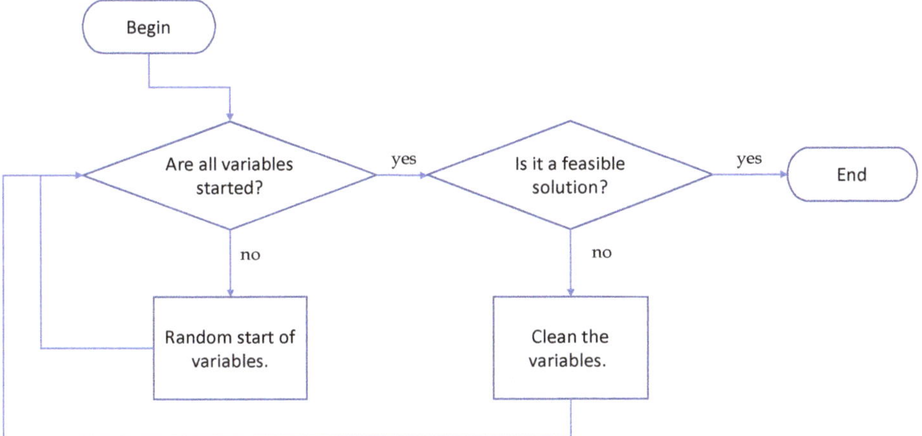

Figure 5. Solution initiation procedure [10].

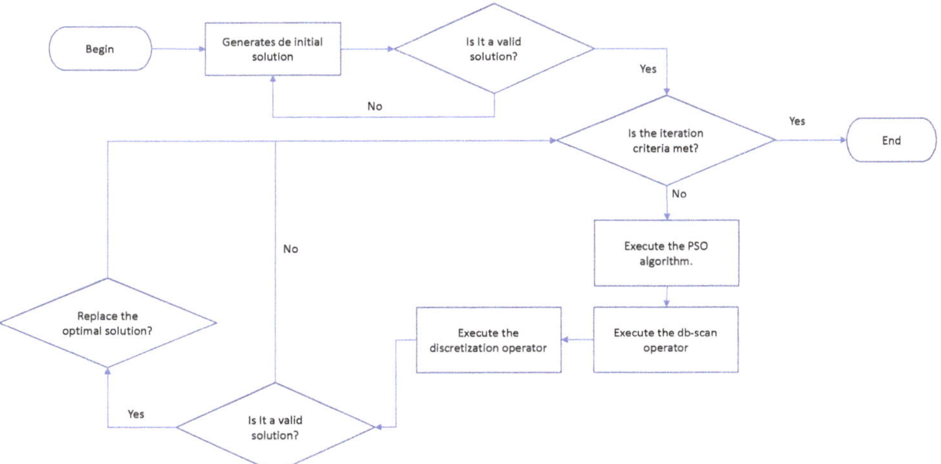

Figure 6. The discrete db-scan algorithm flow chart.

4.1. Particle Swarm Optimization

For the proper functioning of the PSO algorithm, the concepts of population that are usually called a swarm, and each of these solutions is called a particle. The essence of the algorithm is that each particle is guided by a combination of the best value particle obtained so far in the search space (maximum global) together with the best result obtained by the particle (local maximum). The optimization process is iterative until some termination condition is met.

Formally let $f : \mathbb{R}^n \rightarrow \mathbb{R}$ corresponds to the fitness function to be optimized. This function considers a candidate solution that is represented by a vector in $\mathbb{R}^n$ and generates a real value as output. This obtained value, represents the value of the objective function for the given candidate solution. The goal is to find a solution for which $f(a) \leq f(b)$ for all b in the search space, which would mean that a is the global minimum. The algorithm pseudo-code is shown in Procedure 1.

Algorithm 1 Particle swarm optimization algorithm

Objective function f(s)
Generate initial solutions of n particles.
Get the particle's best known position to its initial position: $p_i \leftarrow s_i$.
if $f(p_i) < f(g)$ **then**
  Update the swarm's best known position: $g \leftarrow p_i$
end if
Initialize the particle's velocity: v_i
while stop criterion are meet **do**
  for each particle and dimension **do**
    Pick random numbers: r_p, r_g
    Update the particle's velocity: $v_{i,d} \leftarrow \omega v_{i,d} + \phi_p r_p (p_{i,d} - s_{i,d}) + \phi_g r_g (g_d - s_{i,d})$
    Update the particle's position: $s_i \leftarrow s_i + v_i$
  end for
  if $f(s_i) < f(p_i)$ **then**
    Update the particle's best known position: $p_i \leftarrow s_i$.
    if $f(p_i) < f(g)$ **then**
      Update the swarm's best known position: $p_i \leftarrow s_i$.
    end if
  end if
end while

4.2. db-Scan Operator

The solutions resulting from the execution of the PSO algorithm are grouped by the db-scan operator. We should note that the db-scan operator can be applied to any swarm intelligence continuous metaheuristics. The spatial clustering technique based on noise density of applications (db-scan), requires for the clustering, a set of points S within a vector space, and a metric, usually, the metric is the Euclidean. Db-scan groups the points of S that meet a minimum density criterion. The rest of the points are considered outliers. As input parameters db-scan requires a radius ϵ and the minimum number of neighbors δ. The main steps of the algorithm are shown below:

- Find the points in the ϵ neighborhood of every point and identify the core points with more than δ neighbors.
- Find the connected components of core points on the neighbor graph, ignoring all non-core points.
- Assign each non-core point to a nearby cluster if the cluster is an ϵ neighbor; otherwise, assign it to noise.

In the db-scan (dbscanOp) operator, the db-scan grouping technique is used to make groups of points to which we will assign a probability of transition. This probability of transition will subsequently allow the discretization operator to discretize continuous solutions. The grouping proposal uses the movements obtained by PSO in each dimension for all the particles. Suppose $s(t)$ is a solution in iteration t, then $\Delta^i(s(t))$ represents the magnitude of the offset $\Delta(s(t))$ in the i-th position, considering the iterations t and $t+1$. After all the displacements were obtained, the grouping is carried out. To obtain the groups, the displacement module will be used, which is denoted by $|\Delta^i(s(t))|$. This grouping is done using the db-scan technique. Finally, a generic function P_{tr} is used, which is shown in Equation (2) with the objective of assigning a probability of transition to each group and therefore to each displacement.

Then using the function P_{tr}, a probability is assigned to each group obtained from the clustering process. In this article, we use the linear function given in Equation (2), where Clust (x^i) indicates the location of the group to which $\Delta^i(s)$ belongs. The coefficient α represents the initial transition coefficient and β models the transition separation for the different groups. Both parameters must be estimated. The pseudo-code of the discretization procedure is shown in Algorithm 2.

$$P_{tr}(x) = \alpha + \gamma x \tag{2}$$

where x represents the value of $Clust(s^i)$. Also, because $s^i \in Clust(s^i)$, each element of $Clust(s^i)$, is assigned the same value P_{tr}. That is, $P_{tr}(s^i) = P_{tr}(Clust(s^i))$. On the other hand, $\gamma = \alpha * \beta$ are constants that will be determined in the tuning of parameters, where α corresponds to the initial transition coefficient and β represents the transition probability coefficient.

Algorithm 2 db-scan operator

Function dbscanOp($ls(t)$,$ls(t+1)$)
Input $s(t)$, $s(t+1)$
Output lTransitionProbability$(t+1)$
$l\Delta^i(s(t+1)) \leftarrow$ getDelta($ls(t)$, $ls(t+1)$)
$Clust \leftarrow$ getClusters($l\Delta^i(s(t))$)
lTransitionProbability$(t+1) \leftarrow$ getTranProb($Clust$, $ls(t)$)–Equation (2)
return (lTransitionProbability$(t+1)$)

4.3. Discretization Operator

The discretization operator receives the list lTransitionProbability $(t+1)$. This list contains the values of the transition probabilities which were the result delivered by the db-scan operator. Then given a solution $s(t) \in ls(t)$, we select each of its components $s^i(t)$ and we proceed to determine through the transition probability if this component should be modified. In the case of large transition probabilities, there is a greater possibility of modification. Then a random number is obtained at [0, 1] and this number is compared with the value of the transition probability assigned to the component. In cases where the modification condition is satisfied, it can increase the value by 1 or decrease it by 1. Finally, the selected value is compared with the best value obtained by the algorithm and remains with the minimum of both. The pseudocode of the discrete procedure is shown in Algorithm 3.

Algorithm 3 Discretization operator.

Function DiscOperator(lTransitionProbability$(t+1)$, $ls(t)$)
Input lTransitionProbability$(t+1)$
Output $s(t+1)$–where $s(t+1)$ is discrete.
movement = 0
for $s^i \in s(t) \in ls(t)$ **do**
    if $r_1 > 0.5$ **then**
        movement = 1
    else
        movement = -1
    end if
    s^i = max(1,min(s^i_{best},s^i+movement))
end for

5. Results and Discussion

The experiments developed with the objective of determining the performance of our hybrid algorithm applied to the counterfort retaining wall problem will be detailed in this section. In Section 5.1, we will explain the strategy used to perform the tuning of the parameters. Then, in the first experiment detailed in Section 5.2, we will study the contribution of the db-scan operator to the discretization process. This study will be carried out through a comparison with a random operator. Subsequently, in the second experiment, the db-scan technique will be compared with the k-means clustering technique. This comparison is detailed in Section 5.3. Finally, in Section 5.4 our db-scan PSO proposal will be compared with another algorithm in the literature that solved the same problem. Regarding the scenario of the experiments, each problem will be run 30 times. The value 30 is the minimum number appropriate to be able to obtain statistical conclusions on a population [64], in addition, the signed-rank Wilcoxon test [65] will be incorporated to determine if the difference between the results is statistically significant. For this test, the p-value used was 0.05. The software was built in Python 3.6 and run on a computer with the following configuration: Intel Core i7-4770 CPU and 16 GB of RAM.

5.1. Parameter Settings

In the methodology used to perform the adjustment of the algorithm parameters, heights of 8 and 12 m were considered. The selection of these heights was motivated by their difference in complexity. 8 represents walls of small size and 12 represents walls of a greater size where the satisfaction of stability restrictions has a greater difficulty. After defining the instances to use, each of the defined configurations was resolved 5 times for each height. The set of configurations used are detailed in Table 4. The value 5 was chosen with the intention of being able to execute all the combinations in a limited time. The range column in Table 4 represents the scanned values to perform the PSO adjustment. Obtaining these ranges was based on previous studies where the db-scan technique was applied to other combinatorial problems. For more details on the method, see the reference [26].

In order to find the best configuration, the method proposes using four measurements. The best solution (bs), the worst solution (ws), the average solution (as) and the convergence time (ct). These measures are defined in Equations (3)–(6). The best global value (bgv) corresponds to the best value obtained in all the configurations executed. Furthermore, the best local value (blv) represents the best value obtained by a particular configuration. The worst local value (wlv) represents the worst moment obtained for a given configuration. The average local value (alv) is the average value obtained by a configuration. The minimum global time (mgt) accounts for the minimum convergence time resulting from all evaluated configurations and the convergence local time (clt) the minimum time for a particular configuration. The minimum global time (mgt) corresponds to the best value getting for all configuration. Finally, the worst global value (wgv) accounts for the worst value obtained by all the configurations. Each of the measures defined, the closer to 1, the better performance that indicator has. On the other hand, the closer to 0, the worse performance. Since there are 4 measurements to be able to carry out the evaluation, we incorporate them into a radar graph and calculate their area. As a consequence of the measurement definition, the largest area corresponds to the configuration that has the best performance considering the 4 defined measurements.

Definition 1 ([10,26]). *Measure definitions:*

1. *The deviation of the best local value obtained in five executions compared with the best global value:*

$$bs = 1 - \frac{bgv - blv}{bgv} \tag{3}$$

2. *The deviation of the worst value obtained in five executions compared with the best global value:*

$$ws = 1 - \frac{bgv - wlv}{bgv} \tag{4}$$

3. *The deviation of the average value obtained in five executions compared with the best global value:*

$$as = 1 - \frac{bgv - alv}{bgv} \tag{5}$$

4. *The convergence time for the average value in each experiment is normalized according to Equation (6).*

$$ct = 1 - \frac{clt - mgt}{wgv - mgt} \tag{6}$$

For PSO, the coefficients c_1 and c_2 are set to 2. The parameter ω linearly decreases from 0.9 to 0.4. For the parameters used by db-scan, the minimum number of neighbors ($minPts$) is estimated as a percentage of the number of particles (N). The parameter settings are shown in Table 4 . In the table, the column labeled "Value" represents the selected value, and the column labeled "Range" corresponds to the set of scanned values.

Table 4. Parameter setting for the PSO algorithm.

Parameters	Description	Value	Range
α	Initial transition coefficient	0.1	[0.08, 0.1, 0.12]
β	Transition probability coefficient	0.4	[0.3, 0.4, 0.5]
N	Number of particles	20	[10, 20, 30]
ϵ	ϵ db-scan parameter	0.25	[0.2, 0.25, 0.3]
$minPts$	Point db-scan parameter	10%	[10, 12, 14]
Iteration Number	Maximum iterations	900	[800, 900, 1000]

5.2. db-Scan and Random Operators Comparison

In this first experiment, the contribution of the db-scan operator in optimizing costs and emissions for the design of the wall will be studied. To properly determine the contribution of the db-scan operator, a random operator was designed to replace the discretization performed by db-scan. In particular, the execution of the db-scan operator in Figure 6 is replaced by a random operator. This random operator assigns a fixed probability of 0.5 instead of assigning probabilities per group. Different heights of the wall were considered, starting at 6 (m) and ending at 14 (m) with increments of one meter. For a proper statistical comparison, 30 runs are executed for each height and operator. The results are then documented in tables and boxplots. Finally, the Wilcoxon signed-rank test is used to determine the significance of the results.

Table 5, Figures 7 and 8 detail the results obtained in experiment 1. When comparing the values in Figure 7, in the case of the best value indicator, the operator that uses db-scan was Superior at all heights for both cost and emissions. Additionally, we observe the difference between the operators increases as the height increases. In the case of 6 and 7 m in the case of cost, the difference is 2.2% and 4.6% respectively. In the case of heights of 13 and 14 m, this difference is 30.0% and 34.5% respectively. When carrying out this same analysis for the emissions of CO_2, we find that for the heights of 6 and 7 the differences are 3.0% and 1.2% respectively and at the heights of 13 and 14 we obtain 15.9% and 14.2% respectively. In the comparison of the average indicator, we see a very similar result to that reported in the best value analysis. The average indicator of the db-scan operator is higher for all heights than the random operator in both optimizations. Like the best value case, the difference increases as the height increases. In the case of averages, because the number of values is greater than in the case of the best value, we apply the Wilcoxon significance test. The result indicates that the difference is statistically significant in both cases, costs, and emissions. In Figure 7 we show the boxplots for the cost optimization results. In both cases, db-scan and random it is observed that the dispersion and the Inter-quartile range of the values obtained in the optimization increases as the height increases. However, from height 9 onwards this dispersion is more noticeable in the case of the random operator. When analyzing the results of the optimization of CO_2, emissions, which are shown in Figure 8, the behavior is very similar to that of cost. In the case of emissions, the increase in dispersion and in the inter-quartile range is more noticeable in the random operator from the height of 11 m.

Table 5. Comparison between random and db-scan operators in cost and emission optimization.

Height (m)	Best Cost Value db-Scan	Avg Cost db-Scan	Best Emissions Value db-Scan	Avg Emissions db-Scan	Best Cost Value Random	Avg Cost Random	Best Emissions Value Random	Avg Emissions Random
6	595	598.38	1249	1254.24	608	622.64	1287	1302.92
7	680	691.53	1452	1460.05	711	735.21	1470	1498.69
8	780	793.57	1662	1679.95	791	827.26	1717	1783.61
9	920	937.40	1997	2069.05	1022	1094.61	2190	2215.22
10	1095	1126.22	2462	2546.37	1188	1300.18	3050	3131.50
11	1302	1363.15	3061	3148.37	1526	1694.54	3614	3816.46
12	1528	1581.43	3715	3900.52	1870	1994.33	4299	4547.35
13	1780	1864.57	4470	4643.26	2315	2401.84	5180	5473.56
14	2049	2245.28	5294	5502.76	2756	3032.15	6048	6468.52
Wilcoxon p-value						4.31×10^{-6}		1.87×10^{-5}

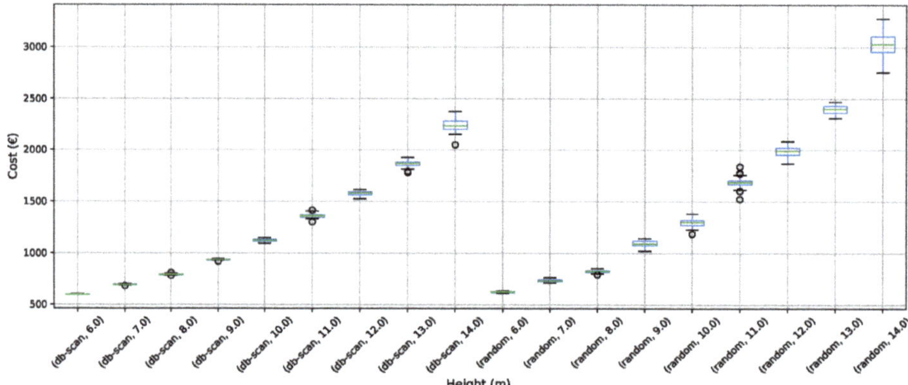

Figure 7. Cost box-plots, comparison between db-scan and random operators.

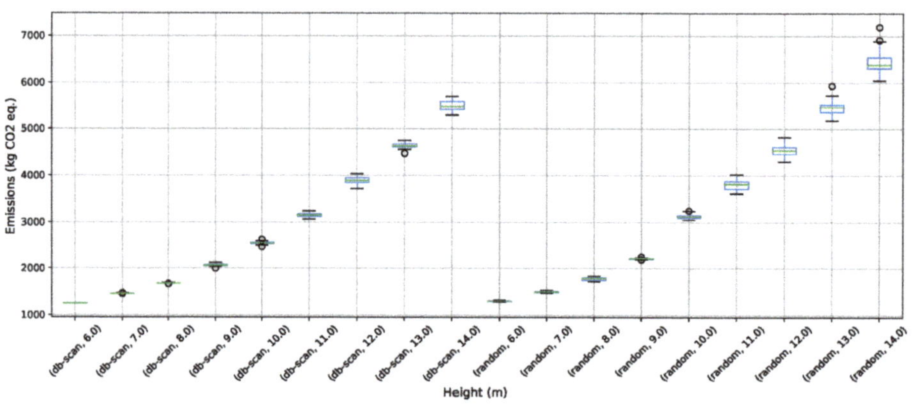

Figure 8. Emission box-plots, comparison between db-scan and random operators.

The convergence diagrams for heights 6, 9, 12 and 14 are shown in Figure 9a,b. These diagrams correspond to the results obtained in cost optimization using the random and db-scan operators. From the db-scan convergence chart, we see that height 6 has the best convergence, followed closely by 9. On the other hand, heights 12 and 14 have a similar convergence, being slower than the case of 6 and 9. At higher wall heights, complying with constraints becomes more complicated and therefore the optimization problem is more difficult to solve. In the case of the random operator, it shows a correspondence between the height and the speed of convergence. The lower the height, the better its convergence. On the other hand, we see that for the random case, already in the 500 iterations the slope tends to stabilize unlike in the db-scan case where for the most difficult problems the slope stabilizes after the 600 iterations.

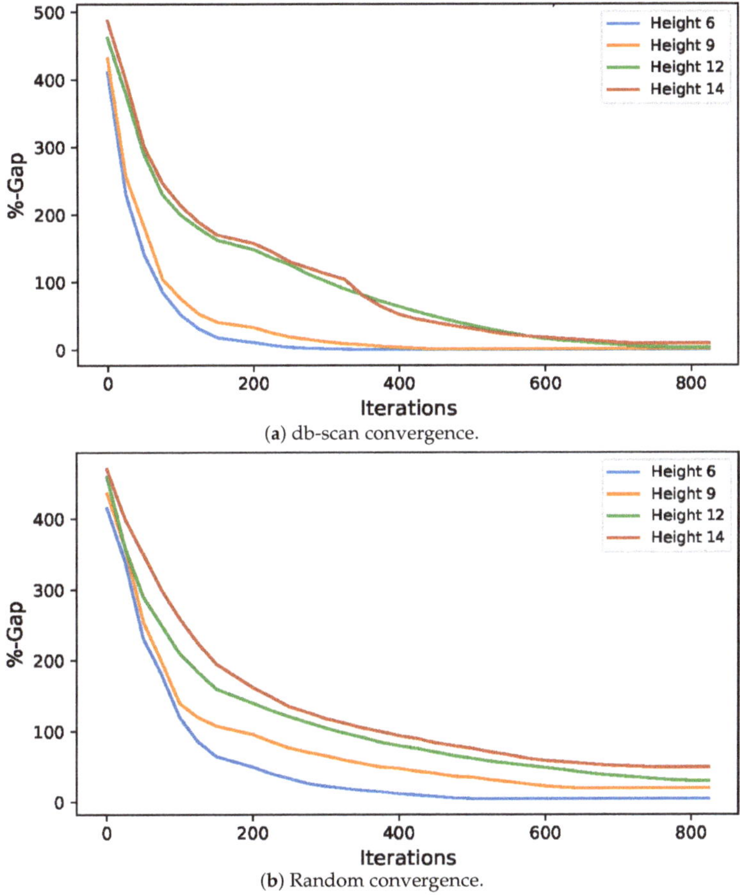

(**a**) db-scan convergence.

(**b**) Random convergence.

Figure 9. Cost optimization convergence plots for db-scan and random operators.

5.3. db-Scan and k-Means Operators Comparison

This second experiment aims to compare the performance of the algorithm that uses db-scan, with another algorithm that uses k-means as a discretization method. This experiment is inspired by the fact that both techniques correspond to unsupervised learning algorithms that aim to cluster points. In the case of k-means, unlike db-scan, the number of clusters must be defined a priori. In this experiment, suggested by the articles [10,26], k was configured with the value 5. The Equation (2) was used to set the values of the probability of transition for each cluster. As in experiment 1, the db-scan module in Figure 6 is replaced in this case by k-means leaving the rest of the modules unchanged.

The results of this experiment are shown in Table 6 and Figures 10 and 11. When analyzing the results of the best value indicator shown in Table 6, we observed in the case of cost optimization, the results are similar, with k-means being higher in 5 of the 9 heights. In the case of optimization of CO_2 emissions, something similar occurs, with very close values between the two algorithms.

In the last case, k-means was higher in 3 cases, db-scan in 1, and in 4 heights their values coincided. When analyzing the average indicator, in the case of optimizing the cost of the wall, we observe that for small heights very close values are obtained in both algorithms, being k-means higher than db-scan. Particularly in the case of heights 6, 7, 8 the difference in percentage was 0.34%, 1.16%, and 1.83% respectively. On the other hand, when we analyze the values obtained in the highest walls, we find that db-scan is superior to k-means. Particularly for heights 12, 13, and 14, we have that the difference is 3.63%, 5.04%, and 3.46% respectively. Wilcoxon's statistical test when analyzing the total population indicates that the difference between both algorithms is not significant. In the case of emission optimization, the behavior of the average indicator is different from that of cost optimization. The db-scan algorithm is superior in 6 of the 9 heights in this indicator. Heights 13 and 14 stand out, achieving differences of 3.3% and 5.78% respectively. When analyzing the total distribution of points, the Wilcoxon test indicates that the difference is significant in favor of db-scan.Finally, when analyzing Figures 10 and 11 both algorithms have a similar behavior between heights 6 and 11. From height 12 onwards, the increase in dispersion and in the Inter-quartile range becomes much more notorious in the case of k-means. This last result shows that for more difficult problems, db-scan behaves more robust than k-means.

Table 6. Comparison between db-scan and k-means operators in cost and emission optimization.

Height (m)	Best Cost Value db-Scan	Avg Cost db-Scan	Best Emissions Value db-Scan	Avg Emissions db-Scan	Best Cost Value k-Means	Avg Cost k-Means	Best Emissions Value k-Means	Avg Emissions k-Means
6	595	598.38	1249	1254.24	591	596.35	1242	1284.92
7	680	691.53	1452	1460.05	678	683.62	1440	1465.52
8	780	793.57	1662	1679.95	775	779.34	1659	1685.83
9	920	937.40	1997	2069.05	911	938.24	1997	2046.26
10	1095	1126.22	2462	2546.37	1095	1107.80	2470	2523.35
11	1302	1363.15	3061	3148.37	1302	1327.16	3061	3139.20
12	1528	1581.43	3715	3900.52	1528	1638.84	3715	3902.61
13	1780	1864.57	4470	4643.26	1775	1958.62	4470	4798.79
14	2049	2245.28	5294	5502.76	2049	2322.87	5294	5821.08
Wilcoxon *p*-value						0.42		4.08×10^{-4}

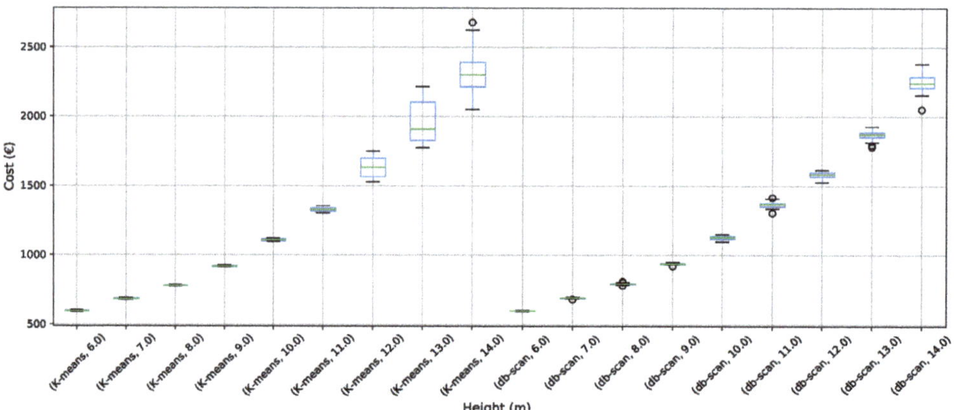

Figure 10. Cost box-plots, comparison between db-scan and k-means operators.

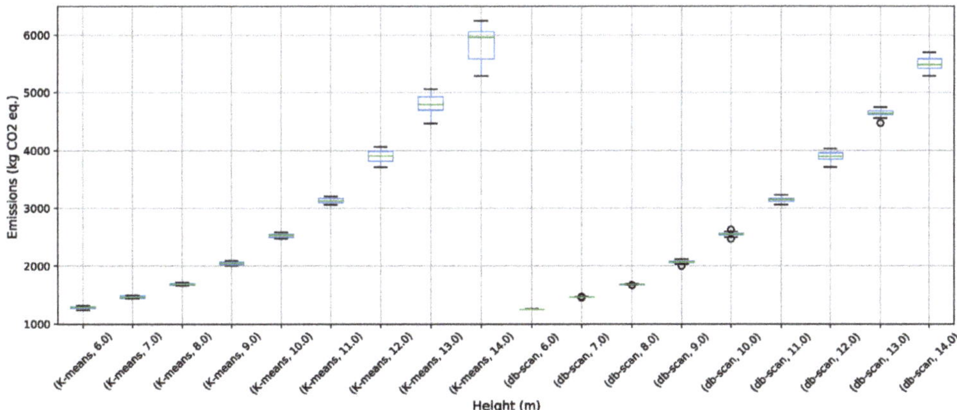

Figure 11. Emission box-plots, comparison between db-scan and k-means operators.

5.4. db-Scan PSO and HS Comparison

The third experiment aims to compare the hybrid db-scan PSO algorithm with results published in the literature. We particularly consider the results published in [18,58]. In these articles, a variant of the HS algorithm for the optimization of buttress retaining walls was developed. To carry out the evaluation, the same procedure as the previous experiments will be followed, i.e., through the best value and average indicators supplemented with boxplots and the Wilcoxon test for the statistical significance of the results.

The comparison of both algorithms is documented in Table 7 and in Figures 12 and 13. When analyzing the best value indicator, the db-scan PSO algorithm is superior in 8 of the 9 instances to HS. In optimizing emissions of CO_2, in 9 of the 9 instances, db-scan performs better. When evaluating the average indicator, the situation is similar to reported by the best value indicator. In 8 of 9 heights of the wall, db-scan PSO has better performance for cost optimization and in the 9 heights, it is superior for emissions. Wilcoxon's test indicates that in both cases the difference is significant. When analyzing the Figures 12 and 13 we observe that from height 12 onwards, the dispersion and the inter-quartile range of HS solutions grow significantly concerning db-scan PSO.

Table 7. Comparison between db-scan PSO and HS algorithms in cost and emission optimization.

Height (m)	Best Cost Value db-Scan	Avg Cost db-Scan	Best Emissions Value db-Scan	Avg Emissions db-Scan	Best Cost Value HS	Avg Cost HS	Best Emissions Value HS	Avg Emissions HS
6	595	598.38	1249	1254.24	595	600.15	1250	1289.14
7	680	691.53	1452	1460.05	689	694.98	1478	1511.53
8	780	793.57	1662	1679.95	784	788.38	1699	1731.54
9	920	937.40	1997	2069.05	934	941.29	2050	2097.45
10	1095	1126.22	2462	2546.37	1130	1143.64	2560	2617.81
11	1302	1363.15	3061	3148.37	1354	1381.50	3124	3201.45
12	1528	1581.43	3715	3900.52	1590	1707.24	3865	4046.95
13	1780	1864.57	4470	4643.26	1840	2067.37	4650	4955.95
14	2049	2245.28	5294	5502.76	2154	2348.71	5550	6241.00
Wilcoxon p-value						4.17×10^{-7}		2.24×10^{-8}

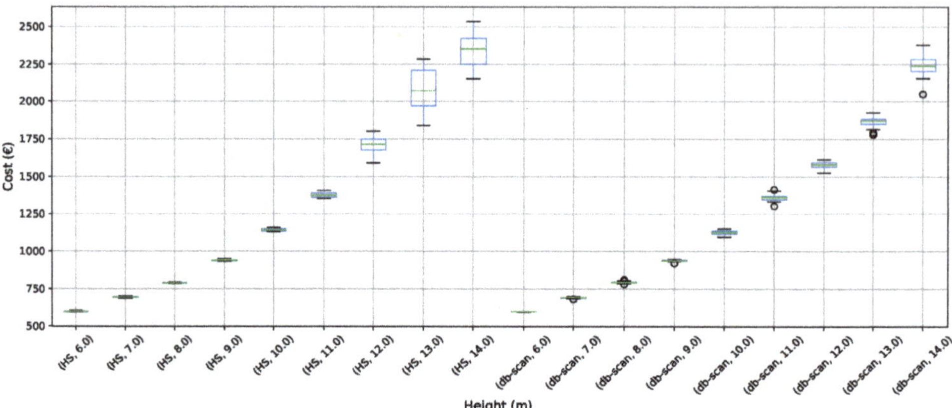

Figure 12. Cost box-plots, comparison between db-scan PSO and HS algorithms.

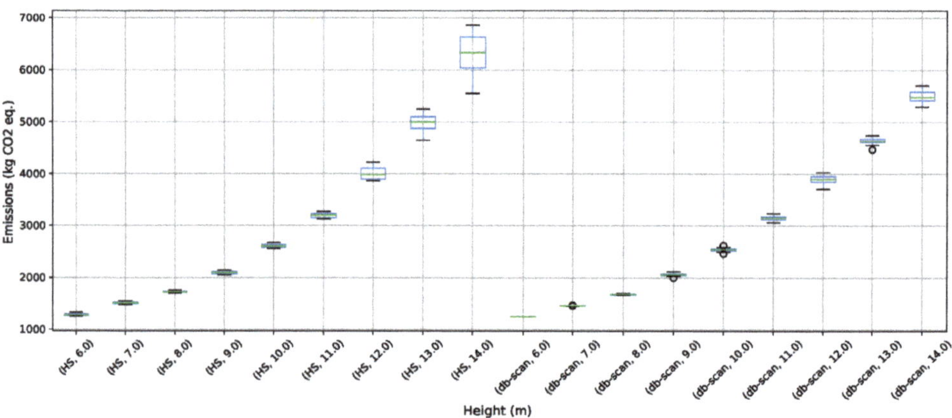

Figure 13. Emission box-plots, comparison between db-scan PSO and HS algorithms.

6. Conclusions

To address the buttressed walls problem, a hybrid algorithm based on the db-scan clustering technique and the PSO optimization algorithm was proposed in this article. This hybridization was necessary because PSO works naturally in continuous spaces and the problem studied is combinatorial. A Db-scan was used as a discretization mechanism. The optimization functions considered were cost and emission of CO_2. To measure the robustness of our proposal, three experiments were designed. The first evaluates the performance of the hybrid algorithm with respect to a random operator. Later in the second experiment, the performance was compared with respect to the k-means clustering technique. Finally, in the last experiment, we studied the performance of the hybrid algorithm when compared to an HS adaptation described in the literature. The first experiment concludes that the db-scan operator contributes to the quality of the solutions, obtaining better values than the random operator, in addition to reducing the dispersion of these. In comparison with k-means mixed results are obtained, in some cases, k-means is superior to db-scan and in other db-scan improves the solutions obtained by k-means. Regarding the dispersion in the different instances, we observed that from height 12 onwards db-scan obtained much smaller dispersions than k-means. Lastly, in comparison with HS, in general, db-scan surpass HS obtaining better results, especially in the instances where the height was over 12 m.

From the results obtained in this study, several lines of research emerge. The first line is related to population management. In the present work, the population was a static parameter. By analyzing the results generated by the algorithm as it iterates, it is possible to identify regions where search needs to be intensified or regions where further exploration is needed. This would allow for dynamic population management. Another interesting research line is related to the parameters used by the algorithm. According to what is detailed in Section 5.1, a robust method was used to get the most suitable configuration. However, this configuration is a static one and is not necessarily the best configuration throughout all execution. Proposing a framework that allows adapting the parameters based on the results obtained by the algorithm as it is executed, would allow generating even more robust methods than the current one.

Author Contributions: Conceptualization, V.Y., J.V.M. and J.G.; methodology, V.Y., J.V.M. and J.G.; software, J.V.M. and J.G.; validation, V.Y., J.V.M. and J.G.; formal analysis, J.G.; investigation, J.V.M. and J.G.; data curation, J.V.M.; writing—original draft preparation, J.G.; writing—review and editing, V.Y., J.V.M. and J.G.; funding acquisition, V.Y. and J.G. All authors have read and agreed to the published version of the manuscript.

Funding: The first author was supported by the Grant CONICYT/FONDECYT/INICIACION/11180056, the other two authors were supported by the Spanish Ministry of Economy and Competitiveness, along with FEDER funding (Project: BIA2017-85098-R).

Conflicts of Interest: The authors declare no conflict of interest.

References

1. Carbonell, A.; González-Vidosa, F.; Yepes, V. Design of reinforced concrete road vaults by heuristic optimization. *Adv. Eng. Softw.* **2011**, *42*, 151–159.
2. Yepes, V.; Alcala, J.; Perea, C.; González-Vidosa, F. A parametric study of optimum earth-retaining walls by simulated annealing. *Eng. Struct.* **2008**, *30*, 821–830. [CrossRef]
3. García, J.; Lalla-Ruiz, E.; Voß, S.; Droguett, E.L. Enhancing a machine learning binarization framework by perturbation operators: Analysis on the multidimensional knapsack problem. *Int. J. Mach. Learn. Cybern.* **2020**. [CrossRef]
4. García, J.; Moraga, P.; Valenzuela, M.; Pinto, H. A db-Scan Hybrid Algorithm: An Application to the Multidimensional Knapsack Problem. *Mathematics* **2020**, *8*, 507. [CrossRef]
5. García, J.; Crawford, B.; Soto, R.; Astorga, G. A clustering algorithm applied to the binarization of Swarm intelligence continuous metaheuristics. *Swarm Evol. Comput.* **2019**, *44*, 646–664.
6. García, J.; Moraga, P.; Valenzuela, M.; Crawford, B.; Soto, R.; Pinto, H.; Pe na, A.; Altimiras, F.; Astorga, G. A Db-Scan Binarization Algorithm Applied to Matrix Covering Problems. *Comput. Intell. Neurosci.* **2019**, *2019*. [CrossRef]
7. Saeheaw, T.; Charoenchai, N. A comparative study among different parallel hybrid artificial intelligent approaches to solve the capacitated vehicle routing problem. *Int. J. Bio-Inspir. Comput.* **2018**, *11*, 171–191. [CrossRef]
8. Crawford, B.; Soto, R.; Astorga, G.; García, J. Constructive metaheuristics for the set covering problem. In *International Conference on Bioinspired Methods and Their Applications*; Springer: Berlin, Germany, 2018; pp. 88–99.
9. García, J.; Altimiras, F.; Pe na, A.; Astorga, G.; Peredo, O. A binary cuckoo search big data algorithm applied to large-scale crew scheduling problems. *Complexity* **2018**, *2018*. [CrossRef]
10. García, J.; Martí, J.; Yepes, V. A Hybrid k-Means Cuckoo Search Algorithm Applied to the Counterfort Retaining Walls Problem. *Mathematics* **2020**, *8*, 555. [CrossRef]
11. Marti-Vargas, J.R.; Ferri, F.J.; Yepes, V. Prediction of the transfer length of prestressing strands with neural networks. *Comput. Concr.* **2013**, *12*, 187–209. [CrossRef]
12. Penadés-Plà, V.; García-Segura, T.; Yepes, V. Robust Design Optimization for Low-Cost Concrete Box-Girder Bridge. *Mathematics* **2020**, *8*, 398.
13. García-Segura, T.; Yepes, V.; Frangopol, D.M.; Yang, D.Y. Lifetime reliability-based optimization of post-tensioned box-girder bridges. *Eng. Struct.* **2017**, *145*, 381–391. [CrossRef]
14. Sierra, L.A.; Yepes, V.; García-Segura, T.; Pellicer, E. Bayesian network method for decision-making about the social sustainability of infrastructure projects. *J. Clean. Prod.* **2018**, *176*, 521–534.

15. Yepes, V.; Martí, J.V.; García-Segura, T. Cost and CO_2 emission optimization of precast–prestressed concrete U-beam road bridges by a hybrid glowworm swarm algorithm. *Autom. Constr.* **2015**, *49*, 123–134. [CrossRef]
16. Yepes, V.; Gonzalez-Vidosa, F.; Alcala, J.; Villalba, P. CO_2-optimization design of reinforced concrete retaining walls based on a VNS-threshold acceptance strategy. *J. Comput. Civ. Eng.* **2012**, *26*, 378–386. [CrossRef]
17. Yepes, V.; Martí, J.V.; García, J. Black Hole Algorithm for Sustainable Design of Counterfort Retaining Walls. *Sustainability* **2020**, *12*, 2767. [CrossRef]
18. Molina-Moreno, F.; Martí, J.V.; Yepes, V. Carbon embodied optimization for buttressed earth-retaining walls: Implications for low-carbon conceptual designs. *J. Clean. Prod.* **2017**, *164*, 872–884. [CrossRef]
19. Kaveh, A.; Biabani Hamedani, K.; Zaerreza, A. A set theoretical shuffled shepherd optimization algorithm for optimal design of cantilever retaining wall structures. *Eng. Comput.* **2020**. [CrossRef]
20. Mergos, P.E.; Mantoglou, F. Optimum design of reinforced concrete retaining walls with the flower pollination algorithm. *Struct. Multidiscipl. Optim.* **2020**, *61*, 575–585.
21. Pons, J.J.; Penadés-Plà, V.; Yepes, V.; Martí, J.V. Life cycle assessment of earth-retaining walls: An environmental comparison. *J. Clean. Prod.* **2018**, *192*, 411–420. [CrossRef]
22. Zastrow, P.; Molina-Moreno, F.; García-Segura, T.; Martí, J.V.; Yepes, V. Life cycle assessment of cost-optimized buttress earth-retaining walls: A parametric study. *J. Clean. Prod.* **2017**, *140*, 1037–1048. [CrossRef]
23. Caserta, M.; Voß, S. Matheuristics: Hybridizing Metaheuristics and Mathematical Programming. In *Metaheuristics: Intelligent Problem Solving*; Springer: Berlin, Germany, 2009; pp. 1–38.
24. Talbi, E.G. Combining metaheuristics with mathematical programming, constraint programming and machine learning. *Ann. Oper. Res.* **2016**, *240*, 171–215.
25. Juan, A.A.; Faulin, J.; Grasman, S.E.; Rabe, M.; Figueira, G. A review of simheuristics: Extending metaheuristics to deal with stochastic combinatorial optimization problems. *Oper. Res. Perspect.* **2015**, *2*, 62–72. [CrossRef]
26. García, J.; Crawford, B.; Soto, R.; Castro, C.; Paredes, F. A k-means binarization framework applied to multidimensional knapsack problem. *Appl. Intell.* **2018**, *48*, 357–380. [CrossRef]
27. Crawford, B.; Soto, R.; Astorga, G.; García, J.; Castro, C.; Paredes, F. Putting continuous metaheuristics to work in binary search spaces. *Complexity* **2017**, *2017*, 8404231.
28. Voß, S. Meta-heuristics: The state of the art. In *Workshop on Local Search for Planning and Scheduling*; Springer: Berlin, Germany, 2000; pp. 1–23.
29. Bishop, C.M. *Pattern Recognition and Machine Learning*; Springer: Berlin, Germany, 2006.
30. Calvet, L.; de Armas, J.; Masip, D.; Juan, A.A. Learnheuristics: Hybridizing metaheuristics with machine learning for optimization with dynamic inputs. *Open Math.* **2017**, *15*, 261–280. [CrossRef]
31. Zhou, Y.K.; Zheng, S.Q. Machine learning-based multi-objective optimisation of an aerogel glazing system using NSGA-II-study of modelling and application in the subtropical climate Hong Kong. *J. Clean. Prod.* **2020**, *253*, 119964.
32. Abasi, A.K.; Khader, A.T.; Al-Betar, M.A.; Naim, S.; Makhadmeh, S.N.; Alyasseri, Z.A.A. Link-based multi-verse optimizer for text documents clustering. *Appl. Soft Comput.* **2020**, *87*, 106002.
33. Martin, S.; Ouelhadj, D.; Beullens, P.; Ozcan, E.; Juan, A.A.; Burke, E.K. A multi-agent based cooperative approach to scheduling and routing. *Eur. J. Oper. Res.* **2016**, *254*, 169–178. [CrossRef]
34. García, J.; Crawford, B.; Soto, R.; Astorga, G. A percentile transition ranking algorithm applied to binarization of continuous swarm intelligence metaheuristics. In *International Conference on Soft Computing and Data Mining*; Springer: Johor, Malaysia, 2018; pp. 3–13.
35. Crepinsek, M.; Ravber, M.; Mernik, M.; Kosar, T. Tuning Multi-Objective Evolutionary Algorithms on Different Sized Problem Sets. *Mathematics* **2019**, *7*, 824.
36. Ries, J.; Beullens, P. A semi-automated design of instance-based fuzzy parameter tuning for metaheuristics based on decision tree induction. *J. Oper. Res. Soc.* **2015**, *66*, 782–793. [CrossRef]
37. Poikolainen, I.; Neri, F.; Caraffini, F. Cluster-based population initialization for differential evolution frameworks. *Inf. Sci.* **2015**, *297*, 216–235. [CrossRef]
38. Li, Z.Q.; Zhang, H.L.; Zheng, J.H.; Dong, M.J.; Xie, Y.F.; Tian, Z.J. Heuristic evolutionary approach for weighted circles layout. In *International Symposium on Information and Automation*; Springer: Berlin, Germany, 2010; pp. 324–331.
39. Yalcinoz, T.; Altun, H. Power economic dispatch using a hybrid genetic algorithm. *IEEE Power Eng. Rev.* **2001**, *21*, 59–60.

40. Kaur, H.; Virmani, J.; Kriti.; Thakur, S. A genetic algorithm-based metaheuristic approach to customize a computer-aided classification system for enhanced screen film mammograms. In *U-Healthcare Monitoring Systems*; Advances in Ubiquitous Sensing Applications for Healthcare; Dey, N., Ashour, A.S., Fong, S.J., Borra, S., Eds.; Academic Press: Cambridge, MA, USA, 2019; pp. 217–259. [CrossRef]
41. Santucci, V.; Milani, A.; Caraffini, F. An Optimisation-Driven Prediction Method for Automated Diagnosis and Prognosis. *Mathematics* **2019**, *7*, 1051. [CrossRef]
42. Bacanin, N.; Bezdan, T.; Tuba, E.; Strumberger, I.; Tuba, M. Optimizing Convolutional Neural Network Hyperparameters by Enhanced Swarm Intelligence Metaheuristics. *Algorithms* **2020**,*13*, 67. [CrossRef]
43. Sun, K.; Tian, P.X.; Qi, H.N.; Ma, F.Y.; Yang, G.K. An Improved Normalized Mutual Information Variable Selection Algorithm for Neural Network-Based Soft Sensors. *Sensors* **2019**, *19*, 5368.
44. De Rosa, G.H.; Papa, J.P.; Yang, X.S. Handling dropout probability estimation in convolution neural networks using meta-heuristics. *Soft Comput.* **2018**, *22*, 6147–6156.
45. Chou, J.S.; Thedja, J.P.P. Metaheuristic optimization within machine learning-based classification system for early warnings related to geotechnical problems. *Autom. Constr.* **2016**, *68*, 65–80. [CrossRef]
46. Pham, A.D.; Hoang, N.D.; Nguyen, Q.T. Predicting compressive strength of high-performance concrete using metaheuristic-optimized least squares support vector regression. *J. Comput. Civ. Eng.* **2015**, *30*, 06015002.
47. Göçken, M.; Özçalıcı, M.; Boru, A.; Dosdoğru, A.T. Integrating metaheuristics and artificial neural networks for improved stock price prediction. *Expert Syst. Appl.* **2016**, *44*, 320–331.
48. Chou, J.S.; Nguyen, T.K. Forward Forecast of Stock Price Using Sliding-Window Metaheuristic-Optimized Machine-Learning Regression. *IEEE Trans. Ind. Inform.* **2018**, *14*, 3132–3142. [CrossRef]
49. Li, M.W.; Geng, J.; Hong, W.C.; Zhang, Y. Hybridizing chaotic and quantum mechanisms and fruit fly optimization algorithm with least squares support vector regression model in electric load forecasting. *Energies* **2018**, *11*, 2226. [CrossRef]
50. Yeoh, J.M.; Caraffini, F.; Homapour, E.; Santucci, V.; Milani, A. A Clustering System for Dynamic Data Streams Based on Metaheuristic Optimisation. *Mathematics* **2019**, *7*, 1229. [CrossRef]
51. Mann, P.S.; Singh, S. Energy efficient clustering protocol based on improved metaheuristic in wireless sensor networks. *J. Netw. Comput. Appl.* **2017**, *83*, 40–52. [CrossRef]
52. De Alvarenga Rosa, R.; Machado, A.M.; Ribeiro, G.M.; Mauri, G.R. A mathematical model and a Clustering Search metaheuristic for planning the helicopter transportation of employees to the production platforms of oil and gas. *Comput. Ind. Eng.* **2016**, *101*, 303–312. [CrossRef]
53. Faris, H.; Mirjalili, S.; Aljarah, I. Automatic selection of hidden neurons and weights in neural networks using grey wolf optimizer based on a hybrid encoding scheme. *Int. J. Mach. Learn. Cybern.* **2019**, *10*, 2901–2920. [CrossRef]
54. Tuba, M.; Alihodzic, A.; Bacanin, N. Cuckoo search and bat algorithm applied to training feed-forward neural networks. In *Recent Advances in Swarm Intelligence and Evolutionary Computation*; Springer: Berlin, Germany, 2015; pp. 139–162.
55. Rere, L.; Fanany, M.I.; Arymurthy, A.M. Metaheuristic algorithms for convolution neural network. *Comput. Intell. Neurosci.* **2016**, *2016*, 1537325. [CrossRef]
56. Rashid, T.A.; Hassan, M.K.; Mohammadi, M.; Fraser, K. Improvement of variant adaptable LSTM trained with metaheuristic algorithms for healthcare analysis. In *Advanced Classification Techniques for Healthcare Analysis*; IGI Global: Hershey, PA, USA, 2019; pp. 111–131.
57. Catalonia Institute of Construction Technology. *BEDEC PR/ PCT ITEC Materials Database*; Catalonia Institute of Construction Technology: Barcelona, Spain, 2009.
58. Molina-Moreno, F.; García-Segura, T.; Martí, J.V.; Yepes, V. Optimization of buttressed earth-retaining walls using hybrid harmony search algorithms. *Eng. Struct.* **2017**, *134*, 205–216. [CrossRef]
59. Ministerio de Fomento. *EHE: Code of Structural Concrete*; Ministerio de Fomento: Madrid, Spain, 2008.
60. Ministerio de Fomento. *CTE. DB-SE. Structural Safety: Foundations*; Ministerio de Fomento: Madrid, Spain, 2008. (In Spanish)
61. Huntington, W.C. *Earth Pressures and Retaining Walls*; Literary Licensing, LLC: Whitefish, MT, USA, 1957.
62. Calavera, J. *Muros de Contención y Muros de Sótano*; INTEMAC: Madrid, Spain, 2001. (In Spanish)
63. CEB-FIB. *Model Code. Design Code*; Thomas Telford Services Ltd.: London, UK, 2008.

64. Hays, W.L.; Winkler, R.L. *Statistics: Probability, Inference, and Decision*; Holt, Rinehart, and Winston: New York, NY, USA, 1971.
65. Wilcoxon, F. Individual comparisons by ranking methods. In *Breakthroughs in Statistics*; Springer: Berlin, Germany, 1992; pp. 196–202.

Article

The Double Traveling Salesman Problem with Multiple Stacks and a Choice of Container Types

Lars Magnus Hvattum [1,*], Gregorio Tirado [2,3,*] and Ángel Felipe [4]

1 Faculty of Logistics, Molde University College, 6410 Molde, Norway
2 Department of Financial and Actuarial Economics & Statistics, Complutense University of Madrid, 28223 Madrid, Spain
3 Interdisciplinary Mathematics Institute, Complutense University of Madrid, 28040 Madrid, Spain
4 Department of Statistics and Operational Research, Complutense University of Madrid, 28040 Madrid, Spain; felipe@mat.ucm.es
* Correspondence: hvattum@himolde.no (L.M.H.); gregoriotd@ucm.es (G.T.)

Received: 26 May 2020; Accepted: 11 June 2020; Published: 16 June 2020

Abstract: The double traveling salesman problem with multiple stacks involves the transportation of goods between two regions. In one region, a vehicle carrying a container visits customers, where pallets of goods are loaded into the container. The container is then shipped to a different region, where another vehicle visits another set of customers where the pallets are unloaded. Pallets are loaded in several rows inside the container, where each row follows the last-in-first-out principle. The standard test instances for the double traveling salesman problem with multiple stacks implies the use of a 45-foot pallet wide container to carry EUR-1 pallets. This paper investigates the effect on transportation costs if an open side container could be used when transporting the pallets. Computational experiments show savings in transportation costs of up to 20%. Moreover, by using a container loaded from the side, rather than from the rear, the defining attributes of the double traveling salesman problem seem to be lost.

Keywords: intermodal transportation; vehicle routing; loading; variable neighborhood search

---

1. Introduction

Petersen and Madsen [1] introduced an optimization problem referred to as the double traveling salesman problem with multiple stacks, based on a prospective customer of a company producing computer software systems for transportation companies. The problem is set in two different regions. A vehicle carrying a container must visit customers to make pickups in one region. The container is then transported to a different region, whereupon a different vehicle carries the container while visiting customers to make deliveries. It is not possible to repack the container en route, and an opening in one end of the container provides the only access to its contents.

The items transported are standardized pallets and each container can fit a given number of R rows with a limited number of L pallets each. This information is crucial, due to the inability to repack the container. Each row of pallets forms a stack that must be loaded and unloaded based on a first-in-last-out principle. This provides a set of difficult linking constraints that must be taken into consideration when routing the vehicle in both the pickup region and the delivery region. For a given total capacity of $R * L$, the loading constraints are most severe when R is low and L is high, whereas having a high value of R provides more flexibility.

Since its introduction, the double traveling salesman problem with multiple stacks has received significant attention from researchers, examining both exact and heuristic solution methods, as well as performing studies on computational complexity and polyhedral analysis. However, it seems that the initial assumptions of the underlying problem structure have not been examined. In this work, one of the hidden assumptions of the problem is questioned: the type of container available to carry out the transportation. By replacing the type of container used in the transportation, the defining characteristics of the problem seem to be lost. This suggests that a technical solution of modifying the container technology used would have cancelled out the need for advancing the frontier of operations research.

Petersen and Madsen [1] provided test instances where the container could contain three rows with 11 pallets each. Standardized Euro Pallets (EPAL, or EUR-1) have dimensions of 1200 by 800 millimetres. This provides a good match with the internal dimensions of a 45-foot pallet wide container, which has an internal width of 2400 millimetres and an internal length of about 13,600 millimetres, with the exact dimensions varying between different manufacturers. Additional test instances were presented with three rows of 22 pallets each, based on situations where the height of each pallet is less than half of the height of the container, and where a pallet can be put on top of another pallet. However, the same loading constraints could also arise from the transportation of another standardized pallet, the EUR-6 pallet, which has dimensions of 600 by 800 millimetres.

The basic original instances have $R = 3$ rows of pallets, each row of length $L = 11$, providing an overall capacity of $R * L = 3 * 11 = 33$ pallets. However, there is an alternative configuration given the dimensions of the container. Placing the EUR-1 pallets with their long side towards the short side of the container, only two rows of pallets will fit. However, there is enough capacity for either 16 or 17 pallets in each row, depending on the exact length of the container. Therefore, the longest editions of the 45-foot containers have enough space for $2 * 17 = 34$ pallets in total. When pallets are vertically stackable, the pattern of three rows provides a capacity of $3 * 22 = 66$ pallets, whereas the pattern of two rows provides a maximum capacity of $2 * 34 = 68$. Transporting EUR-6 pallets provides another possibility. Instead of using $R = 3$ rows of length $L = 22$ and a total capacity of $3 * 22 = 66$, it is possible to use $R = 4$ rows of length $L = 17$ for a total capacity of $4 * 17 = 68$.

The above is true for containers loaded from the back, that is, using one of the short sides. Open side versions also exist for many container sizes, where goods can be loaded from one of the long sides of the container. The existence of such capabilities would open up a wide range of options in the context of the double traveling salesman problem with multiple stacks. For any of the previously mentioned loading options, the number of rows R and their length L can be swapped, producing instances with many rows of relatively low length. Table 1 provides examples of values for R and L that may appear, depending on the type of pallet and container used. Figure 1 illustrates two of the combinations.

Table 1. Selected values for the number of stacks R and their capacity L based on different pallets and container technologies. The first three rows correspond to values that are covered by existing instances.

R	L	Configuration of Pallets and Container
3	11	EUR-1 pallets, 45 ft container loaded from the back
3	22	EUR-6 pallets, 45 ft container loaded from the back
3	44	Vertically stacked EUR-6 pallets, 45 ft container loaded from the back
2	17	EUR-1 pallets, 45 ft container loaded from the back
2	34	Vertically stacked EUR-1 pallets, 45 ft container loaded from the back
4	17	EUR-6 pallets, 45 ft container loaded from the back
4	34	Vertically stacked EUR-6 pallets, 45 ft container loaded from the back
11	3	EUR-1 pallets, 45 ft container loaded from the side
11	6	Vertically stacked EUR-1 pallets, 45 ft container loaded from the side
17	2	EUR-1 pallets, 45 ft container loaded from the side
17	4	EUR-6 pallets, 45 ft container loaded from the side
17	8	Vertically stacked EUR-6 pallets, 45 ft container loaded from the side
22	3	EUR-6 pallets, 45 ft container loaded from the side
22	6	Vertically stacked EUR-6 pallets, 45 ft container loaded from the side

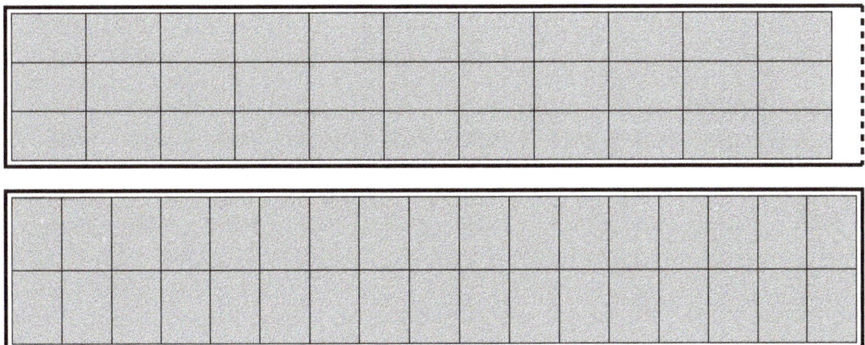

Figure 1. Top: a 45-foot container loaded from the back with three rows of 11 pallets each, resulting in a problem with $R = 3$ and $L = 11$. Bottom: a 45-foot container loaded from the side with two rows of 17 pallets each, resulting in a problem with $R = 17$ and $L = 2$.

The rest of this paper is structured, as follows. Section 2 reviews the related literature, while Section 3 presents the mathematical formulation of the double traveling salesman problem with multiple stacks and the solution methods used in our analysis. Section 4 contains a computational study, with the aim of determining the importance of considering different container types in the considered problem. The paper is concluded with Section 5, where some conclusions and additional thoughts about underlying assumptions of the problem are presented.

2. Literature

The double traveling salesman problem with multiple stacks was first introduced by Petersen and Madsen [1], who presented several heuristic algorithms for the problem, based on iterated local search, tabu search, simulated annealing, and large neighborhood search. Variable neighborhood search for the problem was examined by Felipe et al. [2], with improvements in [3], whereas Urrutia et al. [4] presented a dynamic programming based local search method.

Several mathematical formulations of the double traveling salesman problem with multiple stacks with corresponding solution strategies were presented by Petersen et al. [5]. A specialized algorithm was given by [6] and improved by Lusby and Larsen [6], based on combining separate traveling salesman problems for the pickup region and the delivery region. Branch-and-cut was used by Alba Martínez et al. [7], whereas Carrabs et al. [8] presented a branch-and-bound algorithm for the problem with only two stacks. Two stacks, with infinite capacity each, was also solved by Barbato et al. [9] using a set covering approach, and in [10] using a branch-and-cut algorithm.

Some of the theoretical properties of the problem were investigated by Casazza et al. [11], leading to a simple heuristic method. Given a route for the pickup region and a route for the delivery region, Toulouse and Calvo [12] showed that it can be decided in polynomial time whether or not a feasible stacking exists. Furthermore, Toulouse and Calvo [12] also showed that, given a stacking, optimal routes conditional on the stacking can be determined in polynomial time. Bonomo et al. [13] discussed similar complexity results.

After the introduction of the double traveling salesman problem with multiple stacks, many variants have also been considered in the literature: Iori and Riera-Ledesma [14] presented a generalization with multiple vehicles, and gave three mathematical formulations with corresponding exact solution methods. Heuristics based on iterated local search, simulated annealing, and variable neighborhood descent were proposed for this problem by da Silveira et al. [15]. Another simulated annealing implementation was provided by Chagas et al. [16] and a variable neighborhood search by Chagas et al. [17].

Another variation arises when considering a pickup and delivery traveling salesman problem with multiple stacks, where the nodes to visit are not necessarily split into two separate regions. A large neighborhood search was proposed for this generalization by Côté et al. [18], whereas both Pereira and Urrutia [19] and Sampaio and Urrutia [20] presented branch-and-cut algorithms.

The pickup and delivery problem with time windows and multiple stacks was investigated by [21]. Other types of loading constraints have been considered in the literature as well. Doerner et al. [22] considered the transportation of wood products, while using both a tabu search and an ant colony optimization method. Gendreau et al. [23] used tabu search and Iori et al. [24] a branch-and-cut algorithm for vehicle routing problems with two-dimensional loading constraints. Fuellerer et al. [25] used ant colony optimization for a vehicle routing problem with three-dimensional loading constraints, and Chagas et al. [26] considered partial last-in-first-out loading constraints. Iori [27] presented a survey on combined routing and loading problems and is recommended for further details about related problems.

Even though a wide variety of problems with loading and capacity constraints leading to stacks of items have been approached in the literature through different methodologies, as far as the authors are aware, none of them performed an analysis of the practical consequences that the choice of container types and packing patterns used for transportation may have on the final costs. This is the research gap that this paper tries to fill, focusing on the case of the original double traveling salesman problem with multiple stacks.

3. Background

In this section, we first provide a mathematical model to formally define the studied problem. Next, we describe the solution methods used in the analysis.

3.1. Mathematical Model

The double traveling salesman problem with multiple stacks was modelled for the first time by Petersen and Madsen [1]. Let $G^1 = (V^1, A^1)$, $G^2 = (V^2, A^2)$ be two complete graphs representing, respectively, the pickup and delivery networks of the problem. For every $\omega \in \{1,2\}$, edge $(i,j) \in A^\omega$ of G^ω has a certain weight c_{ij}^ω, representing the travel cost between nodes i and j. Let n be the number of orders and the node sets $V^\omega = \{v_0^\omega, v_1^\omega, \cdots, v_n^\omega\}$, $\omega \in \{1,2\}$, where v_0^ω is the depot and $v_1^\omega, \ldots, v_n^\omega$ represent the n orders. The configuration of the container used for transportation is given by the number of available stacks, being denoted by R, and their maximum capacity, denoted by L.

In addition, we have set $V_*^\omega = V^\omega \backslash \{v_0^\omega\}$, containing all nodes, but the depot in each graph $\omega \in \{1,2\}$, and set $P = \{1, \cdots, R\}$, representing the R available stacks. The set of orders is $D = \{1, \cdots, n\}$, in a way that the item associated to each order $i \in D$ must be picked up at $v_i^1 \in V_*^1$ of G^1, loaded into a certain stack $p \in P$, and delivered at $v_i^2 \in V_*^2$ of G^2.

The problem is modelled as a binary program through three sets of binary variables. The routes of the solution are given by variables $\{x_{ij}^\omega\}$, which determine directly the values of variables $\{y_{ij}^\omega\}$, that indicate precedence between pickups and deliveries. The assignment of orders to stacks is given by variables $\{z_{ip}\}$. They are defined in what follows, where $\omega \in \{1,2\}$.

$$x_{ij}^\omega = \begin{cases} 1 & \text{if } j \text{ is visited immediately after } i \text{ in network } \omega \\ 0 & \text{otherwise} \end{cases} \qquad \forall i,j \in V^\omega$$

$$y_{ij}^\omega = \begin{cases} 1 & \text{if } j \text{ is visited after } i \text{ in network } \omega \\ 0 & \text{otherwise} \end{cases} \qquad \forall i,j \in V_*^\omega$$

$$z_{ip} = \begin{cases} 1 & \text{if order } i \text{ is assigned to stack } p \\ 0 & \text{otherwise} \end{cases} \qquad \forall i \in D, \forall p \in P$$

The model is thus given by Equations (1)–(10).

$$\min f = \sum_{\substack{i,j \in V^{\omega} \\ \omega \in \{1,2\}}} c_{ij}^{\omega} \cdot x_{ij}^{\omega} \tag{1}$$

$$\sum_{i \in V^{\omega}} x_{ij}^{\omega} = 1 \qquad \forall j \in V^{\omega}, \forall \omega \tag{2}$$

$$\sum_{j \in V^{\omega}} x_{ij}^{\omega} = 1 \qquad \forall i \in V^{\omega}, \forall \omega \tag{3}$$

$$y_{ij}^{\omega} + y_{ji}^{\omega} = 1 \qquad \forall i,j \in V_*^{\omega}, i \neq j, \forall \omega \tag{4}$$

$$y_{ik}^{\omega} + y_{kj}^{\omega} \leq y_{ij}^{\omega} + 1 \qquad \forall i,j,k \in V_*^{\omega}, \forall \omega \tag{5}$$

$$x_{ij}^{\omega} \leq y_{ij}^{\omega} \qquad \forall i,j \in V_*^{\omega}, \forall \omega \tag{6}$$

$$y_{ij}^{1} + z_{ip} + z_{jp} \leq 3 - y_{ij}^{2} \qquad \forall i,j \in V_*^{\omega}, \forall p \in P \tag{7}$$

$$\sum_{p \in P} z_{ip} = 1 \qquad \forall i \in D \tag{8}$$

$$\sum_{i \in D} z_{ip} \leq L \qquad \forall p \in P \tag{9}$$

$$x, y, z \in \{0,1\} \tag{10}$$

The objective function (1), to be minimized, is the sum of all pickup and delivery costs. The flow conservation constraints are given in (2) and (3), while the right definition of $\{y_{ij}^{\omega}\}$ variables is ensured by constraints (4)–(6) and the LIFO order to be followed in each stack is imposed by Equation (7). Finally, constraints (8) indicate that each order must be assigned to one, and only one, stack and constraints (9) make sure that the maximum capacity of the stacks is not exceeded.

3.2. Solution Method

Metaheuristics are problem-independent algorithmic frameworks that describe strategies for developing powerful heuristic optimization methods [28]. Therefore, they can be applied to a wide range of problems, such as optimizing non-linear functions of continuous variables [29] or linear functions on binary variables [30]. For the double traveling salesman problem with multiple stacks, to analyze the choice of container types and packing patters, this paper uses a variable neighborhood search [3].

Variable neighborhood search was introduced by Mladenović and Hansen [31]. It is based on the performance of several consecutive local search procedures by changing the neighborhood structure used to define neighbors every time that the search gets stuck in a local optimum. This is a simple idea that has produced very good results in a wide variety of hard optimization problems. The algorithm we use to solve the double traveling salesman problem with multiple stacks is based on an enhanced variable neighborhood search with some additional elements that are specifically adapted to the problem at hand [3]. Six different neighborhood structures or operators, each of which define a local search procedure, are used: Route Swap (swaps the positions of two consecutive orders in one of the routes of the solution), Complete Swap (swaps the stack positions of two orders that are assigned to different stacks), In-Stack Swap (swaps the stack positions of two orders that are assigned to the same stack), Reinsertion (moves one order to a different position in both routes of the solution and reassigns it to a different stack), r-Route Permutation (r orders that are assigned to different stacks and visited consecutively in one route are permuted), and r-Stack Permutation (r orders that are loaded consecutively into the same stack are permuted). The core of the solution method is a variable neighborhood descent (VND) algorithm in which several local search procedures using these neighborhood structures are concatenated. The VND takes as input an initial feasible solution S and the set $\Delta = \{\Delta_k,\ k = 1, \cdots, n_\Delta\}$ of neighborhood structures to be used for the local search. A pseudocode of the VND algorithm is given in what follows.

1. *Initialization:* Do $k = 1$.
2. *Search start:* Do $\hat{S} = S$, $improve = false$.
3. *Local Search:* Find the best solution $\bar{S} \in \Delta_k(\hat{S})$ belonging to the kth neighborhood of $\hat{S}$.
4. If $f(\bar{S}) < f(\hat{S})$, do $\hat{S} = \bar{S}$, $improve = true$ and go back to step 3.
5. *Change of neighborhood structure:*
   - If $improve$ do $k = 1$.
   - Otherwise do $k = k + 1$.
6. *Stopping condition:*
   - If $k \leq n_\Delta$ go back to step 2.
   - If $k > n_\Delta$ and $improve$, do $k = 1$ and go back to step 2.
     Otherwise, END: the best solution found is S.

VND outputs an improved solution which is a local optimum with respect to all neighborhood structures in Δ. Every time a VND execution finishes, a perturbation phase consisting in the performance of a certain number of random moves is applied, in order to escape from the current local optimum, and the VND is applied again to the perturbed solution. In addition to this, several enhancing features are introduced into this basic algorithm in order to improve its performance: more than one initial solution are generated and a different search is performed from each of them; an additional intensification phase is carried out, starting from the best improved initial solution; different orderings of the neighborhood structures used in the VND are considered, choosing randomly among them according to certain probabilities; if the current solution could not be improved after many iterations or it is too far from the best known solution, the search process is restarted; to avoid undoing perturbation moves, tabu lists for the different involved operators are used; the temporal relaxation of some precedence and capacity constraints, controlled through several infeasibility measures and correction procedures to ensure feasibility, is considered for diversification purposes.

The algorithm is fed initial solutions generated in two ways. The first one consists in solving the particular case with only one stack, whose solutions are always feasible for any loading plan with any container configuration. The advantage of this problem is that it reduces to a standard traveling salesman problem in a network whose arc weights are the sum of the weights of the two networks of the original problem. The second one is by using a simple randomized procedure, properly guided in order to ensure the feasibility of the obtained solutions. This second procedure is important to generate a wide variety of initial solutions.

In addition, the two traveling salesman problems induced by dropping all loading constraints on both the pickup and the delivery region of any instance are solved to optimality independently in order to obtain a lower bound: no container configuration or loading pattern can possibly improve the sum of the costs of the two optimal traveling salesman problem solutions. This has been done by using the TSP function of the optimizer OPTMODEL NETWORK from SAS software. This function implements a variant of the branch-and-cut algorithm by Applegate et al. [32], which is one of the most efficient exact methods available in the literature. Default parameters regarding the use of cutting planes, heuristics, node, and branching variable selection, identified as AUTOMATIC by SAS, were used.

4. Computational Study

Test instances for the double traveling salesman problem with multiple stacks were introduced by Petersen and Madsen [1], consisting of twenty instances with 33 customers, and twenty instances with 66 customers. These instances are used in four different experiments to evaluate how the choice of containers and packing patterns influences the transportation costs. The experimental design is simply to solve each of the forty basic instances based on different container configurations and compare the results, rather than relying on more complex experimental designs [33]. The results are obtained by applying the heuristic described in Section 3.2.

The heuristic was run on an Intel Core i5-3210M CPU 2.50 GHz with 6 GB RAM, with a running time of 10 min. for the instances with 33 customers, and 30 min. for the instances with 66 customers.

A lower bound for each instance has been calculated by solving separately the two independent traveling salesman problems that arise when dropping the precedence constraints that are associated with the last-in-first-out principle imposed on the loading of the container. These lower bounds, obtained by using the SAS software, represent the minimum costs that must be covered, even in the absence of loading constraints.

The first experiment considers a container loaded from the back with EUR-1 pallets. The standard packing pattern is $R = 3$ and $L = 11$, for a total capacity of 33 pallets. Assume that the company needs the full capacity to serve regular customers. Now, if the company has an extra demand for transportation of one pallet from the depot of the pickup region to the depot of the unloading region, one solution is to use a different packing pattern, such as $R = 2$ and $L = 17$, for a total capacity of 34 pallets.

Table 2 provides results showing how the total transportation cost changes for 20 standard test instances for selected packing patterns. The table shows, for each instance, the lower bound from solving traveling salesman problems of each region separately ("LB"), the transportation cost ("Cost"), and the relative deviation between the lower bound and the cost ("Dev"). For the first experiment, the average deviation from the lower bound increases from 14.1% to 28.4% when switching packing patterns to facilitate one extra pallet.

The second experiment tests the effect of introducing an open side container, instead of a container loaded from the back. Suitable open side containers may require additional customization, as compared to the more standard containers loaded from the back. To balance this, one would expect that the transportation costs can be reduced, as an alternative packing pattern with many short rows is possible, yielding much flexibility in the routing decisions. Table 2 shows the results for the most flexible alternative packing pattern with $R = 17$ and $L = 2$. For each of the 20 test instances, the heuristic is able to find a solution that matches the lower bound, providing a reduction in the transportation costs of 12% when compared to the use of a standard container loaded from the back ($R = 3$, $L = 11$). In the case where an additional pallet is transported between the depots, the saving in transportation cost amounts to 22% when using $R = 17$ and $L = 2$ instead of $R = 2$ and $L = 17$.

In the third experiment, the transportation of EUR-6 pallets is considered. Assuming that 66 pallets are to be transported in a regular 45-foot pallet wide container, the standard packing pattern of the test instances could be used, with $R = 3$ and $L = 22$. However, for EUR-6 pallets, an alternative pattern is available, with $R = 4$ and $L = 17$, while using the same type of container. This is expected to lead to lower transportation costs, as the added row provides more flexibility. Table 3 confirms this, as the average deviations to the lower bound are 29.7% and 18.7% for $R = 3$ and $R = 4$, respectively.

The fourth experiment deals with vertically stacked EUR-1 pallets and a demand of 66 pallets to be transported. For a standard container, the standard packing pattern has $R = 3$ stacks with $L = 22$ pallets, where pallets in each row are placed in two layers on top of each other. If an open side container is available, however, a packing pattern with $R = 11$ stacks of length $L = 6$ can be used. Table 3 shows that the open side container again leads to significantly reduced transportation costs, saving almost 21% when compared to the container loaded from the back. However, for this setting, there is still a 2.8% gap from the solutions obtained to the lower bound, meaning that it is still important to consider loading decisions when optimizing the routes.

Table 2. Results for instances with 33 customers and three different container configurations, given as $(R * L)$.

Instance	LB	Baseline $(3 * 11)$ Cost	Dev	Experiment 1 $(2 * 17)$ Cost	Dev	Experiment 2 $(17 * 2)$ Cost	Dev
R00	911	1063	16.7%	1237	35.8%	911	0.0%
R01	875	1032	17.9%	1152	31.7%	875	0.0%
R02	935	1065	13.9%	1194	27.7%	935	0.0%
R03	961	1100	14.5%	1270	32.2%	961	0.0%
R04	937	1068	14.0%	1186	26.6%	937	0.0%
R05	900	1008	12.0%	1132	25.8%	900	0.0%
R06	998	1110	11.2%	1256	25.9%	998	0.0%
R07	963	1105	14.7%	1250	29.8%	963	0.0%
R08	978	1123	14.8%	1258	28.6%	978	0.0%
R09	976	1095	12.2%	1211	24.1%	976	0.0%
R10	901	1016	12.8%	1150	27.6%	901	0.0%
R11	892	1001	12.2%	1115	25.0%	892	0.0%
R12	984	1111	12.9%	1241	26.1%	984	0.0%
R13	956	1085	13.5%	1200	25.5%	956	0.0%
R14	879	1039	18.2%	1181	34.4%	879	0.0%
R15	985	1142	15.9%	1287	30.7%	985	0.0%
R16	967	1093	13.0%	1227	26.9%	967	0.0%
R17	946	1073	13.4%	1211	28.0%	946	0.0%
R18	1008	1126	11.7%	1269	25.9%	1008	0.0%
R19	938	1091	16.3%	1211	29.1%	938	0.0%
Average			14.1%		28.4%		0.0%

Table 3. Results for instances with 66 customers and three different container configurations, given as $(R * L)$.

Instance	LB	Baseline $(3 * 22)$ Cost	Dev	Experiment 3 $(4 * 17)$ Cost	Dev	Experiment 4 $(11 * 6)$ Cost	Dev
R00	1237	1653	33.6%	1491	20.5%	1263	2.1%
R01	1257	1641	30.5%	1474	17.3%	1301	3.5%
R02	1295	1666	28.6%	1513	16.8%	1318	1.8%
R03	1290	1640	27.1%	1553	20.4%	1315	1.9%
R04	1295	1626	25.6%	1511	16.7%	1338	3.3%
R05	1204	1545	28.3%	1406	16.8%	1215	0.9%
R06	1294	1702	31.5%	1551	19.9%	1355	4.7%
R07	1307	1669	27.7%	1525	16.7%	1329	1.7%
R08	1297	1649	27.1%	1501	15.7%	1348	3.9%
R09	1276	1617	26.7%	1509	18.3%	1328	4.1%
R10	1339	1734	29.5%	1559	16.4%	1383	3.3%
R11	1268	1624	28.1%	1477	16.5%	1313	3.5%
R12	1295	1671	29.0%	1558	20.3%	1338	3.3%
R13	1275	1635	28.2%	1519	19.1%	1292	1.3%
R14	1245	1655	32.9%	1516	21.8%	1297	4.2%
R15	1228	1623	32.2%	1515	23.4%	1254	2.1%
R16	1356	1758	29.6%	1602	18.1%	1390	2.5%
R17	1274	1711	34.3%	1521	19.4%	1315	3.2%
R18	1328	1761	32.6%	1577	18.8%	1375	3.5%
R19	1256	1651	31.4%	1529	21.7%	1272	1.3%
Average			29.7%		18.7%		2.8%

The experiments show that the choice of container configuration is important. Using a standard container loaded from the rear leads to transportation routes that are much longer and more expensive that what could be achieved with a container loaded from the side. It seems that existing heuristic solution methods can cope very well with the high number of stacks that result from an open side container. One might conjecture that certain exact methods, such as those developed by Lusby et al. [34], may benefit significantly when solving the problem with open side containers.

5. Concluding Remarks

The double traveling salesman problem with multiple stacks, introduced by Petersen and Madsen [1], has received substantial interest from researchers. The problem was based on real-world transportation requirements, which represented a novel challenge in the way that routing decisions were combined with packing decisions. As the problem provided ample venues for research on theoretical aspects of the problem, as well as on the development of efficient heuristic and exact solution methods, it might be valid to ask whether additional aspects of the real-world application may have been overlooked.

5.1. Main Conclusions

The implied choice of container technology in the double traveling salesman problem with multiple stacks was examined in this paper. Additional test instances can be derived, depending on the packing pattern used, and on whether the container is loaded from the back or from the side. It was demonstrated that the total transportation cost is highly dependent on the container choice, in some cases resulting in savings of more than 20% by allowing for the use of open side containers.

5.2. Limitations

The current study might not have exhausted all relevant variations of container technologies. As further examples, one could consider double door containers, where doors are available on both short sides of the container. This means that the container can be loaded while using one of the doors in the pickup-region and unloaded using either of the doors in the delivery-region, implying that there is a choice between last-in-first-out loading and first-in-first-out loading. Yet, other variants can be derived using open top containers, where goods are loaded and unloaded from the top side of the container, given that suitable equipment for loading and unloading is available at each customer location.

Different loading decisions may require the consideration of load stability, depending on the weight of the goods transported. Consider as an example Figure 2: if each pallet has a significant weight, a load as indicated might lead to dangerous situations on the road, as the content of the container is much heavier on one side. The issue of stability might be more critical when using open side containers, but the increased flexibility might nevertheless be economically attractive, at the expense of dealing with an optimization problem that has additional constraints to ensure stability at each leg of the routes.

Figure 2. Illustration of stacking with $R = 3$ and $L = 11$ that may violate stability restrictions if heavy goods is transported.

Another issue not tackled in current research is when some customers have more than one pallet to be transported. The customer might require that only one visit is made, but might also require that different pallets are delivered to different locations.

Author Contributions: Conceptualization, L.M.H. and G.T.; methodology, G.T. and Á.F.; software, G.T. and Á.F.; validation, G.T.; formal analysis, G.T. and L.M.H.; investigation, L.M.H., G.T. and Á.F.; resources, G.T. and Á.F.; data curation, G.T.; writing—original draft preparation, L.M.H.; writing—review and editing, G.T. and Á.F.; visualization, L.M.H.; supervision, L.M.H.; project administration, L.M.H. and G.T.; funding acquisition, L.M.H. All authors have read and agreed to the published version of the manuscript.

Funding: The first author was supported by the AXIOM project, partially funded by the Research Council of Norway. The second author was supported by the Government of Spain, grant MTM2015-65803-R, and the local Government of Madrid, grant S2013/ICE-2845 (CASI-CAM-CM).

Acknowledgments: The authors wish to thank the three anonymous reviewers that contributed with useful inputs to help improve the manuscript.

Conflicts of Interest: The authors declare no conflict of interest.

References

1. Petersen, H.; Madsen, O. The double travelling salesman problem with multiple stacks—Formulation and heuristic solution approaches. *Eur. J. Oper. Res.* **2009**, *198*, 139–147. [CrossRef]
2. Felipe, Á.; Ortuño, M.; Tirado, G. The double traveling salesman problem with multiple stacks: A variable neighborhood search approach. *Comput. Oper. Res.* **2009**, *36*, 2983–2993. [CrossRef]
3. Felipe, Á.; Ortuño, M.; Tirado, G. Using intermediate infeasible solutions to approach vehicle routing problems with precedence and loading constraints. *Eur. J. Oper. Res.* **2011**, *211*, 66–75. [CrossRef]
4. Urrutia, S.; Milanés, A.; Løkketangen, A. A dynamic programming based local search approach for the double traveling salesman problem with multiple stacks. *Int. Trans. Oper. Res.* **2015**, *22*, 61–75. [CrossRef]
5. Petersen, H.; Archetti, C.; Speranza, M. Exact solutions to the double travelling salesman problem with multiple stacks. *Networks* **2010**, *56*, 229–243. [CrossRef]
6. Lusby, R.; Larsen, J. Improved exact method for the double TSP with multiple stacks. *Networks* **2011**, *58*, 290–300. [CrossRef]
7. Alba Martínez, M.; Cordeau, J.F.; Dell'Amico, M.; Iori, M. A branch-and-cut algorithm for the double traveling salesman problem with multiple stacks. *INFORMS J. Comput.* **2013**, *25*, 41–55. [CrossRef]
8. Carrabs, F.; Cerulli, R.; Speranza, M.G. A branch-and-bound algorithm for the double TSP with two stacks. *Networks* **2013**, *61*, 58–75. [CrossRef]
9. Barbato, M.; Grappe, R.; Lacroix, M.; Calvo, R.W. A set covering approach for the double traveling salesman problem with multiple stacks. In *International Symposium on Combinatorial Optimization*; Springer: Berlin/Heidelberg, Germany, 2016; pp. 260–272.
10. Barbato, M.; Grappe, R.; Lacroix, M.; Calvo, R. Polyhedral results and a branch-and-cut algorithm for the double traveling Salesman problem with multiple stacks. *Discret. Optim.* **2016**, *21*, 25–41. [CrossRef]
11. Casazza, M.; Ceselli, A.; Nunkesser, M. Efficient algorithms for the double traveling salesman problem with multiple stacks. *Comput. Oper. Res.* **2012**, *39*, 1044–1053. [CrossRef]
12. Toulouse, S.; Calvo, R. On the complexity of the multiple stack TSP, kSTSP. In *International Conference on Theory and Applications of Models of Computation*; Springer: Berlin/Heidelberg, Germany, 2009; pp. 360–369.
13. Bonomo, F.; Mattia, S.; Oriolo, G. Bounded coloring of co-comparability graphs and the pickup and delivery tour combination problem. *Theor. Comput. Sci.* **2011**, *412*, 6261–6268. [CrossRef]
14. Iori, M.; Riera-Ledesma, J. Exact algorithms for the double vehicle routing problem with multiple stacks. *Comput. Oper. Res.* **2015**, *63*, 83–101. [CrossRef]
15. da Silveira, U.E.F.; Benedito, M.P.L.; dos Santos, A.G. Heuristic approaches to double vehicle routing problem with multiple stacks. In Proceedings of the 2015 15th International Conference on Intelligent Systems Design and Applications (ISDA), Marrakesh, Morocco, 14–16 December 2015; pp. 231–236.
16. Chagas, J.B.; Silveira, U.E.; Benedito, M.P.; Santos, A.G. Simulated annealing metaheuristic for the double vehicle routing problem with multiple stacks. In Proceedings of the 2016 IEEE 19th International Conference on Intelligent Transportation Systems (ITSC), Rio de Janeiro, Brazil, 1–4 November 2016; pp. 1311–1316.
17. Chagas, J.; Silveira, U.; Santos, A.G.; Souza, M. A variable neighborhood search heuristic algorithm for the double vehicle routing problem with multiple stacks. *Int. Trans. Oper. Res.* **2020**, *27*, 112–137. [CrossRef]
18. Côté, J.F.; Gendreau, M.; Potvin, J.Y. Large neighborhood search for the pickup and delivery traveling salesman problem with multiple stacks. *Networks* **2012**, *60*, 19–30. [CrossRef]

19. Pereira, A.H.; Urrutia, S. Formulations and algorithms for the pickup and delivery traveling salesman problem with multiple stacks. *Comput. Oper. Res.* **2018**, *93*, 1–14. [CrossRef]
20. Sampaio, A.H.; Urrutia, S. New formulation and branch-and-cut algorithm for the pickup and delivery traveling salesman problem with multiple stacks. *Int. Trans. Oper. Res.* **2017**, *24*, 77–98. [CrossRef]
21. Cherkesly, M.; Desaulniers, G.; Irnich, S.; Laporte, G. Branch-price-and-cut algorithms for the pickup and delivery problem with time windows and multiple stacks. *Eur. J. Oper. Res.* **2016**, *250*, 782–793. [CrossRef]
22. Doerner, K.F.; Fuellerer, G.; Hartl, R.F.; Gronalt, M.; Iori, M. Metaheuristics for the vehicle routing problem with loading constraints. *Netw. Int. J.* **2007**, *49*, 294–307. [CrossRef]
23. Gendreau, M.; Iori, M.; Laporte, G.; Martello, S. A Tabu search heuristic for the vehicle routing problem with two-dimensional loading constraints. *Netw. Int. J.* **2008**, *51*, 4–18. [CrossRef]
24. Iori, M.; Salazar-González, J.J.; Vigo, D. An exact approach for the vehicle routing problem with two-dimensional loading constraints. *Transp. Sci.* **2007**, *41*, 253–264. [CrossRef]
25. Fuellerer, G.; Doerner, K.F.; Hartl, R.F.; Iori, M. Metaheuristics for vehicle routing problems with three-dimensional loading constraints. *Eur. J. Oper. Res.* **2010**, *201*, 751–759. [CrossRef]
26. Chagas, J.B.; Toffolo, T.A.; Souza, M.J.; Iori, M. The double traveling salesman problem with partial last-in-first-out loading constraints. *arXiv* **2019**, arXiv:1908.08494.
27. Iori, M. An annotated bibliography of combined routing and loading problems. *Yugosl. J. Oper. Res.* **2016**, *23*, 782–793. [CrossRef]
28. Sörensen, K.; Glover, F. Metaheuristics. In *Encyclopedia of Operations Research and Management Science*, 3rd ed.; Gass, S., Fu, M., Eds.; Springer: Boston, MA, USA, 2013; pp. 960–970.
29. Jäntschi, L.; Bolboacă, S.; Bălan, M.; Sestras, R.; Diudea, M. Results of Evolution Supervised by Genetic Algorithms. *Not. Sci. Biol.* **2010**, *2*, 12–15. [CrossRef]
30. Hvattum, L.; Løkketangen, A.; Glover, F. Adaptive Memory Search for Boolean Optimization Problems. *Discret. Appl. Math.* **2004**, *142*, 99–109. [CrossRef]
31. Mladenović, N.; Hansen, P. Variable neighborhood search. *Comput. Oper. Res.* **1997**, *24*, 1097–1100. [CrossRef]
32. Applegate, D.L.; Bixby, R.E.; Chvatal, V.; Cook, W.J. *The Traveling Salesman Problem: A Computational Study*; Princeton University Press: Princeton, NJ, USA, 2006.
33. Bolboacă, S.; Jäntschi, L. Design of Experiments: Useful Orthogonal Arrays for Number of Experiments from 4 to 16. *Entropy* **2007**, *9*, 198–232. [CrossRef]
34. Lusby, R.; Larsen, J.; Ehrgott, M.; Ryan, D. An exact method for the double TSP with multiple stacks. *Int. Trans. Oper. Res.* **2010**, *17*, 637–652. [CrossRef]

Article

Supported Evacuation for Disaster Relief through Lexicographic Goal Programming

Inmaculada Flores [1,2,*], M. Teresa Ortuño [1,2] and Gregorio Tirado [2,3] and Begoña Vitoriano [1,2]

1 Faculty of Mathematical Sciences, Complutense University of Madrid, 28040 Madrid, Spain; mteresa@ucm.es (M.T.O.); bvitoriano@mat.ucm.es (B.V.)
2 Interdisciplinary Mathematics Institute, Complutense University of Madrid, 28040 Madrid, Spain; gregoriotd@ucm.es
3 Faculty of Economics and Business, Complutense University of Madrid, 28223 Madrid, Spain
* Correspondence: inmacufl@ucm.es

Received: 31 March 2020; Accepted: 19 April 2020; Published: 22 April 2020

Abstract: Disasters have been striking human-beings from the beginning of history and their management is a global concern of the international community. Minimizing the impact and consequences of these disasters, both natural and human-made, involves many decision and logistic processes that should be optimized. A crucial logistic problem is the evacuation of the affected population, and the focus of this paper is the planning of supported evacuation of vulnerable people to safe places when necessary. A lexicographic goal programming model for supported evacuation is proposed, whose main novelties are the classification of potential evacuees according to their health condition, so that they can be treated accordingly; the introduction of dynamism regarding the arrival of potential evacuees to the pickup points, according to their own susceptibility about the disaster and the joint consideration of objectives such us number of evacuated people, operation time and cost, among which no trade-off is possible. The performance of the proposed model is evaluated through a realistic case study regarding the earthquake and tsunami that hit Palu (Indonesia) in September 2018.

Keywords: humanitarian logistics; evacuation; multi-criteria decision making; goal programming; disaster relief

---

1. Introduction and Literature Review

Disaster management, understood as the planning, organization and management of all that is needed to deal with the humanitarian aspects of emergencies, disasters or catastrophes, in order to lessen their impact on population and infrastructures, has always been in the focus of the international community. In recent decades, this global concern has given birth to a growing literature on disaster management and in humanitarian logistics, defined in the Humanitarian Logistics Conference, 2004, as the process of planning, implementing and controlling the efficient, cost-effective flow and storage of goods and materials as well as related information, from the point of origin to the point of consumption for the purpose of meeting the end requirements of beneficiaries and alleviate the suffering of vulnerable people. See the survey of Özdamar and Ertem [1], and the books of Tomasini and Van Wassenhove [2] and Vitoriano et al. [3], among others, for optimization problems addressed within the Operational Research community regarding disaster management and humanitarian logistics.

The disaster management cycle comprises four successive phases, see [2]: mitigation, preparedness, response and recovery. Each one of them comprises important logistic operations that must be planned in the most effective and efficient way. Research on these phases may focus on specific types of disasters, such as hurricanes, typhoons and cyclones [4], earthquakes [5], floods [6] or wildfires [7]; or it can address specific problems along the cycle, such as location [6], emergency mitigation [5,8], prepositioning of aid distribution centers [9], transportation and last mile distribution [10–12] or evacuation problems [13,14].

Evacuation is an activity long time considered in humanitarian logistics. According to UNDRR [15], an evacuation consists in moving people and assets temporarily to safer places before, during or after the occurrence of a hazardous event, in order to protect them. Evacuation may be required in buildings, regions or transportation means, such as trains, ships or airplanes. Due to the increasing height of buildings and the expansion of transport travel in the last century, initially, a large collection of works related with buildings, trains, airplanes and ships evacuation arose. Some examples can be found in [13,16–18]. In 2001, a preliminary state of the art review appeared, mainly based on building evacuation, [19]. Some years later, on the other hand, a systematic collection of network flow models applied to regional emergency evacuation and their applications came up, see [14].

Different types of evacuation are stated by the London Resilience Partnership in [20], comprising self-, assisted and supported evacuation. Self-evacuation is understood as individuals moving from an unsafe place to a safer one by their own means, without any kind of assistance. In assisted evacuation the individuals are capable of traveling by themselves but require certain support (for example, information) from agencies. Finally, supported evacuation is needed if individuals require greater and specific support (for example, an ambulance) to be transported from the unsafe areas to safer ones. Self-evacuation and assisted evacuation (both often referred as car-based evacuation) have been vastly studied in the literature, facing problems like traffic congestion, clogging, bottlenecks, density waves, oscillations, patterns and panic, usually solved by nonlinear techniques and queuing theory [18,21–25]. However, and according to Houston et al. [26], sometimes self evacuation is not an option. An example can be found at Hurricane Katrina (New Orleans, 2005), where over 40% of victims did not evacuate, either due to physical disability or because they were caring for a person with one. Support can be needed for different reasons, from people with no access to an adequate vehicle to those that require special transport means (mainly ambulances). In addition, each type of disaster often requires a different evacuation process. For example, in Hamacher and Tjandra [19] two evacuation scenarios are considered, precautionary and life-saving operations. Furthermore, Amideo et al. [27] stablished that hurricanes and wildfires require preventive evacuation while earthquakes and floods demand immediate post-disaster evacuation. Pyakurel et al. [28] established that an evacuation optimizer looks for a plan on an evacuation network for an efficient transfer of the maximum quantity of evacuees from the dangerous points (sources) to safer ones (sinks) as quickly as possible. Other analyses related with evacuation planning in order to save human lives and support humanitarian relief can be found in [29,30].

In terms of regional evacuation, and according to Amideo et al. [27], the related literature prior to 2011 used to be focused on shelter location-allocation and self-evacuation. The first paper on regional supported (or bus-based) evacuation is [31], where this type of evacuation is presented as a variant of the vehicle routing problem. After this paper, few works can be found until 2017, when Shahparvari and Abbasi [32] proposed a stochastic model for a supported evacuation to determine the required vehicles, scheduling and routing under uncertainty, with time windows and bushfire propagation and considering road availability and disruptions. Besides that, Shahparvari et al. [33] developed a capacitated vehicle routing solution to evacuate under short-notice. Finally, the identification of the required number of vehicles and the safest routes and schedules for late evacuees it approached in [34].

In this work, a new model for supported evacuation of vulnerable people after a disaster is presented. The model seeks to evacuate people from pick-up points to shelters with limited capacity, and it is multimodal, considering different types of vehicles; dynamic, because people reach the pick up points at different periods along the time horizon; and multi-objective. It is based on the coordination of network flows of vehicles and people, as stated in [35], where there is a collection of evacuation models based in networks flows with discretized time. We also consider shelters with limited capacity, as in the location-allocation model presented by Sherali et al. [6], that selects a set of candidate shelters according to the available resources and minimizes the total congestion in a car based evacuation.

Capacity constraints in arcs and nodes can also be found in [36]. Regarding the optimization criteria, a wide variety of them have been employed in the humanitarian logistics literature, see [37]. It is important to note that the use of multi-criteria optimization methods has experienced a significant growth in the last years, mainly because they are able to help the decision maker select the best option between several alternatives, evaluating criteria that may generate conflict. An interesting example that takes into consideration multiple criteria such as cost, time, equity, reliability, security or priority can be found in [38]. In particular, for evacuation models, different criteria have been considered in the literature. Shahparvari et al. [7] highlighted the fact that most existing studies have focused on minimizing just the total evacuation time, and they consider two additional objectives: maximize the amount of evacuated people and minimize the usage of resources. Mejia-Argueta et al. [39] proposed to minimize the maximum evacuation and distribution time and the total cost of relief operations, while Alçada-Almeida et al. [40] minimized the total distance required for all the population to reach a shelter, the fire risk faced while traveling, the risks associated with staying in the shelter and the total time from the shelters to the hospital. Our model considers the maximization of the number of evacuees of different types regarding health condition and the minimization of the evacuation time and operation costs.

The main contributions of our research are the following:

- The introduction of dynamism into the supported evacuation model regarding the arrival of potential evacuees to the pick up points, since we consider their arrival may happen at different of points of time during the planning horizon.
- The classification of affected people requiring evacuation according to their particular situation and/or medical condition, leading to several priority levels, so that each group can be treated accordingly.
- The joint consideration of objectives such us number of evacuated people, operation time and cost within a lexicographic approach, in such a way that there is no trade off among them.
- The validation of the proposed model on a realistic case study based on the earthquake and tsunami that hit Indonesia in September 2018.

The structure of the remainder of this paper is as follows. Section 2 presents the detailed description of the supported evacuation problem considered, whose mathematical formulation is proposed in Section 3 through a lexicographic goal programming model. Section 4 introduces the case study used to evaluate the performance of the model and analyzes the computational results provided by its resolution. Finally, Section 5 draws some conclusions from this work.

2. Problem Description

A detailed definition of the problem approached in this paper, focused on regional supported evacuation, will be given in what follows. As stated in the previous section, to mitigate the possible effects of some disasters, agencies may determine certain areas that are necessary to evacuate for security reasons and some other areas that are safe and available to host the evacuated population. It happens frequently that a certain amount of people, for a variety of reasons, cannot evacuate from the compromised areas to safe ones by their own means, requiring additional support from public agencies to be able to evacuate. The way these vulnerable people is evacuated is the main focus of our model.

The inhabitants that cannot self-evacuate by their own means go directly to previously designated pickup points, if they are able to, according to their own feeling regarding the disaster, or are taken there by police or search and rescue teams. According to Özdamar and Ertem [1], many factors influence the feeling of the population regarding the catastrophic situation, including demographic characteristics such as age, health status or gender, and socioeconomic ones, such as homeownership, financial conditions, prior experiences and awareness. In addition, these factors may vary significantly along time. As a result, the arrival of potentially vulnerable evacuees to the designated pick up points is clearly dynamic, and this is a key aspect of the problem.

The analysis of real cases of disasters shows that classifying the population in need of supported evacuation is necessary to prioritize appropriately the attention from the involved agencies. Houston et al. [26] indicate that people with disabilities, medical conditions, homeless or pregnant women should be considered as high priority population and be evacuated with a higher urgency, due to their special vulnerability. Other population groups are usually classified as normal priority evacuees. However, the particular classification considered for the design of the evacuation plans may vary significantly from one case to another, so that it represents the existing situation as faithfully as possible.

Safe areas contain hospitals, for population requiring medical care, and temporary shelters, for the rest of the population, that are associated to the nodes of the network. An additional distinction between different types of shelters, as suggested in [26], into general population shelters and shelters for people with special needs or medical needs shelters, may also be performed. All hospital and shelters have a certain capacity to accommodate evacuees that cannot be exceeded.

Transportation to safe areas during emergencies is critical in order to evacuate people who either have specific mobility issues or do not have access to transportation. For this purpose, different types of vehicles with diverse characteristics will be available. Vehicles may traverse the arcs individually or forming convoys and they can visit one or more locations several times, even though not all of them must be visited mandatorily. Each vehicle type has a certain capacity for the transportation of people that cannot be exceeded. Other characteristics, different for each vehicle type, are the fixed cost by distance unit, the variable cost depending on the amount of people transported, the possibility to traverse certain arcs depending on their conditions and the travel time required to traverse these arcs.

The operational area is a dynamic network composed by a region to be evacuated and a safe region. The streets or roads connecting the nodes are represented by arcs/edges and may be classified, depending on the requisites of the particular case, as paths/unpaved roads, local roads or within towns, highways, freeways, etc., according to their type; and as blocked or shattered, seriously damaged, partially damaged, usable, etc., according to their state and the environmental or traffic conditions. These elements determine whether a certain vehicle can traverse a certain arc or not and its maximum speed when travelling. There may also be a certain flow capacity associated to each arc that represents the maximum number of evacuees or vehicles that can traverse it per time unit.

In short, a solution of the problem must comprise a set of itineraries for the available vehicles and the corresponding flow of people moving from the affected area to the safe one, in such a way that they are compatible with each other and verify all capacity constraints given by the network. The resulting evacuation plans can be evaluated according to different criteria and, as a result, several objectives will be considered. The number of people evacuated successfully, that is to be maximized, seems the most important one. However, the evacuation time, to be minimized, is also crucial. Additionally, the operational cost, even though may not be the main focus of the decision maker, should also be minimized. All these criteria will be considered jointly within a lexicographical goal programming model, as detailed in the next section.

3. The Proposed Evacuation Model

The network of the problem is represented by a graph $\mathcal{G}(\mathcal{N},\mathcal{E})$, where $\mathcal{N}$ is the set of nodes, which may represent areas, cities or locations within them, and $\mathcal{E}$ is the set of arcs (if directed) or edges (if undirected), which correspond to roads or streets that communicate the nodes. Pick up nodes from where the compromised population will be transferred to the safe areas, $\mathcal{NA}$, are located at known places at the unsafe area. Meanwhile, temporary shelters nodes, which usually correspond to schools, universities or government buildings, $\mathcal{NS}$; and hospitals or medical centers, $\mathcal{NH}$, are located at the safe areas. Finally, some transit nodes, $\mathcal{NT}$, may also exist in the network. As stated in the previous section, edges or arcs may be classified according to the requirements of the particular case study or disaster.

Evacuees are classified into different categories depending on the criticality of their health state and the space they require on vehicles or nodes at the secured area, since they may require different transport and assistance conditions. The population is to be transported by fleets of different types of vehicles with different characteristics. All evacuation operations must be carried out within a certain monetary budget and a given time span.

3.1. Parameters of the Model and Decision Variables

Sets and indices

$\mathcal{N}$: Set of nodes or involved areas ($i, j \in \mathcal{N}$: $\mathcal{N} = \mathcal{NA} \cup \mathcal{NS} \cup \mathcal{NH} \cup \mathcal{NT}$).
$\mathcal{E}$: Set of edges or arcs between nodes ($(i,j) = (ij) \in \mathcal{E}$).
$\mathcal{T}$: Discretized time periods ($t, t' \in \{1, ..., \mathcal{T}\}$).
$\mathcal{H}$: Set of types of people to be evacuated ($h \in \mathcal{H}$).
$\mathcal{K}$: Set of types of vehicles available for the evacuation ($k \in \mathcal{K}$).
$\mathcal{M}$: Set of goals to be achieved ($m \in \mathcal{M}$).

Parameters

d_{ij}^k : Distance from node $i \in \mathcal{N}$ to node $j \in \mathcal{N}$ (or length of arc/edge $(i,j) \in \mathcal{E}$) when traversed by a vehicle of type $k \in \mathcal{K}$.
av_{ij}^k : Average velocity of arc/edge $(i,j) \in \mathcal{E}$ for a vehicle of type $k \in \mathcal{K}$.
b : Budget limitation.
qp_{it}^h : Amount of people of type $h \in \mathcal{H}$ to be evacuated from node $i \in \mathcal{NA}$ at the beginning of period $t \leq \mathcal{T}$.
typ^h : Priority of people of type $h \in \mathcal{H}$ (high or normal).
wp^h : Space that a person of type $h \in \mathcal{H}$ occupies in a vehicle to be transported or in a node of the secure area.
cp_i^h : Capacity for people of type $h \in \mathcal{H}$ at node $i \in \mathcal{NS} \cup \mathcal{NH}$.
vpc^k : Capacity of a vehicle of type $k \in \mathcal{K}$.
va_i^k : Number of vehicles of type $k \in \mathcal{K}$ that are available at node $i \in \mathcal{N}$ at the beginning of the operation.
fc_{ij}^k : Fixed cost to traverse the arc/edge $(i,j) \in \mathcal{E}$ with a vehicle of type $k \in \mathcal{K}$, by unit of distance.
vc_{ij}^k : Variable cost to traverse the arc/edge $(i,j) \in \mathcal{E}$ with a vehicle of type $k \in \mathcal{K}$, by unit of distance and cargo transported.
vv^k : Maximum velocity that a vehicle of type $k \in \mathcal{K}$ may reach.
τ_{ij}^k : Time required to travel the arc/edge $(i,j) \in \mathcal{E}$ with a vehicle of type $k \in \mathcal{K}$, calculated as follows: $\tau_{ij}^k = max\{\frac{d_{ij}^k}{vv^k}, \frac{d_{ij}^k}{av_{ij}^k}\}$
tg_m : Aspiration level of goal $m \in \mathcal{M}$.

Decision variables

BT_t : 1 if population with high priority has been evacuated in period t and 0, otherwise.
BT'_t : 1 if population with normal priority has been evacuated in period t and 0, otherwise.
T : Total time required to evacuate population with high priority.
T' : Total time required to evacuate the rest of the population.
P_{it}^{h} : Number of people of type $h \in \mathcal{H}$ located at node $i \in \mathcal{N}$ at the beginning of period $t \leq \mathcal{T}$.
PL_{ijt}^{hk} : Number of people of type $h \in \mathcal{H}$ who started to be transported from node $i \in \mathcal{N}$ to node $j \in \mathcal{N}$ in a vehicle of type $k \in \mathcal{K}$ at the beginning of period $t \leq \mathcal{T}$.
V_{it}^{k} : Number of vehicles of type $k \in \mathcal{K}$ located at node $i \in \mathcal{N}$ at the beginning of period $t \leq \mathcal{T}$.
VL_{ijt}^{k} : Number of vehicles of type $k \in \mathcal{K}$ that started to move from node $i \in \mathcal{N}$ to node $j \in \mathcal{N}$ at the beginning of period $t \leq \mathcal{T}$.

Attribute and deviation variables

E : Number of people with high priority that are evacuated depending on the characteristics of the particular instance.
E' : Number of people with normal priority that are evacuated depending on the characteristics of the particular instance.
TM : Total time required to evacuate all the population (max. of T and T').
$Cost$: Total cost of the operation.
DV_m : Deviation variable with respect to the aspiration level of goal $m \in \mathcal{M}$.
P_m : Slack variable with respect to the aspiration level of goal $m \in \mathcal{M}$.

The four attributes of the model (number of people with high priority evacuated, number of people with normal priority evacuated, total evacuation time and operation cost) are considered jointly within a lexicographical goal programming approach. Each attribute $m \in \mathcal{M}$ is associated with an aspiration level tg_m that represents a certain value that would be satisfying for the decision maker and we seek to achieve. The unwanted deviations from the aspiration levels tg_m are measured by the deviation variables DV_m, in such a way that a zero value in one of them means the associated aspiration level has been achieved. We use a lexicographical approach because some of the considered attributes are incomparably more important than others and thus no trade offs are possible among them. This way, the objective function consists in minimizing lexicographically the four deviation variables following the order of priority of the considered attributes: DV_E (first level of priority, critical population evacuated), $DV_E{}'$ (second level of priority, non critical population evacuated), DV_{TM} (third level of priority, total evacuation time), DV_C (fourth level of priority, operation cost).

3.2. Mathematical Model

The proposed model is given by Equations (1)–(25): the deviation variables of the objective function (1) are to be minimized lexicographically, subject to constraints (2)–(25), that define feasible evacuation plans.

$$\text{LexMin}\left\{[DV_E],[DV_{E'}],[DV_{TM}],[DV_C]\right\} \tag{1}$$

$$\sum_{j|(ji)\in\mathcal{E}}\sum_{k}\sum_{t'\leq t-\tau_{ji}^{k}} PL_{jit'}^{hk} + \sum_{t'\leq t} qp_{it'}^{h} = \sum_{j|(ij)\in\mathcal{E}}\sum_{k}\sum_{t'\leq t} PL_{ijt'}^{hk} + P_{it}^{h}\ \forall h,i,t \tag{2}$$

$$wp^{h}P_{it}^{h} \leq cp_{i}^{h}\ \forall t,h,i\in\mathcal{NH}\cup\mathcal{NS} \tag{3}$$

$$\sum_{h} wp^{h}PL_{ijt}^{hk} \leq vpc^{k}VL_{ijt}^{k}\ \forall i,j\in\mathcal{N}|(ij)\in\mathcal{E}\ \forall k,t \tag{4}$$

$$E = \sum_{h|typ^{h}=\max_{h}\{typ^{h}\},i\in\mathcal{NH}\cup\mathcal{NS}} P_{i\mathcal{T}}^{h} \tag{5}$$

$$E' = \sum_{h|typ^{h}=\min_{h}\{typ^{h}\},i\in\mathcal{NH}\cup\mathcal{NS}} P_{i\mathcal{T}}^{h} \tag{6}$$

$$\sum_{j|(ji)\in\mathcal{E}}\sum_{t'\leq t-\tau_{ji}^{k}} VL_{jit'}^{k} + va_{i}^{k} = \sum_{j|(ij)\in\mathcal{E}}\sum_{t'\leq t} VL_{ijt'}^{k} + V_{it}^{k}\ \forall i,k,t \tag{7}$$

$$\sum_{i} va_{i}^{k} = \sum_{i} V_{i\mathcal{T}}^{k}\ \forall k \tag{8}$$

$$T = \sum_{t} t(BT_{t} - BT_{t-1}) \tag{9}$$

$$BT_{t} \geq BT_{t-1}\ \forall t \tag{10}$$

$$\sum_{h|typ^{h}=\max_{h}\{typ^{h}\}}\ \sum_{i,j|(ij)\in\mathcal{E}}\sum_{k}\sum_{t'|t'=\lceil t-\tau_{ij}^{k}\rceil} PL_{ijt'}^{hk} \leq (1-BT_{t}) \sum_{h|typ^{h}=\max_{h}\{typ^{h}\}}\ \sum_{i,t'} qp_{it'}^{h}\ \forall t \tag{11}$$

$$T' = \sum_{t} t(BT'_{t} - BT'_{t-1}) \tag{12}$$

$$BT'_{t} \geq BT'_{t-1}\ \forall t \tag{13}$$

$$\sum_{h|typ^{h}=\min_{h}\{typ^{h}\}}\ \sum_{i,j|(ij)\in\mathcal{E}}\sum_{k}\sum_{t'|t'=\lceil t-\tau_{ij}^{k}\rceil} PL_{ijt'}^{hk} \leq (1-BT'_{t}) \sum_{h|typ^{h}=\min_{h}\{typ^{h}\}}\ \sum_{i,t'} qp_{it'}^{h}\ \forall t \tag{14}$$

$$TM \geq T \tag{15}$$

$$TM \geq T' \tag{16}$$

$$Cost = \sum_{h,i,j,k,t|(ij)\in\mathcal{E}} d_{ij}^{k}(fc_{ij}^{k}VL_{ijt}^{k} + vc_{ij}^{k}PL_{ijt}^{hk}) \tag{17}$$

$$Cost \leq b \tag{18}$$

$$Cost \geq 0 \tag{19}$$

$$E + DV_{E} - P_{E} = tg_{E} \tag{20}$$

$$E' + DV_{E'} - P_{E'} = tg_{E'} \tag{21}$$

$$TM - DV_{TM} + P_{TM} = tg_{TM} \tag{22}$$

$$Cost - DV_{C} + P_{C} = tg_{C} \tag{23}$$

$$P_{it}^{h}, PL_{ijt}^{hk}, V_{it}^{k}, VL_{ijt}^{k}, E, E', T, T', TM \in \mathbb{Z}^{+}\ \forall i,j,h,k,t \tag{24}$$

$$BT_{t}, BT'_{t} \in \{0,1\}\ \forall t \tag{25}$$

Equations (2) take care of the conservation of the flow of people along the network: for each i, h and t, the amount of people of type h that have been transported to i up to time period t plus the amount of people of type h that arrived on their own to node i up to time period t must equal the amount of people of type h that leave node i up to time t plus the ones that stay there. Node and vehicle capacities are imposed by (3) and (4), respectively. As a result, the number of people of each type that can be evacuated is calculated by (5) and (6). Analogously, Equations (7) and (8) are the vehicle flow conservation constraints.

The time period at which every high (respectively normal) priority evacuee has been moved to a safe area is given by Equation (9) (respectively (12)), while (10) and (11) (respectively (13) and (14)) ensure binary variables BT (respectively BT') are defined properly. The maximum evacuation time is represented by TM, following (15) and (16), and constraints related to the available budget are (17)–(19), defining the total cost as the sum of the fixed cost of moving an empty vehicle and the variable cost for transporting people.

Since we are considering a multi-objective lexicographical goal programming model, the optimal value obtained at each level leads to new constraints that are added to the following levels to ensure that the goals from previous levels are achieved. Objectives with highest priority are those related to the amount of people that is possible to evacuate. Hence, the first and second levels consist in minimizing the deviations from the stated aspiration levels of the total number of people evacuated with high or normal priority, as given in (20) and (21). The third level considers the prompt evacuation of the population, minimizing the deviation with respect to the given aspiration level, as stated in Equation (22). Finally, the minimization of the deviation of the total cost with respect to its aspiration level (see (23)) is taken into account.

4. Case Study: Palu, Indonesia

This section describes the case study used to evaluate the performance of the proposed model. It is based on the earthquake and tsunami that hit the island of Sulawesi, Indonesia, in September 2018. The Indonesian archipelago is known for its catastrophic earthquakes as well as the occurrence of landslide induced tsunamis. The earthquake and tsunami we are considering were caused by a strike-slip faulting event at shallow depth that occurred within the interior of the Molucca Sea micro-plate, which is part of the Sunda tectonic plate, see [41].

On 28 September 2018, 10:02:44 GMT/UTC hour (local hour, WIB, 18:02:44) a series of strong earthquakes struck mainly Palu, the capital of the Indonesian province of Central Sulawesi, and Donggala. The strongest was a 7.4 M earthquake only 10 km deep with its epicenter close to Palu. The earthquake, according to the Humanitarian Country Team [42], triggered a tsunami, whose waves reached up to three metres, that hit the northern parts of Sulawesi island and resulted in liquefaction and landslides. Furthermore, a total of 170 aftershocks occurred after the main one. As a consequence of the liquefaction and aftershocks, some roads from/to Palu were not accessible for several hours and both the port and airport were servery damaged, highly complicating the Search & Rescue and evacuation management. In particular, the BNPB (National Disaster Management Agency) declared the urgent need for planes with the ability to take off and land in short runway (specifically, Hercules C130) due to the cracks at the airport runway.

Based on the updated information from WHO, see [43], 832 persons died, 580 were injured, more than 16,000 were displaced to temporary shelters and dozens of houses were damaged in the first 24 h. The estimated exposed population was more than 310,00 in Donggala regency and more than 350,000, in Palu.

4.1. Data

The case study has been created by using information obtained from AHA Reports [44–51]. The satellite images and geospatial data of the affected area were obtained from the Copernicus Emergency Management Service website, see [52]. The complete affected area, illustrated in Figure 1, is located in the Sulawesi island and divided into 18 grading maps (named Areas of Interest), covering the cities and villages that suffered the shake of the earthquake and/or the impact along the coast. In particular, we have focused on the two most affected Areas of Interest (AoI), that host the most exposed population (the ones corresponding to the two squares inside the red rectangle of Figure 1), namely AoI 7 (Palu) and AoI 11 (Palu East). Their grading maps are shown in Figure 2, where AoI 7 corresponds to the left-hand-side figure and AoI 11 to the right-hand-side one.

The network of the case study contains 44 nodes representing locations of the city of Palu. According to their characteristics, they are classified as:

- 16 pickup nodes (R) located at the affected area, easily recognized by the local population and the Search & Rescue Teams, such as mosques, schools, cafes or shopping centers in relatively unaffected or lightly affected streets. The total number of affected nodes and their locations have been established according to the data of affected buildings and affected population obtained from Copernicus, [52].
- 23 temporary shelter nodes (S) are real evacuation sites corresponding to real temporary shelters enabled by BASARNAS (the National Search and Rescue Agency of Indonesia), whose locations and capacities have been obtained from reports of the Government of Indonesia [53–55]. Initially, there were only 24 evacuation sites, which amounted to 41 and, finally, to 141 on 4 October 2018. According to the time horizon and the total area of this case study, we have considered 5 large shelters, 6 medium ones and 12 small.
- Hospitals and medical center nodes (H) correspond to real locations of hospitals and medical centers obtained from the same reports used to compile the data regarding temporary shelters. These nodes are the destination of severely injured people and pregnant women, especially.
- Palu airport (A) is a node that corresponds to the real location of Mutiara SIS Al-Jufrie, the airport of the city. For the case study, this airport was a destination point for population to be evacuated by plane to other safer cities, especially for severely injured people that could not be treated at the available hospitals due to a lack of medical supplies or appropriate facilities.

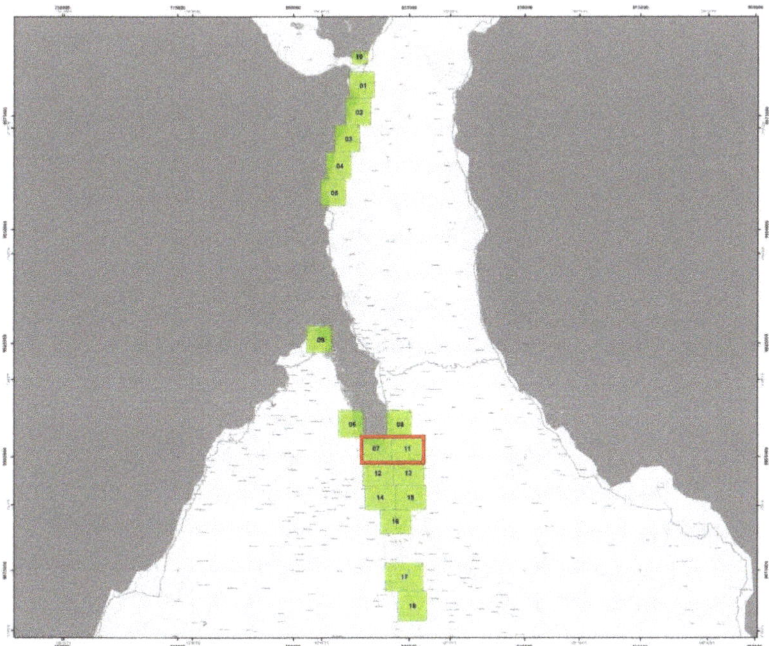

Figure 1. Areas of Interest of the earthquake and tsunami, Indonesia.

(**a**) (**b**)

Figure 2. Considered areas (Source: [52]). (**a**) AoI 7, Palu and (**b**) AoI 11, East Palu.

The nodes are connected by 76 edges, whose lengths were determined by using GoogleMyMaps technology. According to the information obtained from the maps of Copernicus [52], they represent bridges, level crossings, tunnels or roads which suffered different levels of damage, as blocked or shattered, seriously damaged, partially damaged or passable. Furthermore, they may be highways, single carriageways, local, within towns or unpaved roads. The complete graph is shown in Figure 3.

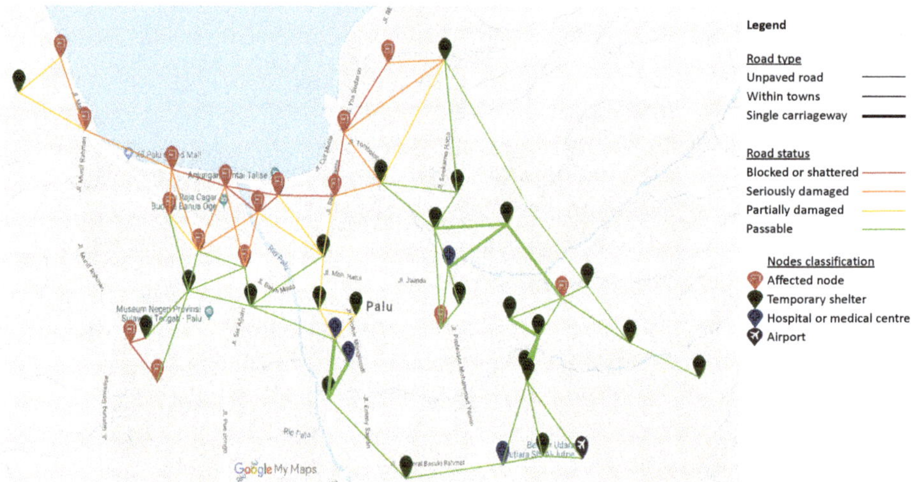

Figure 3. Complete Graph.

The condition of streets and roads, due to the tsunami and the liquefaction of the ground at some areas, made necessary the participation of a diverse number of vehicles for the tasks of people search and evacuation, including helicopters, regular and inflatable boats, army trucks and ambulances. Terrestrial vehicles cannot travel by blocked or shattered arcs and it has been supposed that seriously and partially damaged connections suffer a penalization of 50% and 25% of their maximum velocity, respectively. At the beginning of the operation, ambulances are available at hospitals, while army trucks can be found at several nodes. Moreover, only blocked or shattered roads can be travelled by inflatable boat, because of the floods, water movement and the liquefaction of the terrain. At the beginning of the operation, all inflatable boats are at affected nodes, because they either belong to the neighborhood or the Search & Rescue teams leave them there. Finally, helicopters can travel following a straight line between any pair of points with heliport available and are located at hospitals and at the airport at the beginning of the operation.

Some characteristics of the vehicles such as capacity (number of not badly injured people that can be transported by each type of vehicle), mean velocity and costs are presented in Table 1, together with the maximum amount of vehicles that can be used in the operation. The fixed cost is independent of the load of the vehicle and is paid only per unit of distance; however, the variable cost depends both on the traversed distance and on the cargo transported, being paid per unit of distance and unit of load (u.l.).

Table 1. Characteristics of available vehicles.

Type	Capacity	Mean Velocity	Fixed Cost	Variable Cost	Quantity Available
Helicopter	3	153 km/h	0.50 $/km	1.20 $/(km×u.l.)	8
Truck	18	80 km/h	0.20 $/km	0.80 $/(km×u.l.)	410
Ambulance	9	80 km/h	0.10 $/km	0.50 $/(km×u.l.)	40
Raft	6	3 km/h	0.05 $/km	0.12 $/(km×u.l.)	48

The data regarding affected people and their classification into the different categories considered before have been estimated from several public reports. For example, the data about the number of inhabitants and affected in each AoI have been obtained from the maps of Copernicus [52], as displayed in Table 2. The report of BNPB [55], already mentioned, provides the number of evacuees by 1 October 2018: a total of 48,025 persons at hospitals and temporary shelters out of 60,424, the total affected population. Therefore, only around 20% of the total population evacuated to relative homes or hotels.

Table 2. Population data obtained from the information from Copernicus.

	AoI 7	AoI 11	All AoI
Number of inhabitants	110,265	76,176	608,521
Affected in AoI	37,776	136	60,424

According to the Indonesia Humanitarian Country Team [56], women, unaccompanied minors, adolescent mothers and other vulnerable groups such as people with disabilities, medically homebound, poor, people living with HIV or AIDS and sexual minorities, are at increased risk after the occurrence of a disaster. In fact, it has been estimated that there were around 350,000 cases of abuse, exploitation, forced or early marriage as a consequence of the particular disaster of this case study. In addition, there were cases of food shortages due to the closure of markets, roads and transports. For this reason, besides the classification according to their capacity, shelters have been classified into secure and non secure for vulnerable groups. These groups, together with severely injured people and pregnant women, have been assigned a high priority. In particular, we have considered the following population groups: severely injured adults (SIA), pregnant women (PW), unaccompanied minors (UM), women susceptible to gender violence (GBV) and the rest of the population (RP). According to Stepanov and Smith [57], most of the remaining population who should be able to evacuate on their own usually move to homes of relatives, hotels or other shelters. The Indonesia Humanitarian Country Team [56] established on 10.05.2018 that during the first periods of the response to this disaster, the risk of violence, exploitation or abuse is heightened and pre-existing gender inequalities may be exacerbated with an environment of impunity, where near 100,000 people were stated as under high risk. It has been assumed that at every pick-up node at the affected area, at the beginning of the operation, there was the same number of persons to be evacuated, following an uniform distribution. There are 14 nodes in AoI 7 and 2 in AoI 11. The details and number of people to be evacuated at the beginning of the operation at each AoI are given in Table 3.

Table 3. Population characteristics.

Type	Priority	Required Space	AoI 7	AoI 11
Severely injured adults (SIA)	High	3	171	6
Pregnant women (PW)	High	1.5	85	2
Susc. to gender violence (GBV)	High	1	153	3
Unaccompanied minors (UM)	High	1	98	2
Rest of the population (RP)	Normal	1	476	11

Additionally, people are assumed to be arriving to the pick up nodes dynamically, according to their feeling about the disaster and the Search & Rescue operations. The number of people arriving during each time period is estimated through Equation (26). It is supposed that the feeling regarding the potential danger to the population decreases along time. For this reason, a decreasing exponential function is used. For the first time period, we consider the data of people to be evacuated at the beginning of the operation, displayed in Table 3; in addition, after one sixth of the time span has gone by, we assume that no more people arrive at the pick-up points.

$$qp_{it}^{h} = \lfloor qp_{i0}^{h} e^{ln(qp_{i0}^{h} * \frac{1-t}{1/6 * \mathcal{T} - 1})} \rfloor \quad \forall t \leq \mathcal{T}/6, \forall i \in \mathcal{NA} \tag{26}$$

The capacities of the shelters for different types of people are shown in Table 4. The data was obtained from several reports, such as [53–55]. The authorities prioritized the evacuation of those who were injured or ill. There were flights between Mutiara Sis Al Jufri Airport, in Palu, and nearby airports, mainly at cities like Balikpapan (Sulaiman Seppinggan International Airport), Makassar (Sultan Hasanuddin International Airport), Manado (Sam Ratulangi International Airport) and Jakarta (Halim Perdanakusuma International Airport).

Table 4. Shelter characteristics and capacity distributions. PW: pregnant women. UM: unaccompanied minors. GBV: women susceptible to gender violence. RP: rest of the population.

ID	Name	Capacity	Classif.	Type	PW	UM	GBV	RP
S01	Silae (Miners)	500	medium	normal	0	0	0	500
S02	Makorem (area)	3000	large	normal	0	0	0	3000
S03	Mako Sabhara P.	5000	large	safe	600	1600	2800	0
S04	Perumahan M.R.	150	small	normal	0	0	0	150
S05	GOR Siranindi	200	small	normal	0	0	0	200
S06	Halaman D.	100	small	normal	0	0	0	100
S07	Masjid Raya	300	small	normal	0	0	0	300
S08	Lap. Anoa	100	small	normal	0	0	0	100
S09	Masjid B.A.	1500	large	safe	180	480	840	0
S10	Bundaran STQ	500	medium	normal	0	0	0	500
S11	Lap. Dayodara	700	medium	normal	0	0	0	700
S12	Samping M.A.F.	880	medium	normal	0	0	0	880
S13	Lagarutu	511	medium	normal	0	0	0	511
S14	Lap Vatulemo	1000	large	safe	120	320	560	0
S15	Jalan Swadaya	157	small	normal	0	0	0	157
S16	Malao Atas	150	small	normal	0	0	0	150
S17	Jalan Maleo	100	small	normal	0	0	0	100
S18	Hal. Perkantoran	2000	large	normal	0	0	0	2000
S19	Sepanjang J.G.	250	small	normal	0	0	0	250
S20	BTN Lasoani	300	small	normal	0	0	0	300
S21	Lap. Kawatuna	300	small	normal	0	0	0	300
S22	Sekitar J.B.R.	120	small	normal	0	0	0	120
S23	Lap. Faqih R.	500	medium	normal	0	0	0	500

The time horizon considered corresponds to 72 h, discretized into 24 three-hour periods, from 28 September 2018 10 a.m. to 1 October 2018, 10 a.m. According to the Humanitarian Country Team [42], on 9 October 2018, the HCT's Central Sulawesi Earthquake Response Plan requested US$ 50.5 million for the relief activities after the disaster for the assistance of 191,000 people over three months. As an estimation only for the operations and scope included in our case study, we have considered an available budget of 200,000$.

4.2. Results

This section is devoted to present and analyze the results obtained by solving the proposed model for the base case study introduced in the previous section and several variations of it, carried out to allow for a deeper analysis of the performance of the model. The mathematical models are solved by GAMS 23.7 with a relative gap of 5% on an Intel(R) Core(TM) i5-6300HQ CPU 2.30 GH, 8G RAM running Windows 10.

The 4-level lexicographical goal programming model defined in (1)–(25) will be solved as a way to deal with all criteria in a joint way. Each of the four goals considered requires an aspiration level, as detailed hereunder. The aspiration levels will be: evacuating all people that arrive to the pick-up points along the time horizon, both high (8790 people) and normal (7340 people) priority population (related to the first and second lexicographical levels, respectively), within 36 h, that is 50% of the time span (third lexicographical level), and using up to 50% of the total available budget (fourth lexicographical level), that is, 100,000$. Each level leads to a mixed integer linear programming model whose size (number of constraints, variables, non-zero elements and discrete variables) is given in Table 5, emphasizing the large dimensions of realistic case studies as the one considered here.

Table 5. Size of the resulting models level by level.

	Constraints	Variables	Non-Zeroes	Discrete Var.
Level 1	25,854	87,411	1,960,669	87,409
Level 2	25,856	87,414	1,960,694	87,410
Level 3	25,957	87,467	1,972,821	87,458
Level 4	25,959	87,470	2,050,729	87,458

Solving the four levels lexicographically, ensuring that the deviation from the aspiration level of the previous goal is maintained in the following levels, a solution is obtained in which 98.36% of high priority and 92.89% of normal priority population are evacuated with an operation time of 72 h (twice as desired) and a cost significantly below the aspiration level (54,956.13$). Nearly all the high priority population is evacuated, as well as most of the normal priority population, and the cost is quite restrained; however, the whole time span considered in the model is required to complete the evacuation. Figure 4 illustrates the location of evacuees at the end of the operation in this solution and the detailed information regarding the amount of people of each category that stay at each node can be found in Table 6. Both representations state that most of evacuees are located at safe nodes at the end of the operation. In particular, Mako Sabhara (S03), in the east of Palu river, and Makorem area (S02), in the west, sheltered a very high number of evacuees due to their location and capacity. Additionally, it must be pointed out that S02 is a normal shelter and S03 a safe one, which means that these shelters accommodate people with both high and normal priorities. The next location hosting the highest number of remaining evacuees is the airport, from where they could be transported to other nearby cities. On the other hand, it is important to highlight the fact that, in practice, the information regarding the people that could not be evacuated and remain at unsafe areas (number of people, location, health condition, etc.) is shared with other agencies, in order to coordinate the complete evacuation of all affected population.

Figure 4. Location of evacuees at the end of the operation.

Table 6. Distribution of evacuees at the end of the operation by category. SIA: Severely injured adults. PW: pregnant women. UM: unaccompanied minors. GBV: women susceptible to gender violence. RP: rest of the population.

	SIA	PW	UM	GBV	RP
R09					1
R11	144				521
S01					500
S02					2997
S03		600	1565	2433	
S04					150
S05					200
S06					100
S07					300
S09		180	167	198	
S10					270
S12					863
S14		12			
S18					107
H01	750	250			
H02	750	230			
H03	727	250			
H04	527				
A01	6			1	1333

To provide a more detailed example regarding a particular group of affected people, Figure 5 illustrates how the evacuation of women susceptible to GBV was performed. In particular, it shows the time period (or periods) in which women susceptible to GBV start being transported from one node to another and the type of vehicle used for transportation. Square brackets represent different periods where the same path and type of vehicle are used. It can be seen how trucks is the most widely used type of vehicle, even though helicopters, ambulances and rafts are also used at certain points.

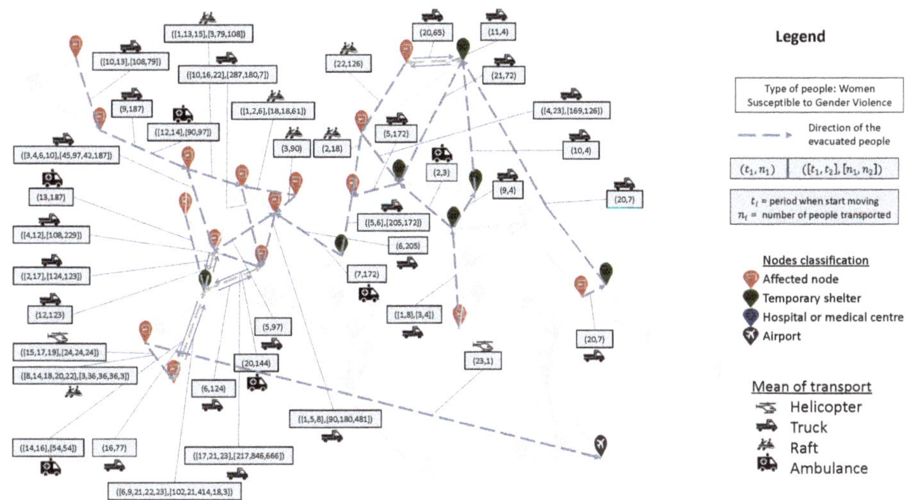

Figure 5. Evacuation of women susceptible to GBV.

In order to further evaluate the performance of the model in different situations, several specific modifications were performed on the case study and the resulting lexicographical goal programming problems were resolved again. For example, as stated by the International Civil Aviation Organization of OCHA [58], the Palu airport suffered a great amount of damage and it could not be used during the first hours of the evacuation operation. As a result, it could be interesting to analyze what may happen if the airport is assumed to be unavailable in the network of the case study. In this situation, 8601 people with high priority (97.85%) and 6824 with normal priority (92.94%) could be evacuated, which is quite similar to what could be achieved originally, and within the same operation time; however, without the airport being available, this can only be achieved at a significantly higher cost of 59,019.69$. A similar situation could arise if the hospitals were unavailable as safe locations for the evacuees. In that case, less high priority population could be evacuated (7764 people, 88.33%), but this would allow the evacuation of more normal priority population (7035 people, 95.84%). The operation time would be unchanged and the cost would be slightly higher (56,204.31$).

Another usual limitation that is often found during or after the occurrence of a disaster is the lack of certain types of vehicles. Since the number and types of vehicles available for the evacuation operation may influence significantly on the results, we have resolved the case study removing some of them. For example, without helicopters, less high priority population (8527, 97.01%), but slightly more normal priority population (6829, 93.01%), can be evacuated, with the same operation time (72 h) and a higher cost (57,792.54$). Meanwhile, the solution without rafts shows that only 8422 (95.81%) of the high priority population and 6827 (92.99%) of the normal priority population could be evacuated. This could be done three hours faster, but with a much higher cost of 70,142.55$. The case in which trucks are not available is especially extreme, because without them very few people could be evacuated: namely, only 40.52% and 2.06% of the high and normal priority population, respectively. Furthermore, 72 h are still necessary to complete the operation, which only costs 5292$. This happens because most of the available vehicles are actually trucks and, in addition, this type of vehicle has the highest capacity.

An additional interesting experiment regards splitting the network of the original problem into two or more subgraphs, in such a way that a solution to the original problem could be obtained from solutions of the subproblems associated to the subgraphs. The solution obtained by merging the solutions obtained independently from each subproblem may not be optimal, but this could simplify the resolution of the original problem by reducing it to simpler subproblems. One way to do this would be to use the partition illustrated in Figure 6, which splits the area of interest into two parts, east and west of the Palu river. In each subproblem, potential evacuees are located only at the pick-up nodes of that particular partition. Similarly, only secure nodes and vehicles of the corresponding partition can be used to perform the evacuation operations. After solving these two subproblems and merging the solutions obtained, the resulting plan allows the evacuation of only 75.92% and 51.23% of the critical and non critical population, respectively, in comparison with 98.36% and 92.89%, that can be achieved with the plan provided by the original model. This shows a very big difference in the effectiveness of the two approaches. On the other hand, the lower number of evacuees that are to be moved allows the operation to be completed in only 39 h and with a significantly lower cost of 34,992.86$ (in comparison with 72 h and 54,956.13$). However, we use a lexicographical approach because we believe there should not be any trade off between the number of evacuated people and other attributes such as time or cost, and as a result, these two solutions are not comparable. This experiment highlights that the coordination of the operations is crucial to obtain effective global evacuation plans.

The previous experiments are all performed with very high aspiration levels for the amount of evacuated population, because this should be the main concern in this type of operations. However, they impose quite strong constraints on the third (operation time) and fourth (cost) lexicographical levels, which cannot vary much. In order to test if it would be possible to obtain significantly faster or cheaper operations at the expense of evacuating less people, we have resolved the model with lower aspiration levels of 75% and 50% for the evacuated people with high and normal priority, respectively (75–50). These aspiration levels are quite similar to the results that could be achieved by solving independently the two sub-problems defined by Figure 6. The results we obtained with these alternative aspiration levels indicate that the stated amount of people could be evacuated much faster, in only 30 h, but with a much higher cost of 151,340.09$. This shows how slight operation time reductions can be extremely costly, even with a reduced number of total evacuees. Additionally, we also solved the case with aspiration levels of 50% and 75% for high and normal priority population, respectively (50–75), obtaining an evacuation plan that could be implemented in only 36 h. However, as it happened earlier, the cost skyrockets to 171,516.01$, stressing the high cost of reducing the operation times in this context.

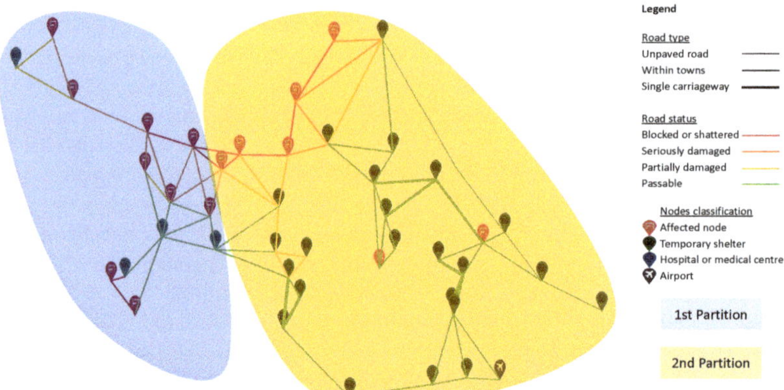

Figure 6. Splitting the network into two parts.

As a summary of the results obtained in this section, the solutions obtained by solving the case study under the different assumptions considered is given in Table 7.

Table 7. Results of the case study.

	% Critical Evacuees	% Non-Critical Evacuees	Evacuation Time (h)	Operation Cost ($)
Base case	98.36	92.89	72	54,956
No Airport	97.85	92.94	72	59,019
No Hospitals	88.33	95.84	72	56,204
No Helicopters	97.01	93.01	72	57,792
No rafts	95.81	92.99	69	70,142
No trucks	40.52	2.06	72	5292
Split in two	75.92	51.23	39	34,992
Asp. Levels 75–50	75.00	50.00	30	151,340
Asp. Levels 50–75	50.00	75.00	36	171,516

5. Concluding Remarks

The main objective of the presented work has been palliating the effects of a disaster by focusing on the supported evacuation of the affected population, in order to protect them. The model provides a feasible evacuation planning to help the authorities facing the consequences of a disaster make decisions about how to proceed operationally. The main novelties introduced with respect to existing approaches have been the consideration of dynamism on the arrival of evacuees to the pick up points, the classification of the population according to their health condition and the joint consideration of conflicting objectives such us the number of people to evacuate, the operation time and the total cost. In order to include all these relevant elements, we have followed a lexicographic goal programming approach, in such a way that there is no possible trade off among the considered criteria: the most important ones are the amount of people evacuated (high priority, first level, and normal priority, second level), followed by the operation time (third level) and finally the cost (last level).

The computational results obtained by solving a realistic case study based on the earthquake and tsunami that hit Indonesia in September 2018 has shown how the proposed model is able to provide detailed feasible evacuation plans that could be implementable in real operations. In addition, the computational experience illustrated how the model could also be used to evaluate the consequences of possible changes in the available resources or infrastructures, as for example if the airport, certain shelters or a hospital were not available, or several vehicles could not be used. Another interesting insight that we have obtained from the results of the case study regards the high conflict among the different criteria considered: in order to be able to evacuate most of the affected population, long operation times are required, even using the whole time span, with reasonably low costs; the only way to obtain faster operations is by reducing the number of people evacuated, and even in that case, the cost increases a lot (up to three times) to achieve this time reduction. If the situation were that of a very high urgency, it could be possible to use the model to design several fast evacuation plans for reduced groups of affected people that could be carried out through the collaboration of several agencies.

One interesting line of future work could be the extension of the proposed evacuation model so that the planning could also take into account other important elements in the response to a disaster, as for example the location of temporal shelters or depots, the pre-positioning of stocks, the distribution of commodities required by the evacuated people, etc. In addition, the introduction of some stochastic variables into the model to represent the arrival of potential evacuees or changes in the status of the infrastructures could also be worth considering.

Author Contributions: Conceptualization, I.F., M.T.O., G.T. and B.V.; Data curation, I.F.; Formal analysis, I.F., M.T.O., G.T. and B.V.; Funding acquisition, B.V.; Investigation, I.F., M.T.O., G.T. and B.V.; Methodology, I.F., M.T.O., G.T. and B.V.; Software, I.F.; Supervision, M.T.O., G.T.; Validation, I.F., M.T.O. and G.T.; Visualization, I.F., M.T.O., G.T. and B.V.; Writing—original draft, I.F., M.T.O. and G.T.; Writing—review & editing, I.F., M.T.O., G.T. and B.V. All authors have read and agreed to the published version of the manuscript.

Funding: This research was funded by Marie Skłodowska-Curie RISE H2020 project GEO-SAFE (691161) and the Government of Spain (MTM2015-65803-R).

Acknowledgments: The authors would like to thank Belén Fernández Fernández, for her valuable collaboration.

Conflicts of Interest: The authors declare no conflict of interest.

References

1. Özdamar, L.; Ertem, M.A. Models, solutions and enabling technologies in humanitarian logistics. *Eur. J. Oper. Res.* **2015**, *244*, 55–65. [CrossRef]
2. Tomasini, R.; Van Wassenhove, L. *Humanitarian Logistics*; Springer: Berlin/Heidelberg, Germany, 2009.
3. Vitoriano, B.; Montero, J.; Ruan, D. *Decision Aid Models for Disaster Management and Emergencies*; Springer Science & Business Media: Berlin/Heidelberg, Germany, 2013.
4. Wu, H.C.; Lindell, M.K.; Prater, C.S. Logistics of hurricane evacuation in Hurricanes Katrina and Rita. *Transp. Res. Part F Traffic Psychol. Behav.* **2012**, *15*, 445–461. [CrossRef]
5. Peizhuang, W.; Xihui, L.; Sanchez, E. Set-valued statistics and its application to earthquake engineering. *Fuzzy Sets Syst.* **1986**, *18*, 347–356. [CrossRef]
6. Sherali, H.D.; Carter, T.B.; Hobeika, A.G. A location-allocation model and algorithm for evacuation planning under hurricane/flood conditions. *Transp. Res. Part B Methodol.* **1991**, *25*, 439–452. [CrossRef]
7. Shahparvari, S.; Chhetri, P.; Abareshi, A.; Abbasi, B. Multi-Objective Decision Analytics for Short-Notice Bushfire Evacuation: An Australian Case Study. *Aust. J. Inf. Syst.* **2015**, *19*. [CrossRef]
8. Dong, W.; Chiang, W.; Shah, H. Fuzzy information processing in seismic hazard analysis and decision making. *Soil Dyn. Earthq. Eng.* **1987**, *6*, 220–226. [CrossRef]
9. Monzón, J.; Liberatore, F.; Vitoriano, B. A Mathematical Pre-Disaster Model with Uncertainty and Multiple Criteria for Facility Location and Network Fortification. *Mathematics* **2020**, *8*, 529. [CrossRef]
10. Vitoriano, B.; Ortuño, M.T.; Tirado, G.; Montero, J. A multi-criteria optimization model for humanitarian aid distribution. *J. Glob. Optim.* **2011**, *51*, 189–208. [CrossRef]
11. Ortuño, M.T.; Tirado, G.; Vitoriano, B. A lexicographical goal programming based decision support system for logistics of Humanitarian Aid. *Top* **2011**, *19*, 464–479. [CrossRef]
12. Tirado, G.; Martín-Campo, F.J.; Vitoriano, B.; Ortuño, M.T. A lexicographical dynamic flow model for relief operations. *Int. J. Comput. Intell. Syst.* **2014**, *7*, 45–57. [CrossRef]
13. Fahy, R.F. EXIT89: An evacuation model for high-rise buildings. *Fire Saf. Sci.* **1991**, *3*, 815–823. [CrossRef]
14. Dhamala, T.N. A survey on models and algorithms for discrete evacuation planning network problems. *J. Ind. Manag. Optim.* **2015**, *11*, 265–289.
15. UN Office for Disaster Risk Reduction. Terminology on Disaster Risk Reduction. 2017. Available online: https://www.unisdr.org/we/inform/terminology (accessed on 1 March 2020).
16. Sharma, S.; Singh, H.; Prakash, A. Multi-agent modeling and simulation of human behavior in aircraft evacuations. *IEEE Trans. Aerosp. Electron. Syst.* **2008**, *44*, 1477–1488. [CrossRef]
17. Shi, J.; Ren, A.; Chen, C. Agent-based evacuation model of large public buildings under fire conditions. *Autom. Constr.* **2009**, *18*, 338–347. [CrossRef]
18. Manley, M.; Kim, Y.S.; Christensen, K.; Chen, A. Modeling emergency evacuation of individuals with disabilities in a densely populated airport. *Transp. Res. Rec.* **2011**, *2206*, 32–38. [CrossRef]
19. Hamacher, H.W.; Tjandra, S.A. *Mathematical Modelling of Evacuation Problems: A State of Art*; Fraunhofer-Institut für Techno- und Wirtschaftsmathematik: Kaiserslautern, Germany, 2001.
20. London Resilience Partnership. Mass Evacuation Framework. 2018. Available online: https://www.london.gov.uk (accessed on 1 March 2020).
21. Smith, J.M. State-dependent queueing models in emergency evacuation networks. *Transp. Res. Part B Methodol.* **1991**, *25*, 373–389. [CrossRef]
22. Sbayti, H.; Mahmassani, H.S. Optimal scheduling of evacuation operations. *Transp. Res. Rec.* **2006**, *1964*, 238–246. [CrossRef]
23. Brown, C.; White, W.; van Slyke, C.; Benson, J.D. Development of a strategic hurricane evacuation–dynamic traffic assignment model for the Houston, Texas, Region. *Transp. Res. Rec.* **2009**, *2137*, 46–53. [CrossRef]
24. Sun, D.; Kang, J.; Batta, R.; Song, Y. Optimization of Evacuation Warnings Prior to a Hurricane Disaster. *Sustainability* **2017**, *9*, 2152. [CrossRef]
25. Özdamar, L.; Aksu, D.T.; Yasa, E.; Ergunes, B. Disaster relief routing in limited capacity road networks with heterogeneous flows. *J. Ind. Manag. Optim.* **2018**, *14*, 1367–1380. [CrossRef]

26. Houston, N.; Easton, A.; Davis, E.; Mincin, J.; Phillips, B.; Leckner, M. *Evacuating Populations with Special Needs: Routes to Effective Planning Primer Series*; US Department of Transportation: Washington, DC, USA, 2009.
27. Amideo, A.E.; Scaparra, M.P.; Kotiadis, K. Optimising shelter location and evacuation routing operations: The critical issues. *Eur. J. Oper. Res.* **2019**, *279*, 279–295. [CrossRef]
28. Pyakurel, U.; Nath, H.N.; Dhamala, T.N. Efficient contraflow algorithms for quickest evacuation planning. *Sci. China Math.* **2018**, *61*, 2079–2100. [CrossRef]
29. Pyakurel, U.; Nath, H.N.; Dempe, S.; Dhamala, T.N. Efficient Dynamic Flow Algorithms for Evacuation Planning Problems with Partial Lane Reversal. *Mathematics* **2019**, *7*, 993. [CrossRef]
30. Pillac, V.; Van Hentenryck, P.; Even, C. A conflict-based path-generation heuristic for evacuation planning. *Transp. Res. Part B Methodol.* **2016**, *83*, 136–150. [CrossRef]
31. Bish, D.R. Planning for a bus-based evacuation. *OR Spectr.* **2011**, *33*, 629–654. [CrossRef]
32. Shahparvari, S.; Abbasi, B. Robust stochastic vehicle routing and scheduling for bushfire emergency evacuation: An Australian case study. *Transp. Res. Part A Policy Pract.* **2017**, *104*, 32–49. [CrossRef]
33. Shahparvari, S.; Abbasi, B.; Chhetri, P. Possibilistic scheduling routing for short-notice bushfire emergency evacuation under uncertainties: An Australian case study. *Omega* **2017**, *72*, 96–117. [CrossRef]
34. Shahparvari, S.; Abbasi, B.; Chhetri, P.; Abareshi, A. Fleet routing and scheduling in bushfire emergency evacuation: A regional case study of the Black Saturday bushfires in Australia. *Transp. Res. Part D Transp. Environ.* **2019**, *67*, 703–722. [CrossRef]
35. Dhamala, T.; Pyakurel, U. Significance of Transportation Network Models in Emergency Planning of Cities. *Cities People Places Int. J. Urban Environ.* **2016**, *2*, 58–76. [CrossRef]
36. Miller-Hooks, E.; Patterson, S.S. On solving quickest time problems in time-dependent, dynamic networks. *J. Math. Model. Algorithms* **2004**, *3*, 39–71. [CrossRef]
37. Gutjahr, W.J.; Nolz, P.C. Multicriteria optimization in humanitarian aid. *Eur. J. Oper. Res.* **2016**, *252*, 351–366. [CrossRef]
38. Ferrer, J.M.; Ortuño, M.T.; Tirado, G. A New Ant Colony-Based Methodology for Disaster Relief. *Mathematics* **2020**, *8*, 518. [CrossRef]
39. Mejia-Argueta, C.; Gaytán, J.; Caballero, R.; Molina, J.; Vitoriano, B. Multicriteria optimization approach to deploy humanitarian logistic operations integrally during floods. *Int. Trans. Oper. Res.* **2018**, *25*, 1053–1079. [CrossRef]
40. Alçada-Almeida, L.; Tralhao, L.; Santos, L.; Coutinho-Rodrigues, J. A multiobjective approach to locate emergency shelters and identify evacuation routes in urban areas. *Geogr. Anal.* **2009**, *41*, 9–29. [CrossRef]
41. Valkaniotis, S.; Ganas, A.; Tsironi, V.; Barberopoulou, A. *A Preliminary Report on the M7.5 Palu Earthquake Co-Seismic Ruptures and Landslides Using Image Correlation Techniques on Optical Satellite Data*; Zenodo: Genève, Switzerland, 2018. [CrossRef]
42. Humanitarian Country Team. *Central Sulawesi Earthquake and Tsunami*; Technical Report 01; Humanitarian Country Team: Jakarta, Indonesia, 2018.
43. World Health Organization/Indonesia. *Situation Report 01 Sulawesi Earthquake and Tsunami, Indonesia*; Technical report; WHO: Geneva, Switzerland, 2018.
44. AHA Centre. *Situation Update No. 1 M 7.4 Earthquake & Tsunami; Sulawesi, Indonesia*; Technical Report 01; AHA Centre: Jakarta, Indonesia, 2018.
45. AHA Centre. *Situation Update No. 2 M 7.4 Earthquake & Tsunami; Sulawesi, Indonesia*; Technical Report 02; AHA Centre: Jakarta, Indonesia, 2018.
46. AHA Centre. *Situation Update No. 3 M 7.4 Earthquake & Tsunami; Sulawesi, Indonesia*; Technical Report 03; AHA Centre: Jakarta, Indonesia, 2018.
47. AHA Centre. *Situation Update No. 4 M 7.4 Earthquake & Tsunami; Sulawesi, Indonesia*; Technical Report 04; AHA Centre: Jakarta, Indonesia, 2018.
48. AHA Centre. *Situation Update No. 5 M 7.4 Earthquake & Tsunami; Sulawesi, Indonesia*; Technical Report 05; AHA Centre: Jakarta, Indonesia, 2018.
49. AHA Centre. *Situation Update No. 6 M 7.4 Earthquake & Tsunami; Sulawesi, Indonesia*; Technical Report 06; AHA Centre: Jakarta, Indonesia, 2018.
50. AHA Centre. *Situation Update No. 7 M 7.4 Earthquake & Tsunami; Sulawesi, Indonesia*; Technical Report 07; AHA Centre: Jakarta, Indonesia, 2018.

51. AHA Centre. *Situation Update No. 9 M 7.4 Earthquake & Tsunami; Sulawesi, Indonesia*; Technical Report 09; AHA Centre: Jakarta, Indonesia, 2018.
52. European Commission (EC); European Space Agency (ESA); European Environment Agency (EEA). Emergency Management Service—Mapping. 2018. Available online: https://emergency.copernicus.eu/mapping (accessed on 1 March 2020).
53. Badan Nasional Penanggulangan Bencana. *Laporan Harian No.1 Gempa M.7,4 dan Tsunami Sulawesi Tengah*; Technical Report 01; Badan Nasional Penanggulangan Bencana: Jakarta, Indonesia, 2018.
54. Badan Nasional Penanggulangan Bencana. *Laporan Harian Penanganan Gempa Bumi dan Tsunami Palu dan Donggala*; Technical Report 02; Badan Nasional Penanggulangan Bencana: Jakarta, Indonesia, 2018.
55. Badan Nasional Penanggulangan Bencana. *Penanganan Bencana Gempabumi M7,4 dan Tsunami di Kota Palu dan Donggala Sulawesi Tengah*; Technical Report 03; Badan Nasional Penanggulangan Bencana: Jakarta, Indonesia, 2018.
56. Indonesia Humanitarian Country Team. *Central Sulawesi Earthquake Response Plan (October 2018–December 2018)*; Technical report; Indonesia Humanitarian Country Team: Jakarta, Indonesia, 2018.
57. Stepanov, A.; Smith, J.M. Multi-objective evacuation routing in transportation networks. *Eur. J. Oper. Res.* **2009**, *198*, 435–446. [CrossRef]
58. International Civil Aviation Organization of OCHA. *Airport Situation Report*; Technical Report 04; International Civil Aviation Organization of OCHA: Geneva, Switzerland, 2018.

Article

Probability-Based Wildfire Risk Measure for Decision-Making

Adán Rodríguez-Martínez * and Begoña Vitoriano

Interdisciplinary Mathematics Institute, Complutense University of Madrid (UCM), 28040 Madrid, Spain; bvitoriano@mat.ucm.es

* Correspondence: adanro01@ucm.es

Received: 6 March 2020; Accepted: 5 April 2020; Published: 10 April 2020

Abstract: Wildfire is a natural element of many ecosystems as well as a natural disaster to be prevented. Climate and land usage changes have increased the number and size of wildfires in the last few decades. In this situation, governments must be able to manage wildfire, and a risk measure can be crucial to evaluate any preventive action and to support decision-making. In this paper, a risk measure based on ignition and spread probabilities is developed modeling a forest landscape as an interconnected system of homogeneous sectors. The measure is defined as the expected value of losses due to fire, based on the probabilities of each sector burning. An efficient method based on Bayesian networks to compute the probability of fire in each sector is provided. The risk measure is suitable to support decision-making to compare preventive actions and to choose the best alternatives reducing the risk of a network. The paper is divided into three parts. First, we present the theoretical framework on which the risk measure is based, outlining some necessary properties of the fire probabilistic model as well as discussing the definition of the event 'fire'. In the second part, we show how to avoid topological restrictions in the network and produce a computable and comprehensible wildfire risk measure. Finally, an illustrative case example is included.

Keywords: wildfire management; risk measure; probability; Bayesian networks; decision-making; prescribed burns; firebreak location

1. Introduction

Forest fires are an annual occurrence in many parts of the world, affecting the population and environment of the adjacent areas with significant economic and ecological losses, and often, human casualties. The number and size of forest fires have increased over time, exceeding the firefighters' response capacity [1]. In the last few years, big fires have taken a global relevance: Portugal (2017) 44,969 ha and 66 fatalities, California (2018) 101,287 ha and 124 fatalities, and Australia (2019) 18,600,000 ha approx. and 34 fatalities (until 6 March 2020).

From the firefighters' perspective, Ref. [2] suggests that the abandonment of farmland and reduced grazing have led to an increase in wildland areas, and they address the issue of how extinguishing small fires can increase the size of future fires.

A considerable amount of literature has been published on the use of applied mathematics to tackle the problem of wildfires. Many of these studies are focusing on subjects such as behavior and spread of fire [3], fire suppression [4], evacuation in case of risk [5–7], and location of firebreaks [8,9], for example. For this work, we focus on those articles that deal with the issue of prevention and mitigation of forest fires. In this sense, Ref. [10] is an analysis of operational research challenges in forestry where 33 open problems are formulated, being Problem 20 formulated as follows:

> *'How can we develop tractable models that can be used to help determine when and where to implement fuel treatments on large flammable forest and wildland landscapes?'*

Fuel treatments include mainly fuel management and location of firebreaks. Fuel management is one of the most important measures for preventing large scale fires. This practice consists of reducing the amount of fuel through the application of treatments such as prescribed fire or mechanical thinning [8,11]. Ref. [12] addresses the question of how fuel management should be used over time to reduce wildfire impact. The problem with spatial and temporal dimensions is solved by using a mixed-integer linear programming model. In a later study, Ref. [13] take additional considerations into account for preserving habitat quality. Finally, to obtain robust models against uncertainty, as can be seen for instance in [14] for other fields, stochasticity is included in the model. The complete study can be found in [15].

Firebreak location is a very extended activity to provide safe and accessible places to fight against fires. Effectiveness of firebreaks has been studied extensively in many territories [9,16]. The right position where those firebreaks should be located is a very complex problem. Ref. [8] shows how different configurations of firebreaks affect fire propagation.

The purpose of this paper is to obtain a spatial probabilistic measure for wildfire risk, suitable to compare different landscape configurations after applying fuel treatments. Paper [17] entitled 'Probability-based models for estimation of wildfire risk', presents a statistical perspective to solve this problem. They use a partition of the landscape for estimating the probability of fire in each 1 km^2 pixel. For every pixel, two different events are defined: ignition or small fire (between 0.04 ha and 40.5 ha) and large fire (greater than 40.5 ha). Finally, Ref. [17] proposes estimating the probability of ignition and the probability that a small fire will become a large fire using a logistic regression based on several fire indicators. The aim of that work is to estimate the number of large fires to support tactical decisions to reduce this number.

We propose a network representation of the landscape considering fire as a phenomenon that can be spread out through the land instead of considering it as a one-off event. In a recent study, Ref. [18] presents a methodology used by Catalan Fire and Rescue Service based on modelling landscape with a network. A case study of Odena (Spain) is shown where the landscape is divided into 23 sectors (Figure 1). Considering two wind scenarios, they define main connections between sectors as 'extreme fire behavior', 'intense fire behavior' or 'low fire behavior'. The resulting scheme is the starting point to plan tactical actions on the landscape.

The aim of this work is to establish a mathematical basis to be able to compute fire risk and to be optimized into decision-making models. We focus on the definition of fire risk as well as its calculations in realistic dimensions.

Ref. [19] models wildfires and [20] models fire spread in buildings; both develop mathematical tools for computing fire probabilities in a network. However, the theoretical analysis presented in this paper shows some non-trivial properties of the probabilistic space that were not taken into account.

The second section of this paper is a theoretical discussion about the definition of fire in a probabilistic model and the issue of computing the probability of fire. In Section 2.1, the working space is defined as the product probabilistic space, and the problem of defining 'fire' in this space is addressed. In Section 2.2, it is shown that under some conditions, the probabilistic behavior of fire conforms to a Bayesian network. In the second part, Section 3, a methodology to generalize the previous theoretical results to be applied in a real case is presented. Finally, we use a simulated case study provided by [21] to illustrate the application of the proposed method for decision-making.

2. Fire Probability

A tool measuring the risk that takes into account fire connectivity can be helpful to decide which changes on the landscape are more efficient to reduce fire risk. Also, an estimation of the effectiveness of every action as an expected value can be used to justify the investment in preparedness.

Risk is a broad concept that can be defined in different ways. In our context, we will define fire risk as the expected losses due to bush fires in a determined landscape for a limited period of time. Computing this value implies being able to compute probabilities of burning, which will be done through a representation of the landscape as a network. Nodes represent homogeneous sectors in which we divide the territory and arcs represent the probability of fire spread among those parts. The first issue is to compute the probability of a node burns, assumed known the ignition probabilities in the nodes and the spread probabilities through the arcs.

1st assumption: Landscape can be divided into homogeneous areas: sectors. A sector is a portion of land where in case of a fire, the only possible way of extinguishing it will be within its boundaries (Figure 1).

This assumption is the cornerstone of the methodology presented. It is a usual simplification taken both by researchers and firefighters [12,13,17–19,22]. Although fire and landscape are naturally seen as a continuous phenomenon, preventive actions used to be taken within finite set of possible places and a network is an appropriate framework for these decisions.

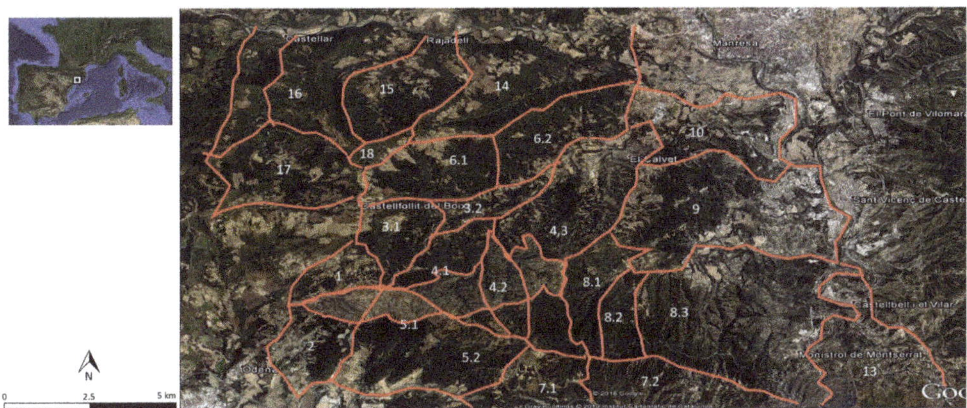

Figure 1. Segmentation of the area of interest to identify tactical objectives [18].

2.1. Probabilistic Network Model

1st assumption allows us to work with a network representing the landscape. We denote the network as $G = (N, A)$, with

$$\begin{aligned} N &= \{n_1, \dots, n_m\} \\ A &= \{a_{ij} = (n_i, n_j) : n_i, n_j \in N\}, \quad \text{with } |A| = p \end{aligned} \tag{1}$$

being N the set of all nodes representing sectors of the landscape and A the set of directed edges connecting adjacent sectors.

Random experiments on nodes will be related to fire ignition (Fire ignition is defined as a fire greater than 0.04 ha [17]). Elementary events for each node will be 'there is ignition' and 'there is **not** ignition', defining the following probability spaces:

$$\begin{array}{llll} \Omega_{n_i} = \{ig_i, ig_i^c\}, & & & i = 1, \dots, m \\ ig_i : \text{ignition in node } n_i, & & & i = 1, \dots, m \\ \mathcal{A}_{n_i} = \mathcal{P}(\Omega_{n_i}), & & & i = 1, \dots, m \\ p_{n_i} : \quad \mathcal{A}_{n_i} & \longrightarrow & \mathbb{R}, & i = 1, \dots, m \\ \quad \varnothing & \longrightarrow & 0 & \\ \quad \{ig_i\} & \longrightarrow & p_{n_i}(\{ig_i\}) & \\ \quad \{ig_i^c\} & \longrightarrow & 1 - p_{n_i}(\{ig_i\}) & \\ \quad \Omega_{n_i} & \longrightarrow & 1 & \end{array} \tag{2}$$

In the same way, elementary events for each edge will be 'a_{ij} is able to spread fire' and 'a_{ij} is **not** able to spread fire':

$$
\begin{array}{llll}
\Omega_{a_{ij}} = \{sp_{ij}, sp_{ij}^c\}, & & & a_{ij} \in A \\
sp_{ij} : \ a_{ij} \text{ is able to spread fire,} & & & a_{ij} \in A \\
\mathcal{A}_{a_{ij}} = \mathcal{P}\left(\Omega_{a_{ij}}\right), & & & a_{ij} \in A \\
p_{a_{ij}} : \quad \mathcal{A}_{a_{ij}} & \longrightarrow & \mathbb{R}, & a_{ij} \in A \\
\qquad \emptyset & \longrightarrow & 0 & \\
\qquad \{sp_{ij}\} & \longrightarrow & p_{a_{ij}}\left(\{sp_{ij}\}\right) & \\
\qquad \{sp_{ij}^c\} & \longrightarrow & 1 - p_{a_{ij}}\left(\{sp_{ij}\}\right) & \\
\qquad \Omega_{a_{ij}} & \longrightarrow & 1 &
\end{array}
\tag{3}
$$

With this definition, if n_i burns and a_{ij} 'is able to spread fire', then n_j also burns. Arc spread capability is related mainly with topographical and meteorological conditions (especially those related with wind directions). Estimation of ig_i and sp_{ij} probabilities will be discussed in Section 3 but it is not part of the current work.

Time period considered will be long enough for characterizing a global risk measure, but short enough so that a cell does not burns twice during the period (no regeneration).

Next assumption is, at this point, a technical necessity to work with the probabilistic space. The compatibility of this assumption with reality will be discussed in Section 3.

2nd assumption: independence between nodes ignitions and arc spread capabilities.

Hence, with this assumption, the probabilistic space on the network model is the product space:

$$
\begin{array}{l}
\Omega = \Omega_{n_1} \times \cdots \times \Omega_{n_m} \times \Omega_{a_{i_1 j_1}} \times \cdots \times \Omega_{a_{i_p j_p}} \approx \{0,1\}^{m+p} \\
\mathcal{A} = \mathcal{A}_{n_1} \times \cdots \times \mathcal{A}_{n_m} \times \mathcal{A}_{a_{i_1 j_1}} \times \cdots \times \mathcal{A}_{a_{i_p j_p}} \\
p : \quad \mathcal{A} \longrightarrow \mathbb{R} \\
\left(w_1, \ldots, w_m, t_{i_1 j_1}, \ldots, t_{i_p j_p}\right) \quad \mapsto \quad \prod_{n_i \in N} p_{n_i}(w_i) \cdot \prod_{a_{ij} \in A} p_{a_{ij}}\left(t_{ij}\right), \quad \forall w_i \in \mathcal{A}_{n_i}, t_{ij} \in \mathcal{A}_{a_{ij}}
\end{array}
\tag{4}
$$

For notational simplicity, we will denote, in the following, the next (non-elementary) events:

$$
\begin{array}{llll}
\Omega_{n_1} \times \cdots \times \{ig_i\} \times \cdots \times \Omega_{n_m} \times \Omega_{a_{i_1 j_1}} \times \cdots \times \Omega_{a_{i_p j_p}} & \subset \ \Omega & \text{as} & ig_i, \quad \text{and} \\
\Omega_{n_1} \times \cdots \times \Omega_{n_m} \times \Omega_{a_{i_1 j_1}} \times \cdots \times \{sp_{i_k j_k}\} \times \cdots \times \Omega_{a_{i_p j_p}} & \subset \ \Omega & \text{as} & sp_{i_k j_k}
\end{array}
\tag{5}
$$

Also, $c_{ij} = \{a_{il_1}, a_{l_1 l_2}, \ldots, a_{l_k j}\} \subset A$ will be a 'path from n_i to n_j' and we denote the set of all paths from n_i to n_j with C_{ij}.

Ones the probability basis for the model has been established, in the following part we will focus on how to define fire event and how to compute its probability. In both cases, we have two options: one simple but computationally intractable and another one non-trivial but computable for real size networks.

Definition 1. *The event F_i: 'n_i burns', can be defined as*

'n_i burns if there is ignition there or if fire is coming from another node where there is ignition through a path connecting it with n_i'.

In terms of elementary events, this is

$$
F_i = ig_i \cup \bigcup_{\substack{j=1 \\ j \neq i}}^{m} \left(ig_j \cap \bigcup_{c_{ji} \in C_{ji}} \left(\bigcap_{a_{kl} \in c_{ji}} sp_{kl} \right) \right). \tag{6}
$$

Please note that paths may not be disjoint, especially in a topography network. From this definition is possible to compute $p(F_i)$ with a method based on the Law of total probability. We consider the partition over $\left(\Omega_{a_{i_1j_1}} \times \cdots \times \Omega_{a_{i_pj_p}}\right)$, i.e.,

$$\Omega^k = \Omega_{n_1} \times \cdots \times \Omega_{n_m} \times \left(\Phi(sp_{a_{i_1j_1}}), \ldots, \Phi(sp_{a_{i_pj_p}})\right), \text{ where } \Phi(sp_{a_{ij}}) = \begin{cases} \{sp_{a_{ij}}\}, \; a_{ij} \in A^k \\ \{sp^c_{a_{ij}}\}, \; a_{ij} \notin A^k \end{cases}, \; A^k \in \mathcal{P}(A) \\ \Omega^k \cap \Omega^{k'} = \emptyset \; \forall k \neq k' \text{ and } \textstyle\bigcup_{k=1}^{2^p} \Omega^k = \Omega \tag{7}$$

Clearly $\{\Omega^k : k = 1, \ldots, 2^p\}$ is a partition of Ω and probability of every Ω^k is:

$$p\left(\Omega^k\right) = p\left(\{sp_{ij} : a_{ij} \in A^k\} \wedge \{sp^c_{ij} : a_{ij} \in A \setminus A^k\}\right) = \prod_{a_{ij} \in A^k} p\left(sp_{ij}\right) \cdot \prod_{a_{ij} \in A \setminus A^k} \left(1 - p\left(sp_{ij}\right)\right) \tag{8}$$

Then, conditional probability of fire in n_i given Ω^k is the probability that there is some ignition in any of its ancestors in A^k, that is

$$\begin{aligned} p\left(F_i|\Omega^k\right) &= p\left(ig_i \cup \bigcup_{j \in \{j : \exists c_{ji} \subset A^k\}} ig_j\right) = 1 - p\left(ig_i^c \cap \bigcap_{j \in \{j : \exists c_{ji} \subset A^k\}} ig_j^c\right) = \\ &= 1 - \left(1 - p\left(ig_i\right)\right) \cdot \left(\prod_{j \in \{j : \exists c_{ji} \subset A^k\}} \left(1 - p\left(ig_j\right)\right)\right) \end{aligned} \tag{9}$$

Finally, applying the Law of total probability

$$p\left(F_i\right) = \sum_{k=1}^{2^p} p\left(\Omega^k\right) \cdot p\left(F_i|\Omega^k\right) \tag{10}$$

This algorithm has its weak point in the partition of the probabilistic space in 2^p parts. This partition implies an exponential computing time with respect to $|A|$.

To explore alternative algorithms, we propose the following recursive definition of the event fire.

Definition 2. *The event F_i: 'n_i burns' can be expressed in a recursive way as*

'n_i burns if there is ignition there or if a neighbor burns and fire spreads to n_i'.

Or expressed in terms of elementary events,

$$F_i = ig_i \cup \bigcup_{\substack{j=1 \\ j \neq i}}^{m} \left(F_j \cap sp_{ji}\right) \tag{11}$$

However, the next example proves that this recursive definition is not always consistent.

Example 1. *Consider the graph*

$$\begin{aligned} G = (N = &\{n_1, n_2\}, \\ A = &\{a_{12} = (n_1, n_2), a_{21} = (n_2, n_1)\}) \end{aligned} \tag{12}$$

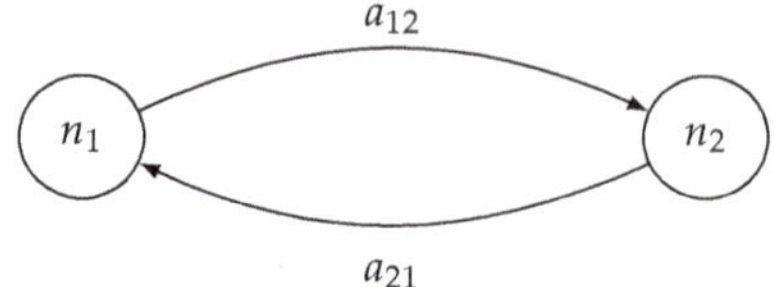

Then, Definition 2 would be:

$$\begin{aligned} F_1 &= ig_1 \cup (sp_{21} \cap F_2) \\ F_2 &= ig_2 \cup (sp_{12} \cap F_1) \end{aligned} \tag{13}$$

Solving the system (13) by substitution

$$F_1 = ig_1 \cup (sp_{21} \cap (ig_2 \cup (sp_{12} \cap F_1))) \tag{14}$$

Using intersection and union sets' properties we obtain

$$F_1 = (ig_1 \cup sp_{21}) \cap (ig_1 \cup ig_2 \cup sp_{12}) \cap (ig_1 \cup ig_2 \cup F_1) \tag{15}$$

And, finally, taking into account that, from definition of F_1, $ig_1 \subset F_1$

$$F_1 = (ig_1 \cup sp_{21}) \cap (ig_1 \cup ig_2 \cup sp_{12}) \cap (ig_2 \cup F_1) \tag{16}$$

This Boolean implicit equation can be solved following [23] work. First, we need to consider the characteristic function of every event:

$$A \to \mathcal{X}^A(\omega) = \begin{cases} 1, \textit{if } \omega \in A \\ 0, \textit{if } \omega \notin A \end{cases} \qquad (\omega \in \Omega) \tag{17}$$

for simplicity we just write $\mathcal{X}^A$.

$$\mathcal{X}^{F_1} = (\mathcal{X}^{ig_1} \vee \mathcal{X}^{sp_{21}}) \wedge (\mathcal{X}^{ig_1} \vee \mathcal{X}^{ig_2} \vee \mathcal{X}^{sp_{12}}) \wedge (\mathcal{X}^{ig_2} \vee \mathcal{X}^{F_1}) \tag{18}$$

Using the resolution method exposed on [23]

$$(\mathcal{X}^{ig_1} \vee \mathcal{X}^{sp_{21}}) \wedge (\mathcal{X}^{ig_1} \vee \mathcal{X}^{ig_2} \vee \mathcal{X}^{sp_{12}}) \wedge \mathcal{X}^{ig_2} \leq \mathcal{X}^{F_1} \leq (\mathcal{X}^{ig_1} \vee \mathcal{X}^{sp_{21}}) \wedge (\mathcal{X}^{ig_1} \vee \mathcal{X}^{ig_2} \vee \mathcal{X}^{sp_{12}}) \tag{19}$$

and grouping terms and expressing it again in sets language, we obtain

$$F_1^m = (ig_1 \cap ig_2) \cup (sp_{12} \cap sp_{21} \cap ig_2) \subseteq F_1 \subseteq ig_1 \cup (sp_{21} \cap (ig_2 \cup sp_{12})) = F_1^M \tag{20}$$

Each set F_1 included in F_1^M (Figure 2a) and containing F_1^m (Figure 2b) is solution of (14).

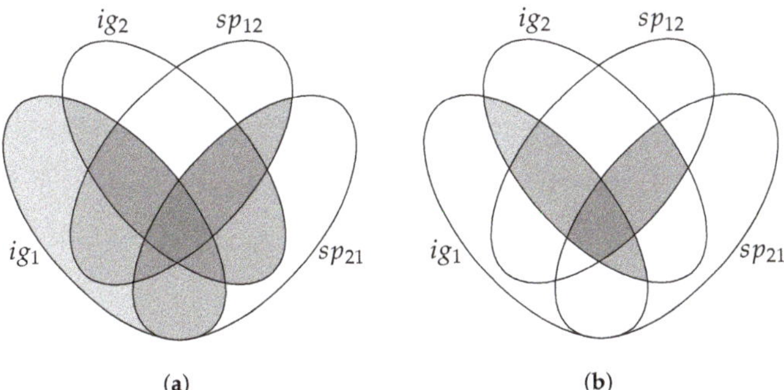

Figure 2. (**a**) Venn diagram for maximal set event solution F_1^M. (**b**) Venn diagram for minimum set event solution F_1^m.

Therefore, there are multiple solution for F_1, specifically there are $2^6 = 64$ different set satisfying inequality (20). If one of them is chosen, F_2 is determined from the second equation of system (13), obtaining so a different solution for each of the solutions of F_1.

Following explicit Definition 1, $F_1 = ig_1 \cup (sp_{21} \cap ig_2)$ (Figure 3).

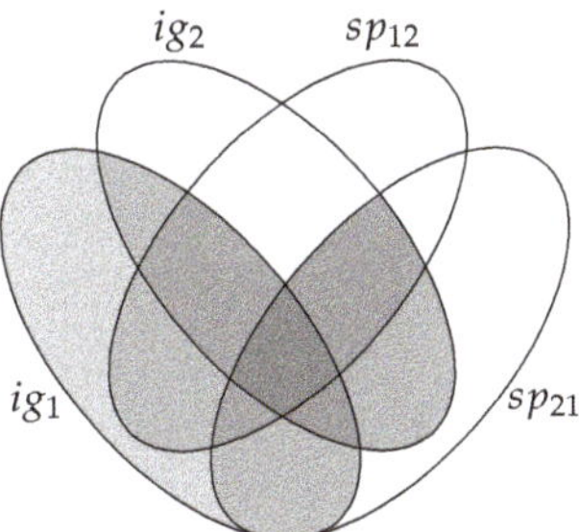

Figure 3. Venn representation of F_1 following explicit Definition 1.

It can be concluded that Definition (2) is not consistent since it does not guarantee unicity. Moreover, regarding the 'maximal solution'

$$\begin{aligned} F_1^M &= ig_1 \cup (sp_{21} \cap (ig_2 \cup sp_{12})) \\ F_2^M &= ig_2 \cup (sp_{12} \cap (ig_1 \cup sp_{21})) \end{aligned} \tag{21}$$

it would imply that if arcs a_{12} and a_{21} are able to spread $(sp_1 \cap sp_2)$ then n_1 and n_2 burn without any ignition.

This example shows that Definition 1 is not consistent for the general case. In the following, it will be proven that this definition is consistent for acyclic networks, proving that Definitions 2 and 1 are equivalent in that case. Next proposition will be used later to prove the result.

Proposition 1. *If $F_1, \ldots, F_{i-1}, F_{i+1}, \ldots, F_m$ are defined by Definition 1 and F_i by Definition 2, then Definitions 2 and 1 are equivalent in F_i:*

$$F_i = ig_i \cup \bigcup_{\substack{j=1 \\ j \neq i}}^{m} (F_j \cap sp_{ji}) = ig_i \cup \bigcup_{\substack{j=1 \\ j \neq i}}^{m} \left(ig_j \cap \bigcup_{c_{ji} \in C_{ji}} \left(\bigcap_{a_{kl} \in c_{ji}} sp_{kl} \right) \right). \tag{22}$$

Proof. This proposition is mainly proved using arithmetic set properties. It has been included the case $m = 1$, which is trivial since both formulas state that $F_i = ig_i$ when there is only one node ($m = 1$). Assume without loss of generality that $i = 1$ and $m > 1$. So, we want to proof that

$$F_1 = ig_1 \cup \bigcup_{j=2}^{m} (F_j \cap sp_{j1}) = ig_1 \cup \bigcup_{j=2}^{m} \left(ig_j \cap \bigcup_{c_{j1} \in C_{j1}} \bigcap_{a_{kl} \in c_{j1}} sp_{kl} \right) \tag{23}$$

$$\begin{aligned} F_1 &= ig_1 \cup \textstyle\bigcup_{j=2}^{m} (F_j \cap sp_{j1}) = \\ &= ig_1 \cup \textstyle\bigcup_{j=2}^{m} \left((ig_j \cap sp_{j1}) \cup \bigcup_{\substack{i=1 \\ i \neq j}}^{m} \left(ig_i \cap sp_{j1} \cap \left(\bigcup_{c_{ij} \in C_{ij}} \bigcap_{a_{kl} \in c_{ij}} sp_{kl} \right) \right) \right) = \\ &= ig_1 \cup \left(\textstyle\bigcup_{j=2}^{m} (ig_j \cap sp_{j1}) \right) \cup \left(\bigcup_{j=2}^{m} \bigcup_{\substack{i=1 \\ i \neq j}}^{m} \left(ig_i \cap sp_{j1} \cap \left(\bigcup_{c_{ij} \in C_{ij}} \bigcap_{a_{kl} \in c_{ij}} sp_{kl} \right) \right) \right) = \\ &= ig_1 \cup \left(\textstyle\bigcup_{j=2}^{m} (ig_j \cap sp_{j1}) \right) \cup \left(\bigcup_{j=2}^{m} \left(ig_1 \cap sp_{j1} \cap \left(\bigcup_{c_{1j} \in C_{1j}} \bigcap_{a_{kl} \in c_{1j}} sp_{kl} \right) \right) \right) \cup \\ &\cup \left(\textstyle\bigcup_{j=2}^{m} \bigcup_{\substack{i=2 \\ i \neq j}}^{m} \left(ig_i \cap sp_{j1} \cap \left(\bigcup_{c_{ij} \in C_{ij}} \bigcap_{a_{kl} \in c_{ij}} sp_{kl} \right) \right) \right) \end{aligned} \tag{24}$$

and using $\bigcup_{j=2}^{m} \left(ig_1 \cap sp_{j1} \cap \bigcup_{c_{1j} \in C_{1j}} \left(\bigcap_{a_{kl} \in c_{1j}} sp_{kl} \right) \right) \subset ig_1$

$$\left.\begin{aligned} F_1 &= ig_1 \cup \left(\textstyle\bigcup_{j=2}^{m} (ig_i \cap sp_{i1}) \right) \cup \left(\bigcup_{j=2}^{m} \bigcup_{\substack{i=2 \\ i \neq j}}^{m} \left(ig_i \cap sp_{j1} \cap \left(\bigcup_{c_{ij} \in C_{ij}} \bigcap_{a_{kl} \in c_{ij}} sp_{kl} \right) \right) \right) = \\ &= ig_1 \cup \left(\textstyle\bigcup_{j=2}^{m} (ig_i \cap sp_{i1}) \right) \cup \left(\bigcup_{i=2}^{m} \bigcup_{\substack{j=2 \\ j \neq i}}^{m} \left(ig_i \cap sp_{j1} \cap \left(\bigcup_{c_{ij} \in C_{ij}} \bigcap_{a_{kl} \in c_{ij}} sp_{kl} \right) \right) \right) = \end{aligned}\right\} \tag{25}$$

Last two equations just differ in the index union order. In both cases, the union is defined over the index set $\{(i,j) \in \{2,\ldots,m\}^2 : i \neq j\}$. This change will be useful for finishing the proof.

$$\begin{aligned} F_1 &= ig_1 \cup \left(\textstyle\bigcup_{j=2}^{m} (ig_i \cap sp_{i1}) \right) \cup \left(\bigcup_{i=2}^{m} ig_i \cap \bigcup_{\substack{j=2 \\ j \neq i}}^{m} \bigcup_{c_{ij} \in C_{ij}} \bigcap_{a_{kl} \in c_{ij}} sp_{kl} \cap sp_{j1} \right) = \\ &= ig_1 \cup \textstyle\bigcup_{i=2}^{m} ig_i \cap \left(\left(\bigcup_{\substack{c_{i1} \in C_{i1} \\ c_{i1} = \{sp_{i1}\}}} sp_{i1} \right) \cup \left(\bigcup_{\substack{j=2 \\ j \neq i}}^{m} \bigcup_{c_{ij} \in C_{ij}} \bigcap_{a_{kl} \in c_{ij}} sp_{kl} \cap sp_{j1} \right) \right) = \\ &= ig_1 \cup \textstyle\bigcup_{i=2}^{m} \left(ig_i \cap \bigcup_{c_{i1} \in C_{i1}} \bigcap_{a_{kl} \in c_{i1}} sp_{kl} \right) \end{aligned} \tag{26}$$

For the last step, note that $\{sp_{i1}\}$ is the set of all paths from n_i to n_1 with length 1, and

$$\{sp_{kl} \cap sp_{j1} : i \neq j = 2, \ldots, m,\ c_{ij} \in C_{ij},\ a_{kl} \in c_{ij}\} \tag{27}$$

is the set of all paths from n_i to n_1 with length ≥ 2. □

Following result is directly derived from Proposition 1. It will complete discussion about fire event definition.

Corollary 1. *If (N, A) is a directed acyclic graph (DAG), then Definitions 2 and 1 are equivalent.*

Proof. In a DAG, there exists a partial order $<_G$ defined as

$$n_i <_G n_j \Leftrightarrow \text{ exists a directed path from } n_i \text{ to } n_j \tag{28}$$

Assuming N is a finite set, we can say that a subgroup of nodes exists $N_0 \subset N$ with no-parents, or in other words

$$\exists \emptyset \neq N_0 \subset N \text{ such as } n_i \nless_G n_j \quad \forall n_i \in N,\ n_j \in N_0 \tag{29}$$

Now, considering $N \setminus N_0$ we can repeat same reasoning and denote this set of minimal elements as N_1. This process can be repeated until $(N \setminus N_0) \setminus \dots) \setminus N_u = \emptyset$, so N can be split on u disjoint sets of nodes, satisfying

$$k \leq k' \Rightarrow n^{k'} \nless_G n^k \quad \forall n^k \in N_k \text{ and } \forall n^{k'} \in N_{k'} \tag{30}$$

This is *there is no path from any node of $N_{k'}$ to any node of N_k if $k \leq k'$* (see Figure 4).

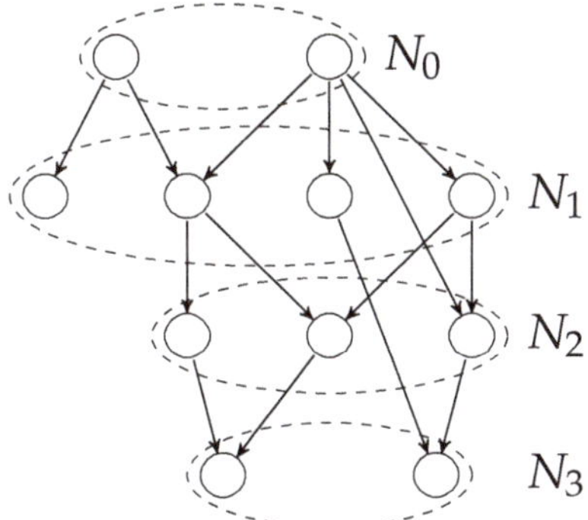

Figure 4. Nodes structure in DAG.

Finally, note that every node of N_{k+1} has its parents on $\bigcup_{i=1}^{k} N_k$, then assuming that nodes of this union are defined by (6) and nodes of N_{k+1} are defined by (11), then all the conditions of Proposition 1 are satisfied and Definitions 1 and 2 are equivalent in N_{k+1}. Moreover, for nodes in N_0, equivalence of (11) and (6) is trivial. Therefore, using complete induction and Proposition 1 we can conclude Definitions 1 and 2 are equivalent for all N. □

2.2. Bayesian Network

In this section, we will prove that random (binary) variables $x_F = \{x_{F_1}, \dots, x_{F_m}\}$ associated with fire events conform a Bayesian network, as long as G is a DAG. This result will allow us to use specific algorithms for computing fire probabilities.

Next result it is a technical property related with mutually independent sets. It will be necessary in the proof of the main result of this section, Proposition 2.

Lemma 1. *Let C_1 and C_2 two mutually independent sets of events, this is*

$$p(A \cap B) = p(A) \cdot p(B), \quad \forall A \in C_1,\ B \in C_2, \tag{31}$$

then σ-algebras

$$\sigma(C_1) \text{ and } \sigma(C_2) \tag{32}$$

are mutually independent, i.e.,

$$p(A \cap B) = p(A) \cdot p(B) \quad \forall A \in \sigma(C_1),\ B \in \sigma(C_2) \tag{33}$$

Proof. This is a particular case of Corollary 10.1 (b) collected in ([24], pp. 284–285). □

Proposition 2. *Random (binary) variables* $x_F = \{x_{F_1}, \ldots, x_{F_m}\}$ *associate with fire events defined as Definition 1 or 2 conform a Bayesian network over a directed acyclic graph* G.

Proof. There are a few ways to define Bayesian networks, we will use the Local Markov property [25]. Given a DAG, there is a natural partial order defined in its nodes:

$$n_i <_G n_j \text{ if exists a directed path from } n_i \text{ to } n_j. \tag{34}$$

With this notation, the Local Markov property consist on

$$\begin{gathered} p\left(x_{F_i} = f_i \mid x_{F_j} = f_j \ : n_j \not>_G n_i \right) = \\ p\left(x_{F_i} = f_i \mid x_{F_j} = f_j \ : n_j <_G n_i \right), \quad f_i, f_j \in \{0,1\} \end{gathered} \tag{35}$$

Using binary random variables instead of events, we can express (11) as

$$x_{F_i} = x_{n_i} \vee \bigvee_{\substack{j=1 \\ j \neq i}}^{m} \left(x_{F_j} \wedge x_{a_{ji}} \right) = x_{n_i} \vee \bigvee_{j \in \{j : a_{ji} \in A\}} \left(x_{F_j} \wedge x_{a_{ji}} \right) \tag{36}$$

and

$$\begin{aligned} p\left(x_{F_i} = f_i\right) &= f_i \cdot p\left(x_{F_i} = 1\right) + (1 - f_i) \cdot p\left(x_{F_i} = 0\right) = \\ &= f_i \cdot p\left(x_{F_i} = 1\right) + (1 - f_i) \cdot \left(1 - p\left(x_{F_i} = 1\right)\right) = \\ &= (2f_i - 1) \cdot p\left(x_{F_i} = 1\right) + (1 - f_i) = \\ &= (2f_i - 1) \cdot p\left(x_{n_i} = 1 \vee \bigvee_{j \in \{j : a_{ji} \in A\}} \left(x_{F_j} = 1 \wedge x_{a_{ji}} = 1\right)\right) + (1 - f_i) \end{aligned} \tag{37}$$

Then,

$$\begin{gathered} p\left(x_{n_i} = 1 \vee \bigvee_{j \in \{j : a_{ji} \in A\}} \left(x_{F_j} = 1 \wedge x_{a_{ji}} = 1\right) \middle| \ x_{F_j} = f_j \ : n_j \not>_G n_i \right) = \\ p\left(x_{n_i} = 1 \vee \bigvee_{j \in \{j : a_{ji} \in A\}} \left(x_{F_j} = 1 \wedge x_{a_{ji}} = 1\right) \middle| \ x_{F_j} = f_j \ : n_j <_G n_i \right) \end{gathered} \tag{38}$$

For the first conditional probability we have

$$\begin{gathered} p\left(x_{n_i} = 1 \vee \bigvee_{j \in \{j : a_{ji} \in A\}} \left(x_{F_j} = 1 \wedge x_{a_{ji}} = 1\right) \middle| \ x_{F_j} = f_j \ : n_j \not>_G n_i \right) = \\ p\left(x_{n_i} = 1 \vee \bigvee_{j \in \{j : a_{ji} \in A \wedge f_j = 1\}} \left(x_{a_{ji}} = 1\right) \middle| \ x_{F_j} = f_j \ : n_j \not>_G n_i \right), \end{gathered} \tag{39}$$

and following same reasoning for second conditional probability

$$\begin{gathered} p\left(x_{n_i} = 1 \vee \bigvee_{j \in \{j : a_{ji} \in A\}} \left(x_{F_j} = 1 \wedge x_{a_{ji}} = 1\right) \middle| \ x_{F_j} = f_j \ : n_j <_G n_i \right) = \\ p\left(x_{n_i} = 1 \vee \bigvee_{j \in \{j : a_{ji} \in A \wedge f_j = 1\}} \left(x_{a_{ji}} = 1\right) \middle| \ x_{F_j} = f_j \ : n_j <_G n_i \right). \end{gathered} \tag{40}$$

Finally, with Lemma 1 we can see that F_k is independent to

$$ig_i \cup \bigcup_{j \in \{j : a_{ji} \in A \wedge f_j = 1\}} ig_j \tag{41}$$

for any k such as $n_k \not>_G n_i$. It is enough to see that $F_k \in \sigma\left(\{ig_i, sp_{ji} : n_i \leq_G n_k\}\right)$ and $\left(ig_i \cup \bigcup_{j \in \{j : a_{ji} \in A \wedge f_j = 1\}} ig_j\right) \in \sigma\left(\{ig_i, sp_{ji} : n_i \not\leq_G n_k\}\right)$ to obtain the result. □

Therefore, we can use plenty of specific Bayesian Networks algorithms that take profit of this structure of $(\Omega, \mathcal{A}, p)$ in order to compute $p\left(F_i\right)$ efficiently.

3. Risk Measure Proposed: Losses Expected Value

Definition 3 (Fire Risk Measure). *Given a probabilistic space $(\Omega, \mathcal{A}, p)$ over a network (N, A), and defined the event fire F_i, we define the Fire Risk Measure (FRM) as the losses expected value due to forest fire*

$$FRM = \sum_{i=1}^{m} \nu(n_i)\, p(F_i), \tag{42}$$

where $\nu(n_i)$ is the value associated with n_i sector that may depends on human, economic and ecological factors.

This definition needs 1st assumption to consider a probabilistic space over a network . 2nd assumption is needed to be able to compute $p(F_i)$ (Equation (10)).

Previous section is intended to explore under which conditions it is possible to tackle the fire probability problem from a Bayesian perspective. In this section, a methodology suitable for computing it in real cases is shown.

Network resulting from representing a landscape will rarely be a Bayesian network. In a general case, we can assume wind scenarios under which the associated graph is a DAG.

3rd assumption: A finite set of meteorological scenarios of disjoint events can be considered and, under each one of them, the network representing landscape is a DAG.

Formally, this assumption means that a set $\mathcal{S}$ and a probabilistic space $(\mathcal{S}, \mathcal{P}(\mathcal{S}), p_{\mathcal{S}})$ exists, such as

$$\begin{array}{l} \mathcal{S} = \{\omega_1, \ldots, \omega_s\}, \\ \omega_i : \text{ meteorological scenario}, \\ \mathcal{A}_{\mathcal{S}} = \mathcal{P}(\mathcal{S}), \\ \begin{array}{rccl} p_{\mathcal{S}}: & \mathcal{A}_{\mathcal{S}} & \longrightarrow & \mathbb{R}, \\ & \emptyset & \longrightarrow & 0 \\ & \{\omega_i\} & \longrightarrow & p_{\mathcal{S}}(\{\omega_i\}), \end{array} \end{array} \tag{43}$$

$$\sum_{i=1}^{s} p_{\mathcal{S}}(\{\omega_i\}) = 1 \text{ and } p_{\mathcal{S}}\left(\{\omega_i\} \cap \{\omega_j\}\right) = 0,\ \forall i \neq j \tag{44}$$

and

$$\begin{array}{l} \tilde{\Omega} = \mathcal{S} \times \Omega_{n_1} \times \cdots \times \Omega_{n_m} \times \Omega_{a_{i_1 j_1}} \times \cdots \times \Omega_{a_{i_p j_p}} \\ \tilde{\mathcal{A}} = \mathcal{P}(\mathcal{S}) \times \mathcal{A}_{n_1} \times \cdots \times \mathcal{A}_{n_m} \times \mathcal{A}_{a_{i_1 j_1}} \times \cdots \times \mathcal{A}_{a_{i_p j_p}} \\ \begin{array}{rlcl} \tilde{p}: & \mathcal{A} & \longrightarrow & \mathbb{R} \\ & \left(\omega, w_1, \ldots, w_m, t_{i_1 j_1}, \ldots, t_{i_p j_p}\right) & \mapsto & p_{\mathcal{S}}(\omega) \cdot \prod_{n_i \in N} p_{n_i}(w_i) \cdot \prod_{a_{ij} \in A} p_{a_{ij}}\left(t_{ij}\right), \\ & & & \forall w \in \mathcal{P}(\mathcal{S})\, w_i \in \mathcal{A}_{n_i}, t_{ij} \in \mathcal{A}_{a_{ij}} \end{array} \end{array} \tag{45}$$

Finally,

$$G^{\omega} = (N, A^{\omega} = \{a_{ij} : \tilde{p}\left(sp_{ij}|\omega\right) \neq 0\}) \text{ is DAG},\ \forall \omega \in \mathcal{S} \tag{46}$$

Definition 3 can be applied with extended probability $\tilde{p}$:

$$\text{FRM} = \sum_{i=1}^{m} \nu(n_i)\, \tilde{p}(F_i) = \sum_{\omega \in \mathcal{S}} \tilde{p}(\omega) \sum_{i=1}^{m} \nu(n_i)\, \tilde{p}(F_i|\omega) \tag{47}$$

The consideration of scenarios has the purpose of being able to apply Bayesian algorithms to compute FRM in real cases. However, 2nd assumption, independence between ignition and spreads, is more realistic under a particular scenario. Also, estimations of $p(ig_i)$ and $p(sp_{ij})$ are usually dependent of meteorological variables, therefore, fixing a meteorological state we can estimate Bayesian network parameters.

Even in a dominant wind scenario, loops are possible. In such cases, a more efficient partition where each G_k is a DAG is possible (see Example 2), being possible computing fire probabilities within a reasonable time.

A relaxation of 3rd assumption can be considered instead of the original one:

3rd relaxed assumption: A finite set of meteorological scenarios of disjoint events can be considered and, under each one of them, the network representing landscape is a quasi-DAG.

Next example shows how to work with quasi-DAG.

Example 2. *Suppose we are interested in studying the non-DAG network represented in Figure 5,* $G = (N, A)$.

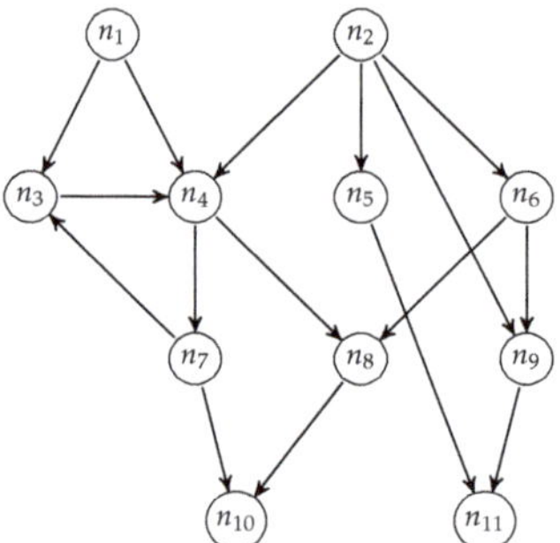

Figure 5. Network with a loop.

Edges $\{a_{34}, a_{47}, a_{73}\}$ *conform the only loop of the graph. With a similar argument of methodology showed in Equation (10), we only need to split the probabilistic model over* $(\Omega_{a_{34}} \times \Omega_{a_{47}} \times \Omega_{a_{73}})$. *All* $8(= 2^3)$ *resulted models represented with 8 different graphs in Figure 6 can be seen as Bayesian networks. For the last model, we can see nodes* n_3, n_4 *and* n_7 *as a unique node since connectivity between all then has probability 1.*

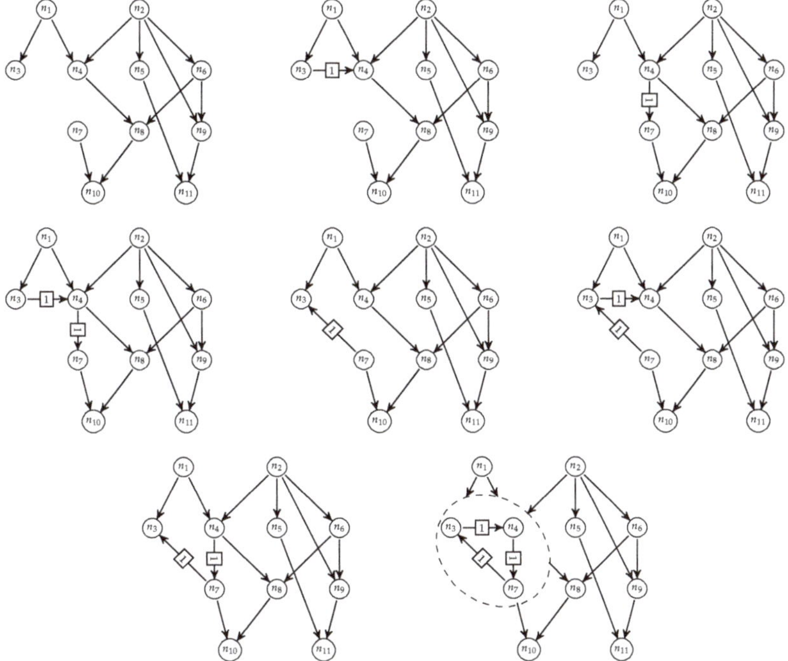

Figure 6. Representation of partition of the probabilistic space.

Consideration of scenarios and partition over some possible loops may complicate the methodology for computing an FRM. Table 1 of computational times comparing partition algorithm and Bayes-based algorithm shows the necessity of the Bayes-based algorithms in networks with more than 20 arcs.

Table 1. Computational time comparative between inference Bayes algorithm and naive algorithm (Equation (10)) with different network size ($m = |N|$, $p = |A|$). Algorithms has been executed on a computer with an Intel Core i7 processor and 8gb RAM.

$(\mathbf{m}, \mathbf{p})$	$(\mathbf{5}, \mathbf{3})$	$(\mathbf{10}, \mathbf{17})$	$(\mathbf{11}, \mathbf{19})$	$(\mathbf{12}, \mathbf{21})$	$(\mathbf{1000}, \mathbf{1750})$
Bayes	0.91×10^{-3}s	5.1×10^{-3}s	5.3×10^{-3} s	5.4×10^{-3} s	410 s
Partition	0.75×10^{-3} s	25.8 s	110 s	500 s	–

4. Case Example

Forests authorities need to evaluate risk of fire of the territory to decide and prioritize preventive actions as prescribed burns and firebreak location. A study of fuel, dominant winds, and orography of sectors allows the obtaining of an estimation of fire's behavior. To obtain accurate results, many types of information must be taken into account.

This section has an illustrative purpose trying to avoid many of the technical issues presented in a real case. It has been simulated an island (Figure 7a) using [21] to apply the methodology exposed in the previous sections. For this example, we suppose that the fire affecting the island is always wind-driven [2] and dominant winds are in the vertical line, it means, two wind direction scenarios $\{\uparrow, \downarrow\}$ referred to South and North wind, respectively. Partition of this example is done considering this assumption (Figure 7b).

(a) (b)

Figure 7. (**a**) Fictional island. (**b**) Sectors considered for a wind-driven fire type.

Ignition probability is considered proportionally to its area fixing the probability of there being any ignition on the island over a year to 0.5 ($p(\bigcup ig_i) = 0.5$, Table 2). In a real case, this number should be estimated with historical data. Probability of spread follows wind direction which depends on the scenario. Figure 8 is the network resulting from scenario '↓'. The network for scenario '↑' will include the same arcs but with opposite direction with the same probabilities that can be seen in Figure 8. Now, it is possible to compute fire probability of each sector and consequently, to obtain the expected burned area. In this case, $\nu(n_i)$ is proportional to the area of the sector and is measured in terms of Monetary Unit [MU]. In a general case, ν should be estimated taking into account several factors, as for instance, the assets located in the sector, including the forest itself.

Table 2. Ignition probability and area considered for each sector.

Sector	1	2	3	4	5	6	7	8	9	10	11
$p(ig_i) \times 10^2$	1.483	1.705	5.239	4.422	2.172	4.216	8.166	2.528	3.433	3.221	1.623
Sector Area (ha)	402.6	462.7	1422	1200	589.5	1144	2216	686.2	931.6	874.3	440.5
Sector	12	13	14	15	16	17	18	19	20	21	22
$p(ig_i) \times 10^2$	0.559	0.595	8.755	0.733	3.373	0.109	1.943	0.823	4.667	2.395	5.466
Sector Area (ha)	151.6	161.5	2376	198.9	915.5	29.62	527.43	223.4	1267	650.0	1483

The network has 43 arcs and with this size, the partition algorithm is untraceable: there are more than 8×10^{12} of subgraphs G_k. Using Bayesian algorithms provided by [26], we calculate FRM: 2453 MU. 13.44% of the landscape is expected to burn, from which 35.59% is due to ignitions.

With the aim of showing how the defined measure can support decision-making, we will compare two different preventive actions: prescribed burns and firebreaks. Prescribed burns usually do not affect trees; controlled fire burns bush and small plants. Thus, we will assume in this example that prescribed burns affect the value of a sector reducing it by 20% and ignition and spread probabilities by 50%. We compare the results of the risk measure considering that prescribed burn can be performed in only one sector, resulting sector 9 the optimal selection (Table 3). In this case, a controlled fire in this sector increases FRM by $0.2 \cdot \nu(n_9)\,(1 - \tilde{p}\,(F_9)) = 167$ MU, but the expected value of losses due to fire in the landscape decreases until 509 MU.

Prescribed burns is not the only option, and even sometimes it is not possible perform it. Land ownership or the protection of natural spaces restrict location and techniques for prescribed burns, being firebreaks an alternative option. A firebreak is a safe place where firefighters can stop fires from spreading out. Considering that only it is allowed locating one firebreak, in our example, the connection that produces the greatest reduction in the risk measure is $(4,9)$ (and $(9,4)$). A firebreak located there, assuming 100% of effectiveness ($p\,(sp_{49}) = p\,(sp_{94}) = 0$) reduces the value of FRM to 2100 MU (Table 4).

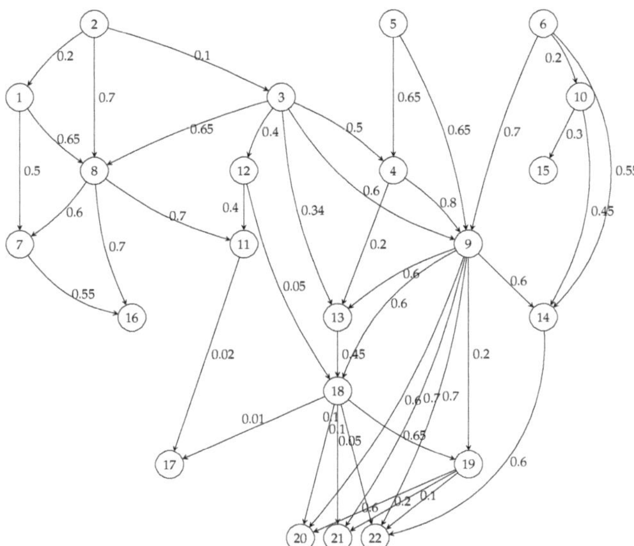

Figure 8. Network representation for scenario '↓'. Network associated with scenario '↑' is equivalent but with arrows inverted.

Table 3. FRM after performing prescribed burns in each sector. Minimum value is highlighted in bold.

Chosen Sector	1	2	3	4	5	6	7	8	9	10	11
FRM (MU)	2491	2501	2412	2378	2465	2497	2671	2424	**1944**	2572	2520
Chosen Sector	12	13	14	15	16	17	18	19	20	21	22
FRM (MU)	2477	2418	2506	2491	2546	2459	2394	2384	2527	2515	2481

Table 4. FRM after locating a firebreak between two sectors. Minimum value is highlighted in bold.

FRM (MU)	1	2	3	4	5	6	7	8	9	10	11	12	13	14	15	16	17	18	19	20	21	22
1	·	2442	·	·	·	·	2432	2420	·	·	·	·	·	·	·	·	·	·	·	·	·	·
2	2442	·	2439	·	·	·	·	2419	·	·	·	·	·	·	·	·	·	·	·	·	·	·
3	·	2439	·	2322	·	·	·	2364	2295	·	·	2446	2442	·	·	·	·	·	·	·	·	·
4	·	·	2322	·	2382	·	·	·	**2100**	·	·	·	2443	·	·	·	·	·	·	·	·	·
5	·	·	·	2382	·	·	·	·	2382	·	·	·	·	·	·	·	·	·	·	·	·	·
6	·	·	·	·	·	·	·	·	2299	2438	·	·	·	2395	·	·	·	·	·	·	·	·
7	2432	·	·	·	·	·	·	2339	·	·	·	·	·	·	·	2392	·	·	·	·	·	·
8	2420	2419	2364	·	·	·	2339	·	·	·	2431	·	·	·	·	2409	·	·	·	·	·	·
9	·	·	2295	**2100**	2382	2299	·	·	·	·	·	·	2387	2195	·	·	·	2330	2423	2353	2391	2319
10	·	·	·	·	·	2438	·	·	·	·	·	·	·	2404	2451	·	·	·	·	·	·	·
11	·	·	·	·	·	·	·	2431	·	·	·	2449	·	·	·	·	2453	·	·	·	·	·
12	·	·	2446	·	·	·	·	·	·	·	2449	·	·	·	·	·	·	2452	·	·	·	·
13	·	·	2442	2443	·	·	·	·	2387	·	·	·	·	·	·	·	·	2379	·	·	·	·
14	·	·	·	·	·	2395	·	·	2195	2404	·	·	·	·	·	·	·	·	·	·	·	2295
15	·	·	·	·	·	·	·	·	·	2451	·	·	·	·	·	·	·	·	·	·	·	·
16	·	·	·	·	·	·	2392	2409	·	·	·	·	·	·	·	·	·	·	·	·	·	·
17	·	·	·	·	·	·	·	·	·	·	2453	·	·	·	·	·	·	2453	·	·	·	·
18	·	·	·	·	·	·	·	·	2330	·	·	2452	2379	·	·	·	2453	·	2340	2436	2444	2444
19	·	·	·	·	·	·	·	·	2423	·	·	·	·	·	·	·	·	2340	·	2360	2438	2437
20	·	·	·	·	·	·	·	·	2353	·	·	·	·	·	·	·	·	2436	2360	·	·	·
21	·	·	·	·	·	·	·	·	2391	·	·	·	·	·	·	·	·	2444	2438	·	·	·
22	·	·	·	·	·	·	·	·	2319	·	·	·	·	2295	·	·	·	2444	2437	·	·	·

A hypothetical decision-maker could prefer a more detailed output. The analysis computes the fire probability of each sector showing the effect of the preventive actions on the fire probabilities of each sector. This information is plotted in greyscale differentiating risk under different wind scenarios in Table 5.

Table 5. FRM with different scenario set and forestry treatments: no treatment, prescribed burn in sector 9, location of firebreak between sector 4 and 9.

Scenario	No Treatment	Prescribed Burns	Firebreak
$\{\uparrow\}$			
$\{\downarrow\}$			
$\{\uparrow, \downarrow\}$			
FRM	2453 MU	1951 MU	2100 MU

5. Conclusions and Future Work

The aim of this work is to develop a methodology to support decision-making in the fight against wildfires. Focused on preventive actions on fuel management, a theoretical framework for a risk measure in a wildfire context has been studied without diverting attention from the main purpose.

A risk measure has been defined through the representation of the landscape as a network and based on the ignition and spread probabilities on it. The measure, defined as the expected value of fire losses, requires an efficient way to compute probabilities of each node burning. Under several assumptions and for a specific wind direction scenario, it has been proved that the network can be conformed to a Bayesian Network. Therefore, efficient algorithms developed for Bayesian Networks can compute the probabilities. The proposed methodology has been applied to an example to illustrate its usefulness for decision-making to minimize the risk of the network, comparing different landscape configurations after fuel management treatments. Moreover, with an accurate estimation of the sectors values, the proposed risk measure can also be used to justify prevention investments.

The limitations of the study are given by the assumptions made:

1st assumption considers that landscape can be divided in homogeneous areas. Researchers and firefighters often make this assumption when working with big scales territories. Usually watersheds are considered to section the landscape, but other methodologies may be explored.

2nd assumption relates to the independence between ignitions and spread capabilities. Correlations between spread and ignition events could arise mainly due to a mutual meteorological scenario affecting them. Consideration of scenarios in the analysis makes this 2nd assumption more acceptable.

3rd assumption, related to acyclic graphs, in its relaxed version, is accomplished considering wind direction scenarios and assuming no spread against wind direction.

Other limitations come from basic input data estimation. Ignition probabilities can be estimated from historical data and a study of influential factors. Fire simulators can be used to estimate spread probabilities. Wind direction scenarios can be obtained from historical data, although scenario generation is a research area under development. Nevertheless, in real cases, expert opinion will remain decisive. Future work will be devoted to obtaining accurate estimations of these input data.

The aim of this paper is to provide the measure to be optimized into decision-making models. Therefore, future research will be devoted to the development of an optimization model to support decisions on the best actions to be implemented to minimize this fire risk measure with limited resources. These limited resources usually will be related to costs and budget, or to time for developing the activities. It must be taken into account that there is limited time window for developing some preventive activities as prescribed burns determined mainly by weather conditions.

Finally, note that tactical planning is crucial in the fight against forest fires. Firefighters and forest guards dedicate effort, time and resources to these tasks. A model to measure the effects of their preventive actions on the landscape will increase their capacity of decision-making and their effectiveness.

Author Contributions: Conceptualization, A.R.-M. and B.V.; data curation, A.R.-M.; formal analysis, A.R.-M. and B.V.; funding acquisition, B.V.; investigation, A.R.-M.; methodology, A.R.-M. and B.V.; project administration, B.V.; resources, B.V.; software, A.R.-M.; validation, A.R.-M.; visualization, A.R.-M.; writing—original draft, A.R.-M. and B.V.; writing—review and editing, A.R.-M. and B.V. All authors have read and agreed to the published version of the manuscript.

Funding: This research has been supported by the H2020 Marie Skłodowska-Curie RISE Action GEO-SAFE 691161, the Government of Spain grant project MTM2015-65803-R and the Complutense University grant CT17/17-CT18/17.

Conflicts of Interest: The authors declare no conflict of interest.

References

1. Westerling, A.L.; Hidalgo, H.G.; Cayan, D.R.; Swetnam, T.W. Warming and earlier spring increase western US forest wildfire activity. *Science* **2006**, *313*, 940–943. [CrossRef]
2. Alcubierre, P.C.; Ribau, M.C.; de Egileor, A.L.O.; Bover, M.M.; Kraus, P.D. *Prevention of Large Wildfires Using the Fire Types Concept*; Generalitat de Catalunya: Barcelona, Spain, 2011.
3. Séro-Guillaume, O.; Margerit, J. Modelling forest fires. Part I: a complete set of equations derived by extended irreversible thermodynamics. *Int. J. Heat Mass Transf.* **2002**, *45*, 1705–1722. [CrossRef]
4. Belval, E.J.; Wei, Y.; Bevers, M. A stochastic mixed integer program to model spatial wildfire behavior and suppression placement decisions with uncertain weather. *Can. J. For. Res.* **2016**, *46*, 234–248. [CrossRef]
5. Cova, T.J.; Dennison, P.E.; Kim, T.H.; Moritz, M.A. Setting wildfire evacuation trigger points using fire spread modeling and GIS. *Trans. GIS* **2005**, *9*, 603–617. [CrossRef]
6. Pultar, E.; Raubal, M.; Cova, T.J.; Goodchild, M.F. Dynamic GIS case studies: Wildfire evacuation and volunteered geographic information. *Trans. GIS* **2009**, *13*, 85–104. [CrossRef]
7. McCaffrey, S.; Rhodes, A.; Stidham, M. Wildfire evacuation and its alternatives: perspectives from four United States' communities. *Int. J. Wildland Fire* **2015**, *24*, 170–178. [CrossRef]
8. Finney, M.A. Design of regular landscape fuel treatment patterns for modifying fire growth and behavior. *For. Sci.* **2001**, *47*, 219–228.
9. Price, O.F.; Edwards, A.C.; Russell-Smith, J. Efficacy of permanent firebreaks and aerial prescribed burning in western Arnhem Land, Northern Territory, Australia. *Int. J. Wildland Fire* **2007**, *16*, 295–305. [CrossRef]
10. Rönnqvist, M.; D'Amours, S.; Weintraub, A.; Jofre, A.; Gunn, E.; Haight, R.G.; Martell, D.; Murray, A.T.; Romero, C. Operations research challenges in forestry: 33 open problems. *Ann. Oper. Res.* **2015**, *232*, 11–40. [CrossRef]
11. King, K.J.; Bradstock, R.A.; Cary, G.J.; Chapman, J.; Marsden-Smedley, J.B. The relative importance of fine-scale fuel mosaics on reducing fire risk in south-west Tasmania, Australia. *Int. J. Wildland Fire* **2008**, *17*, 421–430. [CrossRef]
12. Minas, J.P.; Hearne, J.W.; Martell, D.L. A spatial optimisation model for multi-period landscape level fuel management to mitigate wildfire impacts. *Eur. J. Oper. Res.* **2014**, *232*, 412–422. [CrossRef]
13. León, J.; Reijnders, V.M.; Hearne, J.W.; Ozlen, M.; Reinke, K.J. A Landscape-scale optimisation model to break the hazardous fuel continuum while maintaining habitat quality. *Environ. Model. Assess.* **2019**, *24*, 369–379. [CrossRef]

14. Penadés-Plà, V.; García-Segura, T.; Yepes, V. Robust Design Optimization for Low-Cost Concrete Box-Girder Bridge. *Mathematics* **2020**, *8*, 398. [CrossRef]
15. Caballero, J.L. Mathematical Programming with Uncertainty and Multiple Objectives for Sustainable Development and Wildfire Management. Ph.D. Thesis, Universidad Complutense de Madrid, Madrid, Spain, 2020.
16. Suffling, R.; Grant, A.; Feick, R. Modeling prescribed burns to serve as regional firebreaks to allow wildfire activity in protected areas. *For. Ecol. Manag.* **2008**, *256*, 1815–1824. [CrossRef]
17. Preisler, H.K.; Brillinger, D.R.; Burgan, R.E.; Benoit, J. Probability based models for estimation of wildfire risk. *Int. J. Wildland Fire* **2004**, *13*, 133–142. [CrossRef]
18. Castellnou, M.; Prat-Guitart, N.; Arilla, E.; Larrañaga, A.; Nebot, E.; Castellarnau, X.; Vendrell, J.; Pallàs, J.; Herrera, J.; Monturiol, M.; et al. Empowering strategic decision-making for wildfire management: Avoiding the fear trap and creating a resilient landscape. *Fire Ecol.* **2019**, *15*, 31. [CrossRef]
19. Wei, Y.; Rideout, D.; Kirsch, A. An optimization model for locating fuel treatments across a landscape to reduce expected fire losses. *Can. J. For. Res.* **2008**, *38*, 868–877. [CrossRef]
20. Cheng, H.; Hadjisophocleous, G.V. The modeling of fire spread in buildings by Bayesian network. *Fire Saf. J.* **2009**, *44*, 901–908. [CrossRef]
21. Andrino, D. Three Ways of Generating Terrain with Erosion Features. 2018. Available online: https://github.com/dandrino/terrain-erosion-3-ways (accessed on 14 February 2020).
22. Minas, J.; Hearne, J.; Martell, D. An integrated optimization model for fuel management and fire suppression preparedness planning. *Ann. Oper. Res.* **2015**, *232*, 201–215. [CrossRef]
23. Levchenkov, V. Solution of equations in Boolean algebra. *Comput. Math. Model.* **2000**, *11*, 154–163. doi:10.1007/BF02359182. [CrossRef]
24. Ibarrola, P.; Pardo, L.; Quesada, V. *Teoría de la Probabilidad*; Síntesis: Madrid, Spain, 2010.
25. Russel, S.; Norvig, P. *Artificial Intelligence: A Modern Approach*; EUA, Prentice Hall: Upper Saddle River, NJ, USA, 2003.
26. Schreiber, J. Pomegranate: Fast and flexible probabilistic modeling in python. *J. Mach. Learn. Res.* **2017**, *18*, 5992–5997.

Article

A Hybrid k-Means Cuckoo Search Algorithm Applied to the Counterfort Retaining Walls Problem

José García [1,†], Victor Yepes [2,*,†] and José V. Martí [2,†]

1 Escuela de Ingeniería en Construcción, Pontificia Universidad Católica de Valparaíso, Valparaíso 2362807, Chile; jose.garcia@pucv.cl

2 Institute of Concrete Science and Technology (ICITECH), Universitat Politècnica de València, 46022 València, Spain; jvmartia@cst.upv.es

* Correspondence: vyepesp@cst.upv.es

† These authors contributed equally to this work.

Received: 16 March 2020; Accepted: 3 April 2020; Published: 10 April 2020

Abstract: The counterfort retaining wall is one of the most frequent structures used in civil engineering. In this structure, optimization of cost and CO_2 emissions are important. The first is relevant in the competitiveness and efficiency of the company, the second in environmental impact. From the point of view of computational complexity, the problem is challenging due to the large number of possible combinations in the solution space. In this article, a k-means cuckoo search hybrid algorithm is proposed where the cuckoo search metaheuristic is used as an optimization mechanism in continuous spaces and the unsupervised k-means learning technique to discretize the solutions. A random operator is designed to determine the contribution of the k-means operator in the optimization process. The best values, the averages, and the interquartile ranges of the obtained distributions are compared. The hybrid algorithm was later compared to a version of harmony search that also solved the problem. The results show that the k-mean operator contributes significantly to the quality of the solutions and that our algorithm is highly competitive, surpassing the results obtained by harmony search.

Keywords: CO_2 emission; earth-retaining walls; optimization; k-means; cuckoo search

---

1. Introduction

Combinatorial optimization problems appear naturally in the areas of engineering and science. From the research point of view, these problems present interesting challenges in the areas of operations research, computational complexity, and algorithm theory. Examples of combinatorial problems are found in, scheduling problems [1,2], transport [2], machine learning [3], facility layout design [4], logistics [5], allocation resources [6,7], routing problems [8,9], robotics applications [10], civil engineering problem [11–13], engineering design problem [14], fault diagnosis of machinery [15], and social sustainability of infrastructure projects [16], among others. Combinatorial optimization algorithms should explore the solutions space to find optimal solutions. When the space is large, this cannot be fully explored. Combinatorial optimization algorithms address this difficulty, reducing the effective space size, and exploring the search space efficiently. Among these algorithms, metaheuristics have been a good approximation to obtain adequate solutions. However, addressing large instances and requiring almost real-time solutions for some cases, motivates lines of research that strengthen the methods that address these problems.

One way to classify metaheuristics is according to the search space in which they work. In that sense, we have metaheuristics that work in continuous spaces, discrete and mixed spaces [17]. An important line of inspiration for metaheuristic algorithms are natural phenomena, many of which develop in a continuous space. Examples of metaheuristics inspired by natural phenomena in continuous spaces are: Optimization of particle swarm [18], black hole [19], cuckoo search (CS) [20], Bat algorithm [21], Algorithm of fire fly [22], fruit fly [23], artificial fish swarm [24], gravitational search algorithm [25], among others. The design of discrete versions of these algorithms carries important challenges because they must try to preserve the intensification and diversification properties [17].

With the aim of modifying and improving the results of a metaheuristic algorithm, the main tool available to them is tuning their control parameters. These types of modifications correspond to small modifications of the algorithm since the operating mechanism of the algorithm is not altered. However, many optimization problems cannot be efficiently addressed by an algorithm through modification of its parameters and require deep modifications that alter its mechanism of operation [17,26]. A strategy that has strengthened the results of metaheuristic algorithms has been the hybridization of these with techniques that come from the same and the other areas. Hybridization strategies involve deep changes to the original algorithm. The main hybridization proposals found in the literature are the following: (i) matheuristics, where mathematical programming is combined with metaheuristic techniques [27], (ii) hybrid heuristics, hybridization between different metaheuristic methods [28], (iii) simheuristics, that combine simulation and metaheuristics [29], and (iv) the hybridization between machine learning and metaheuristics. The latest line of research exploring the integration between the areas of machine-learning and metaheuristic algorithms is an emerging line of research in the areas of computing, mathematics, and operational research. In developing the state-of-the-art, we find that hybridization occurs primarily for two purposes. The first, with the goal that metaheuristics help machine-learning algorithms improve their results (for example, [30,31]). In the second intention, machine-learning techniques are used to strengthen metaheuristic algorithms (for example, [32,33]). The details of the hybridization forms are specified in Section 2.

In this article, inspired by the research lines mentioned above. A hybrid algorithm is designed, which explores the application of a machine-learning algorithm in a discrete operator to allow continuous metaheuristics to address combinatorial optimization problems. We will apply our hybrid algorithm to the discrete problem of the design of counterfort retaining walls. The contributions of this work are detailed below:

- A machine-learning algorithm is proposed to allow metaheuristics commonly defined and used in continuous optimization addressing discrete optimization problems simply and effectively. To perform this process, the algorithm uses k-means. This clustering technique has been selected because it has solved other binary combinatorial problems efficiently [5,33]. The selected metaheuristic is CS. Its selection is because it has been frequently used in solving continuous optimization problems and its tuning is relatively simple, which allows focusing on the discretization process.
- A random operator is designed to study the contribution of the k-means operator in the discretization process.
- This hybrid algorithm is applied to the design of the buttresses retaining wall problem. Optimization is carried out for cost and emissions of CO_2. A comparison is made between the proposed hybrid algorithm and an adaptation of the harmony search (HS) proposed in [34]. The design of the counterfort retaining walls will be detailed in Section 3.

The structure of the remaining of the paper is as follows: a state-of-the-art of hybridizing metaheuristics with machine learning is developed in Section 2. In Section 3 the optimization problem, the variables involved, and the restrictions are defined. The discrete k-means algorithm is detailed in Section 4. The experiments and results obtained are shown in Section 5. Finally, in Section 6 the conclusions and new lines of research are summarized.

2. Hybridizing Metaheuristics with Machine Learning

Metaheuristics form a wide family of algorithms. These algorithms are classified as incomplete optimization techniques and are usually inspired by natural or social phenomena [17,35]. The main objective of these is to solve problems of high computational complexity and they own the property of not having to deeply alter their optimization mechanism when the problem to be solved is modified. On the other hand, machine-learning techniques correspond to algorithms capable of learning from a dataset [36]. If we make a classification according to the way of learning, there are three main categories: Supervised learning, unsupervised learning, and learning by reinforcement. Usually, these algorithms are used in problems of regression, classification, transformation, dimensionality reduction, time series, anomaly detection, and computational vision, among others.

In the state-of-the-art of algorithms that integrate machine-learning techniques with metaheuristic algorithms, we have found two main approaches [26]. In the first approach, machine-learning techniques are used in order to improve the quality of the solutions and convergence rates obtained by the metaheuristic algorithms. The second approach uses metaheuristic algorithms to improve the performance of machine-learning techniques [26]. Usually, the metaheuristic is responsible for solving more efficiently an optimization problem related to the machine-learning technique. Adapted and extended from [26], in Figure 1, we have proposed a general scheme of techniques where machine learning and metaheuristics are combined.

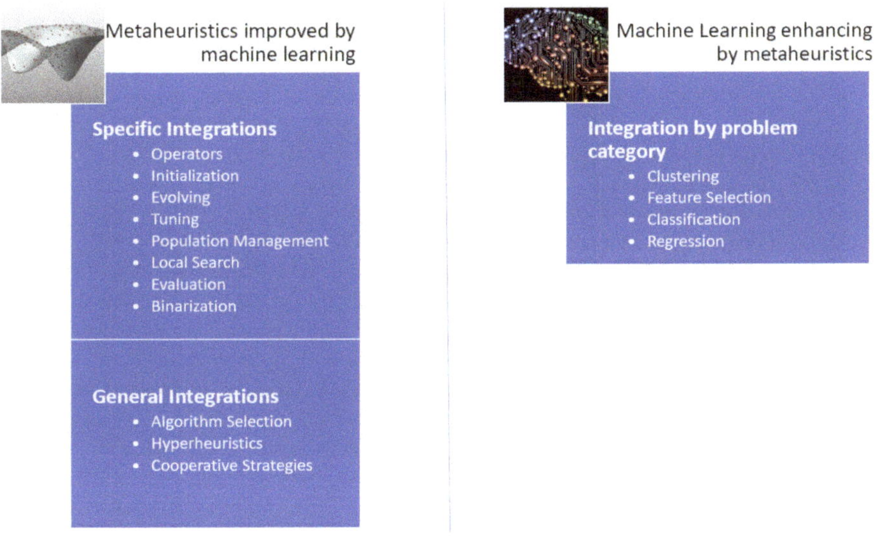

Figure 1. General scheme: Combining Machine Learning and Metaheuristics.

When we analyze the integration mechanisms of the first approach, we identify two lines of research. In the first line, machine-learning techniques are used as metamodels in order to select different metaheuristics, therein choosing the most appropriate for each instance. The second line aims to use specific operators that make use of machine-learning algorithms and subsequently this specific operator is integrated into a metaheuristic [26].

In the general integration of machine-learning algorithms on metaheuristic techniques we find three main groups: algorithm selection, hyper-heuristics, and cooperative strategies. In algorithm selection, the goal is to choose from a set of algorithms and considering the associated characteristics for each instance of the problem, an algorithm that works best for similar instances. Another way to approach the problem is to automate the design of heuristic or metaheuristic methods to tackle a wide range of problems, we call this hyperheuristic approximation. Finally, the objective of cooperative strategies is to mix algorithms in a parallel or sequential way to obtain more robust methods. The cooperation mechanism can be complete, i.e., sharing the complete solution, or partial when only part of the solution is shared. In [32], the problem of docking scheduling in massive terminals was addressed by algorithm selection techniques. In [37], the nurse training problem through a tensor-based hyperheuristic algorithm was addressed. Finally, a distributed framework that uses agents was proposed in [38]. In this framework, each agent corresponds to a metaheuristic and the framework can adapt through direct cooperation. This framework was applied to the problem of permutation flow stores.

Additionally, a metaheuristic incorporates operators that allow strengthening its performance. Examples of these operators are initialization operators, solution perturbation, population management, binarization, parameter configuration, and local search operators, among others [26]. Specific integrations explore the machine-learning application in some of these operators [26]. In the design of binary versions of algorithms that work naturally in continuous spaces, we find binarization operators in [2]. These binary operators use unsupervised learning techniques to perform the binarization process. In [39], the concept of percentile was explored in the process of generating binary algorithms. In addition, in [1], the Apache spark big data framework was applied to manage the population size of solutions to improve convergence times and the quality of results. Another interesting research line was found in the adjustment of metaheuristic parameters. In [40], the parameter setting of a chess classification system was implemented. Based on decision trees, and using fuzzy logic, a semi-automatic parameter setting algorithm was designed in [41]. Usually, in the initiation of solutions, a random mechanism is used. However, using machine-learning techniques, algorithms have been developed that improve the initiation stage of a solution. Applied to the problem of designing a weighted circle in [42], a case-based reasoning technique was used to start a genetic algorithm. In [43] Hopfield neural networks were applied to initiate solutions of a genetic algorithm. This initiation was applied to an economic dispatch problem.

To improve the quality of machine-learning algorithms, we see that metaheuristics have a broad contribution. Contributions are found in problems of feature selection, classification, clustering, extraction of characteristics, among others. In the identification of breast cancer, in [44] a genetic algorithm was designed that improves the performance of support vector machine (SVM) when applied in image analysis. The genetic algorithm was used in the characteristic extraction stage from the images. In [45] a multiverse algorithm was used to adjust the parameters of an SVM classifier. An improved monarch butterfly algorithm was used in [46] with the goal to optimize the weights of a feed-forward neural network. In [47], a geotechnical problem was addressed. In this article, a firefly algorithm was integrated with the least-squares support vector machine technique. When considering regression problems, there are again several cases where metaheuristics have contributed. For example, in predicting the strength of high-performance concrete, in [48] a metaheuristic was used to improve the least-squares technique. In [49], share price prediction was improved by integrating metaheuristics into a neural network. Again, in the prediction of the share price of construction companies in Taiwan, in [30], a sliding-window metaheuristic-optimized machine-learning algorithm was designed, which integrates a metaheuristic in its learning process. Optimization of the parameters of a least-squares regression was improved, in [50] using a firefly algorithm. We also found metaheuristic applications for unsupervised learning techniques. In particular, there are significant contributions from metaheuristics to clustering techniques. For example, in [51], we found an algorithm application that integrates a metaheuristic with a kernel intuitionistic fuzzy c-means. This algorithm was applied to different data sets. Another integration found for clustering techniques is the search for centroids. This problem is to find the set of centroids that best group the data under a certain metric. This is an NP-hard type problem and therefore it is natural to approach it with a metaheuristic. In [52] an algorithm was proposed that uses a bee colony to solve a problem of grouping energy efficiency data obtained from a wireless sensor network. In [53], a metaheuristic was used to find centroids in planning the transportation of employees from an oil platform via helicopters. In the case of neural networks, metaheuristics have contributed to different types of integration. In [54] a metaheuristic algorithm is used to automatically find the appropriate number of neurons in a feed-forward neural network. In [55] a metaheuristic algorithm allows determining the dropout probability in convolutional neural networks. The cuckoo search and bat algorithm algorithms were used in [56] to adjust the weights of neural networks of feed-forward type. In [57] an algorithm was designed using metaheuristics to optimize convolutional neural networks. An application to long term short memory (LSTM) neural network training was studied in [58]. LSTM networks were applied in healthcare analysis.

Inspired by the lines of research detailed above, this work proposes a hybrid algorithm in which the unsupervised k-means learning technique is used to obtain binary versions of the cuckoo search optimization algorithm. This hybrid algorithm was used to solve the problem of the counterfort retaining walls. The k-means technique has been widely used and in recent studies, it has been applied in [59] to bioinformatics for detecting gene expression profile, image segmentation for pest detection [60] in agriculture, and brain tumor identification [61], among others. Particularly the k-means technique has been previously applied in obtaining binary versions of continuous metaheuristics and used to solve the multidimensional knapsack problem [33] and the set covering problem [5] which are NP-hard problems. In the case of the cuckoo search, hybrid versions of this algorithm have been applied in [62] to numerical optimization problems in engineering. In [63] cuckoo search was applied to benchmark functions and engineering design problems. An algorithm with reinforced learning was used in [64] to solve a logistic distribution problem. In [65] an improved version of the cuckoo search was applied to drone location.

3. Problem Definition

In this Section, we will detail the buttressed earth-retaining walls optimization problem. In Section 3.1 we will give the definition of the optimization problem. Later in Section 3.2, we will detail the design variables. Then the design parameters will be described in Section 3.3, and finally, in Section 3.4, we define the constraints.

3.1. Optimization Problem

The optimization problem considers two objective functions, which will be addressed independently. The first function corresponds to the cost (p_i) of the wall construction, expressed in euros. The second one considers the CO_2 equivalent emissions units (e_i). These construction units correspond to formwork, materials, excavation and earth-fill. The cost and emission functions are based on a 1 m wide strip [34]. The emission and cost values were obtained from [34,66] and are shown in Table 1. Then, in a general way, our optimization problem is defined according to Equation (1).

$$O(x) = \sum_{i=1}^{r} a_i x_i \tag{1}$$

where x is the set of decision variables and $a_i \in \{c, e\}$, corresponds to cost or emissions. Additionally, the optimization problem is subject to a set of restrictions determined by the ultimate (ULS) and service (SLS) state limits.

3.2. Problem Design Variables

In the design of the buttressed retaining wall to study, three groups of variables are defined. The geometric variables, the concrete and steel grade, and the passive reinforcement of the wall. In total there are 32 variables where each of these has a discrete amount of possibilities. In the group of concrete and steel grades. Concrete HA-25 to HA-50 is considered in discrete intervals of 5 MPa. In the case of steel, the B500S and B400S types are considered. In the group of geometric variables, there is the thickness of the footing (c), thickness of the stem (b), the length of the toe (p), the thickness of the buttresses (e_c), the length of the heel (t) and the distance between buttresses (d). The last group of variables that correspond to the passive reinforcement of the wall. There are 24 variables shown in Figures 2 and 3. The diameter and the number of bars define the reinforcement. Three reinforcement flexural bars defined as A1, A2 and A3 contribute in the main bending of the stem. The vertical reinforcement of foundation in the rear side of the stem is given by A4, up to a height L1. The secondary longitudinal reinforcement is given by A5 for shrinkage and thermal effects in the stem. The longitudinal reinforcement of the buttress is given by A6. The area of reinforcement bracket from the bottom of the buttress is given by A7 and A8. The upper and bottom heel reinforcement are defined by A9 and A11 and the shear reinforcement in the footing by A12. The longitudinal effects in the toe are defined by A10. The set of combinations of the values for the 32 variables shown in Table 2, constitutes the solution space.

Table 1. Unit breakdown of emissions and cost.

Unit	Emissions (CO_2-eq)	Cost (€)
kg of steel B400	3.02	0.56
kg of steel B500	2.82	0.58
m^3 of concrete HA-25 in stem	224.34	56.66
m^3 of concrete HA-30 in stem	224.94	60.80
m^3 of concrete HA-35 in stem	265.28	65.32
m^3 of concrete HA-40 in stem	265.28	70.41
m^3 of concrete HA-45 in stem	265.91	75.22
m^3 of concrete HA-50 in stem	265.95	80.03
m^2 stem formwork	1.92	21.61
m^3 of backfill	28.79	5.56
m^3 of concrete HA-25 in foundation	224.34	50.65
m^3 of concrete HA-30 in foundation	224.94	54.79
m^3 of concrete HA-35 in foundation	265.28	59.31
m^3 of concrete HA-40 in foundation	265.28	64.40
m^3 of concrete HA-45 in foundation	265.91	69.21
m^3 of concrete HA-50 in foundation	265.95	74.02

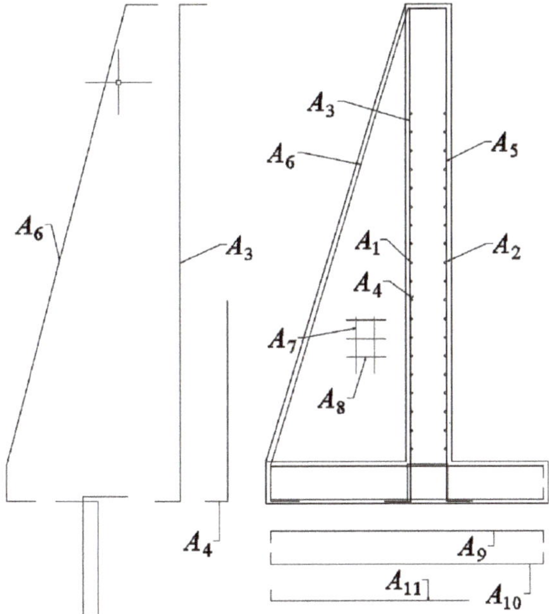

Figure 2. Reinforcement variables for the design of earth-retaining walls. *Source:* [67].

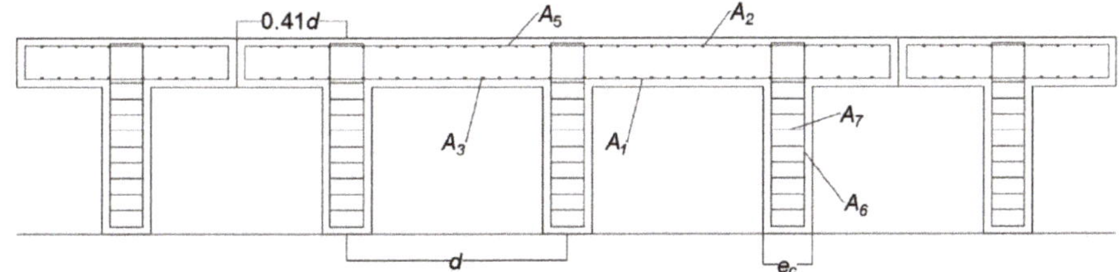

Figure 3. Earth-retaining buttressed wall. Floor cross-section. *Source:* [67].

Table 2. Design variables.

Variables	Lower Bound	Increment	Upper Bound	N of Values
c	H/20	5 cm	H/5	f(H) [1]
b	25 cm	2.5 cm	122.5	40
p	20 cm	10 cm	610	60
t	20 cm	15 cm	905	60
e_c	25 cm	2.5 cm	122.5	40
d	H/5 cm	5 cm	2H/3	f(H) [1]
f_{ck}	25, 20, 25, 40, 45, 50			7
f_{yk}	400, 500			2
A_1 to A_{10}	6, 8, 10, 12, 16, 20, 25, 32			8
	1 steel rebar	2 rebars	12 rebars	6
A_{11} to A_{12}	6, 8, 10, 12, 16, 20, 25, 32		8	
	1 steel rebar	4 rebars	10 rebars	7

(1) Number of values depends on the height.

3.3. Problem Design Parameters

The data that will remain fixed in the optimization process will be called problem design parameters. The main design parameters are shown in Figure 4. The height of the wall and the depth of the soil in front of the wall are denoted by H and H2, respectively. The maximum bearing pressure is the soil foundation ultimate bearing capacity divided by the bearing capacity factor of safety. The maximum support pressure for the operating conditions is represented by σ, the base-friction coefficient is μ and the backfill slope at the top of the rod is β. The density, the internal friction angle and the friction angle that determine the earth's pressure angle are given respectively by P (γ, ϕ, δ). The roughness between the wall and the fill is determined by a fraction of ϕ. The values of the problem design parameters are shown in Table 3.

Table 3. Problem design parameters values.

Parameter Considered	Value
Bearing capacity	0.3 MPa
Fill slope	0
Foundation depth, H2	2 m
Uniform load on top of the fill, γ	10 kN/m^2
Wall-fill friction angle, δ	0°
Base-friction coefficient, μ	tg 30°
Safety coefficient against sliding, γ_{fs}	1.5
Safety coefficient against overturning, γ_{fo}	1.8
EHE safety coefficient for loading	Normal
ULS safety coefficient of concrete	1.5
ULS safety coefficient of steel	1.15
EHE ambient exposure	IIa

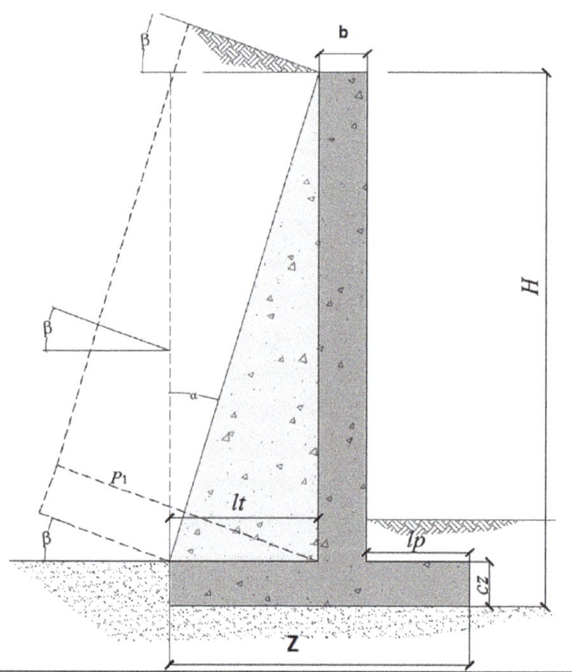

Figure 4. Problem design parameters. *Source:* [67].

3.4. Problem Constraints

The feasibility of the structure is verified in accordance with the Spanish technical standards defined in [68] and the recommendations detailed in [69]. The flexural and shear limit states are checked. The structure is verified in accordance with the approach specified in [70]. To check the structure limit states, a uniform surface load at the top of the fill is considered [71]. For the active earth pressure calculation the surface loads and fill are considered. The major forces in wall analysis consider wall weight, heel backfill load, surface load, earth pressure, weight at the front toe, and passive resistance against the toe. Buttresses receive a load equivalent the product of the distance between the buttresses by the pressure distribution in the stem.

The structural model considers that the upper part of the stem works as a cantilever, while the lower part of the stem is strongly coerced by two elements: the footing and the lower part of the buttress, located at the rear of the stem. Calculation bending moments are taken in the midsection between the buttresses and are given by B_1 and B_2 described in Equations (2) and (3) respectively.

$$B_1 = -0.03p_1d(H-c) \tag{2}$$

$$B_2 = -0.0075p_1d(H-c) \tag{3}$$

B_1 is the bending moment at the connection of the stem to the footing, B_2 is the maximum bending moment on the stem and p_1 represents the pressure over the slab on the upper side of the footing. When the spacing of the buttresses is less than 50% of the height, Equation (4) defines the shear resistance (s) at the connection of the plate to the footing. For an accurate estimate of the moments in each section of the stem, as a result of the vertical bending stress in the stem, the trapezoidal pressure distribution is considered [70]. 50% of the maximum pressure at the top of the foundation is taken as the maximum value. Taking into account the vertical bending moment in the upper quarter part of the stem may be insignificant due to the involvement of [70].

$$s = 0.4p_1d \tag{4}$$

Verification of the bending stress in the T-shaped horizontal cross-section is made considering the effective width, according to [72]. Mechanical bending and shear capacity are evaluated using the equations expressed in [71]. In this manual, the construction limit states of the EHE Structural Concrete Code are considered. The checks against sliding, overturning, and soil stresses, are carried out taking into account the effect of the buttresses, and are given in Equations (5)–(7). In the overturning check, Equation (5) ensures that the favorable overturning moments are sufficiently greater than the unfavorable overturning moments. In Equation (6), B_{of} is the total favorable overturning moment. In Equation (7), B_{ou} is the unfavorable total overturn moment, and the overturning safety factor is γ_{to} and is taken as 1.8 for frequent events. Equation (8) represents the reaction of soil against sliding. As μ is the base-friction coefficient, N' corresponds to the total sum of the ground and wall weights located at the heel and toe, and E_p determines the passive resistance against the toe defined by Equation (9).

$$B_{of} - \gamma_{to}B_{ou} \geq 0 \tag{5}$$

$$B_{of} = N'(\frac{B}{2} - e_p) - E_p(H_t - h') \tag{6}$$

$$B_{ou} = E_h * h_e - E_v(\frac{B}{2} - f) \tag{7}$$

$$R = N'\mu + E_p \tag{8}$$

$$E_p = \frac{1}{2\gamma(H_t^2 - (H_t - c))^2} \frac{(1 + sin(\phi))}{(1 - sin(\phi))} \tag{9}$$

4. The K-Means Discrete Algorithm

The first stage of the algorithm corresponds to generating valid solutions randomly. In the generation procedure, it is first validated if all the variables that make up the solution have been started. If not, the variables are started randomly. Once all the variables are generated, the next step is to validate if the solution obtained is feasible. In the event that it is not feasible, all the variables are cleaned, and they have generated again. The detail of the initiation procedure is shown in Figure 5. Once we have a valid set of solutions, continuous metaheuristics are used in this case CS, to produce a new solution in continuous space. The CS algorithm will be described in Section 4.1. Subsequently, we apply the k-means operator to transform continuous movements into transition probabilities. The k-means operator will be detailed in Section 4.2. The result of the transition probabilities generated by k-means are used by the discretization operator to generate a new solution. The discretization operator is detailed in Section 4.3. Finally, the new solution is validated and, in the case that it complies with the restrictions, it is compared with the best solution obtained. In the case that the new one is superior, it is replaced. The detail flow chart of the solution is shown in Figure 6.

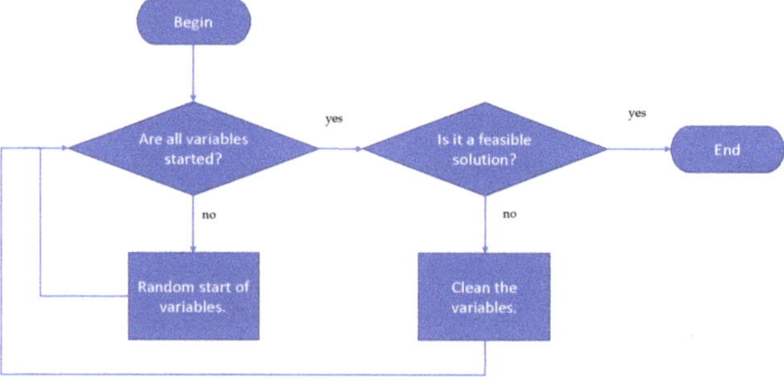

Figure 5. Solution initiation procedure.

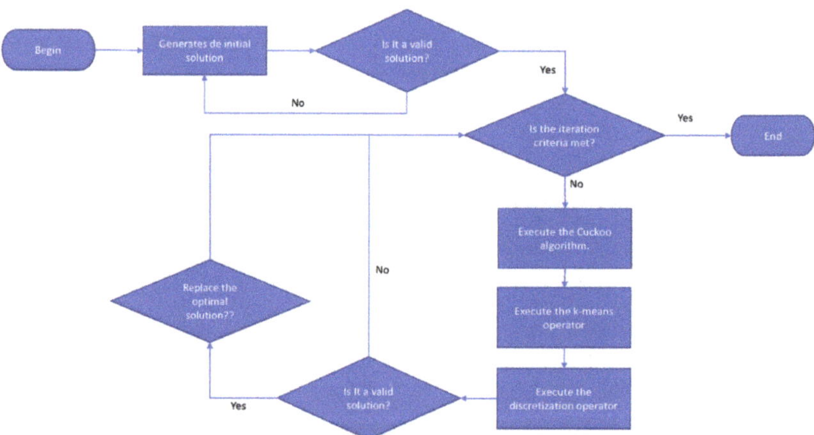

Figure 6. The discrete k-means algorithm flow chart.

4.1. Cuckoo Search Algorithm

The phenomenon of cuckoo species, which lay their eggs in the nests of other bird species, has inspired the CS algorithm. Such is the level of sophistication of cuckoo birds that in some cases even the colors and patterns of the eggs of the chosen host species are mimicked. In the analogy, an egg corresponds to a solution. The concept behind the analogy is to use the best solutions (cuckoos) with the aim of replacing those that do not perform well. The CS algorithm uses three basic rules:

1. Each cuckoo lays one egg at a time and deposits its egg in a randomly chosen nest.
2. The nests with the best results, i.e., with high-quality eggs, will be considered in the next generation.
3. The number of nests available is a fixed parameter. The egg laid by a cuckoo can be discovered by the host bird with a probability $p_a \in (0, 1)$

The algorithm pseudo-code is shown in Algorithm 1.

Algorithm 1 Cuckoo search algorithm

```
Objective function f(x)
Generate initial solutions of n host nests.
while stop criterion are meet do
    Get a cuckoo randomly and replace using Lévy flights.
    Evaluate the fitness.
    Choose in a random way a nest j among n:
    if f_i > f_j then
        replace the solution.
    end if
    portion p_a of the worst nests are eliminated and new ones are created.
    keep best solutions.
    find the current best
end while
```

4.2. k-Means Operator

The goal of the k-means operator is to group the different solutions obtained by executing continuous metaheuristics in this case CS. However, this operator can be applied to any continuous swarm intelligence metaheuristics. When we consider solutions as particles, we understand the position of the particle as the location of the solution in the search space, while the velocity represents the vector of the transition of the solution from the t iteration to the $t + 1$ iteration.

In the k-means operator (kmeanOp), the k-means clustering technique is used to make transitions in a discrete space. The proposal is based on using the movements generated by the CS metaheuristic in each dimension for all the particles. Let $x(t)$ be a solution in iteration t, then $\Delta^i(x(t))$ represents the magnitude of the displacement $\Delta(x(t))$ in the i-th position, considering the iterations t and $t + 1$. Once all the displacements have been calculated, they are grouped using the magnitude of the displacement $|\Delta^i(x(t))|$, we do not consider the sign. This grouping is done using the k-means technique where k represents the number of groups used. Finally, we propose a generic function P_{tr} shown in Equation (10) with the objective of assigning a transition probability to each displacement.

$$P_{tr} : \mathbb{Z}/k\mathbb{Z} \rightarrow [0, 1] \tag{10}$$

Then using the function P_{tr}, a probability is assigned to each group obtained from the clustering process. In this article, we use the linear function given in Equation (11). Where Clust (x^i) indicates the location of the group to which $\Delta^i(x)$ belongs. The coefficient α represents the transition probability and the coefficient β models the transition separation for the different groups. Both parameters must be estimated. The pseudo-code of the discretization procedure is shown in Algorithm 2.

$$P_{tr}(x^i) = P_{tr}(Clust(x^i)) = \alpha + \beta Clust(x^i)\alpha \tag{11}$$

Algorithm 2 k-means operator

Function kmeanOp($lx(t), lx(t+1)$)
Input $x(t)$, $x(t+1)$
Output lTranProb(t+1)
$l\Delta^i(x(t+1)) \leftarrow$ getDelta($lx(t), lx(t+1)$)
$Clust \leftarrow$ getClusters($l\Delta^i(x(t))$,k)
$lTranProb(t+1) \leftarrow$ getTranProb($Clust, lx(t)$)–Equation (11)
return ($lTranProb(t+1)$)

4.3. Discretization Operator

As input the operator receives the transition probabilities ($lTranProb(t+1)$) obtained in the previous stage and the list of solutions $lx(t)$. For each solution $x(t) \in lx(t)$, we consider each component $x^i(t)$ and evaluate if it should be modified according to its transition probability. The transition probability is compared with a random number r_1. In case the change is selected, the movement can increase the value (+1) or decrease it (-1). Finally, the selected value is compared with the best value obtained by the algorithm and remains with the minimum of both. The pseudo-code of the discrete procedure is shown in Algorithm 3.

Algorithm 3 Discretization operator.

Function DiscOp($lTranProb(t+1), lx(t)$)
Input $lTranProb(t+1)$
Output $x(t+1)$–Where x(t+1) is discrete.
movement = 0
for $x^i \in x(t) \in lx(t)$ **do**
  if $r_1 > 0.5$ **then**
    movement = 1
  else
    movement = -1
  end if
  x^i = max(1,min(x^i_{best},x^i + movement))
end for

5. Results and Discussion

This Section aims to describe the results obtained by the hybrid algorithm when applying it to the counterfort retaining walls problem. In the first stage, the contribution of the k-means operator in the discretization process will be determined. This is described in Section 5.2. Later, in a second stage, we will compare our proposal with another algorithm that has solved the problem, Section 5.3. For each problem, we make 30 independent runs. The value 30 is widely accepted for statistical conclusions [73]. The Wilcoxon signed-rank [74] method was the test selected to determine if the difference is statistically significant. The p-value used was 0.05. The selection of the test is based on the methodology proposed in [73]. In this methodology, the Shapiro–Wilk or Kolmogorov—Smirnov-Lilliefors normality test is applied first. In the event that one of the populations is not normal, and both populations have the same number of points, the Wilcoxon signed-rank is suggested to verify the difference. In our case, the Wilcoxon test was applied to the entire population of instances considering all executions and comparing the results of both algorithms. For the execution of the instances, a laptop with a Core i7-4770 as a processor and 16GB in RAM has been used. The algorithm was programmed in Python 3.6.

5.1. Parameter Settings

To calibrate the parameters, heights 8 and 12 were used. These heights were selected with the intention of considering walls of different complexity. 8 represents small size walls and 12 represents larger size walls where the satisfaction of stability restrictions is much more difficult. Subsequently, each configuration was executed 5 times for each height considering all the configurations proposed in Table 4. The value 5 was chosen with the intention of being able to execute all the combinations in a limited time. In Table 4, the range column shows the scanned values to perform the CS adjustment. These ranges were inspired by previous studies in which the k-means technique was applied to solve the multidimensional knapsack problem [33] and the problem of the set covering [5]. The flow chart of the parameter settings is shown in Figure 7. For more detail on the method, reference [33] can be consulted.

To determine the best configuration, the method proposed in [33] was used. In this method, 4 measurements are defined that I know are shown in Equations (12)–(15). The BestGlobalValue represents the best value obtained considering all the configurations. The BestLocalValue, corresponds to the best value obtained in the evaluated configuration. The minGlobalTime is the minimum convergence time considering all settings. Finally, the convergenceLocalTime represents the average convergence time for a configuration. Each of the measures defined, the closer to 1, the better performance that indicator is having. On the other hand, the closer to 0 the worse performance. Since there are 4 measurements to be able to make the evaluation, we incorporate them into a radar chart and calculate their area. As a consequence of the measurement definition, the largest area corresponds to the configuration that has the best performance considering the 4 defined measures.

1. The deviation of the best local value obtained in five executions compared with the best global value:
$$bSolution = 1 - \frac{BestGlobalValue - BestlocalValue}{BestGlobalValue} \quad (12)$$
2. The deviation of the worst value obtained in five executions compared with the best global value:
$$wSolution = 1 - \frac{BestGlobalValue - WorstLocalValue}{BestGlobalValue} \quad (13)$$
3. The deviation of the average value obtained in five executions compared with the best global value:
$$aSolution = 1 - \frac{BestGlobalValue - -- AverageLocalValue}{BestGlobalValue} \quad (14)$$
4. The convergence time for the average value in each experiment is normalized according to Equation (15).
$$nTime = 1 - \frac{convergenceLocalTime - minGlobalTime}{maxGlobalTime - minGlobalTime} \quad (15)$$

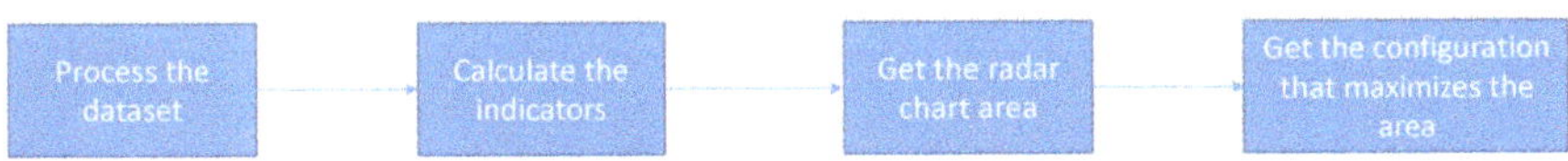

Figure 7. Flow chart of parameter setting.

Table 4. Parameter setting for the Cuckoo search algorithm.

Parameters	Description	Value	Range
N	Number of Nest	5	[5, 10, 15]
k	Number of transition groups K-means Operator	5	[4, 5, 6]
γ	Step Length	0.01	0.01
κ	Lévy distribution parameter	1.5	1.5
Iteration Number	Maximum iterations	800	[800]

5.2. Random Operator

This experiment aims to quantify the contribution of the k-means operator in the process of optimizing the cost of the wall. To achieve this objective, a random operator was designed which replaces the k-means operator described in Section 4.2. This random operator behaves very similarly to that of k-means but instead of assigning transition probabilities per cluster, it assigns a fixed transition probability with a value of 0.5. In the experiment, wall height values between 6 and 14 m are considered. The k-means operator is compared with the random operator through the best value and the average obtained over 30 runs for each height. Later we will use the violin charts to compare both distributions of results and finally the Wilcoxon signed-rank test to validate that the difference is significant.

The results are shown in Table 5 and Figure 8 for cost optimization. In the case of emission optimization, the results are detailed in Figure 9 and in Table 6. In the case of cost optimization when analyzing the best value indicator, we observe that for all heights the k-means operator obtains better results compared to the random operator. In the case of a height of 6 and 7 m, this difference does not reach 2%. However, as the wall grows, the difference increases, reaching in the case of a height of 14 (m), 26%. The average indicator has similar behavior to that previously analyzed. The k-means operator has better performance in all cases. When applying the Wilcoxon test, it indicates that the difference is significant. When analyzing the distributions shown in Figure 8, we observe that the interquartile ranges obtained by the k-means operator are completely displaced to values close to zero with respect to the results obtained by the random operator. On the other hand, the dispersion of the distribution is greater in the solutions obtained by the random operator. In the case of optimization of CO_2 emissions, when analyzing the best value indicator, we observe that the k-means operator obtains better results at all heights. However, the maximum difference from the random operator does not exceed 10%. The average indicator is again higher in the case of k-means, the Wilcoxon test indicating that the difference is significant. When analyzing the violin plots of the emissions shown in Figure 9, we see that the interquartile range obtains better quality results in the case of the k-means discretization. However, the result is not as remarkable as in the case of cost optimization.

Table 5. Comparison between random and k-means operators in cost optimization.

Height (m)	Best Value k-Means	Avg k-Means	Best Value Random	Avg Random	Best Value HS	Avg Random HS
6	591	595.5	600	621.3	595	600.15
7	678	682.8	687	721.4	689	694.98
8	775	778.9	785	851.7	784	788.38
9	911	922.3	981	1094.3	934	941.29
10	1095	1127.0	1184	1296.7	1130	1143.64
11	1302	1384.5	1545	1704.5	1354	1381.50
12	1528	1608.9	1839	1994.3	1590	1707.24
13	1775	1905.3	2241	2510.8	1840	2067.37
14	2049	2301.6	2775	3267.1	2154	2348.71
Wilcoxon p-value				1.6×10^{-5}		1.31×10^{-3}

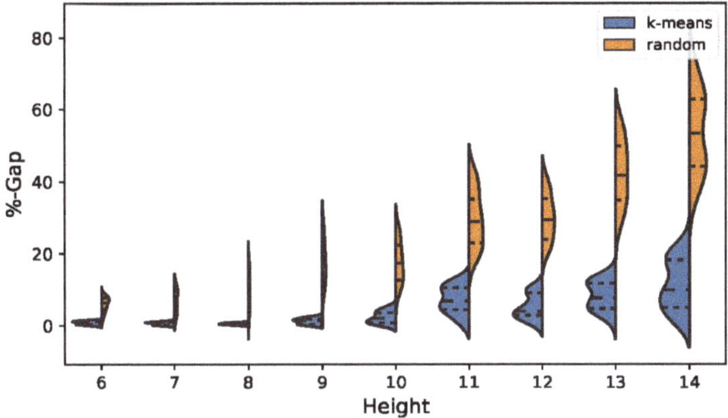

Figure 8. Cost violin plots, comparison between k-means and random operators.

Table 6. Comparison between random and k-means operators in emission optimization.

Height (m)	Best Value k-Means	Avg k-Means	Best Value Random	Avg Random	Best Value HS	Avg Random HS
6	1242	1251.2	1274	1304.3	1250	1289.14
7	1440	1458.5	1467	1501.3	1478	1511.53
8	1659	1683.2	1696	1788.6	1699	1731.54
9	1997	2111.1	2180	2214.6	2050	2097.45
10	2470	2572.1	2975	3132.7	2560	2617.81
11	3061	3206.6	3540	3801.7	3124	3201.45
12	3715	3921.5	4168	4555.7	3865	4046.95
13	4470	4676.4	5039	5484.5	4650	4955.95
14	5294	5621.2	5877	6319.1	5550	6241.00
Wilcoxon p-value				1.2×10^{-7}		2.51×10^{-4}

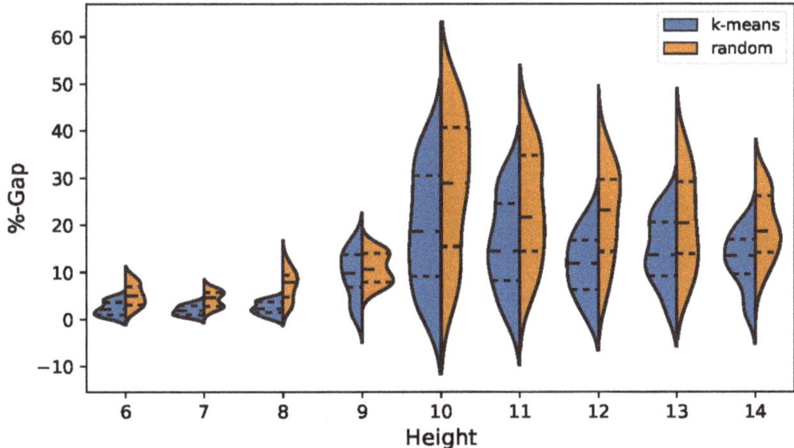

Figure 9. Emission Violin plots, comparison between k-means and random operators.

Figure 10a,b show the convergence diagrams for cost optimization for the algorithm using k-means and random, respectively. The heights chosen were 6, 9, 12 and 14. From Figure 10a it is observed that heights 6 and 9 have a better convergence than in the case of 12 and 14. The same effect is observed for the case of the random operator, Figure 10b, although in the latter case it is not so notorious. When we make a comparison between the graphs, we observe for the random case the slope stabilizes much earlier than in the case of k-means, as a consequence of this the latter obtains better results.

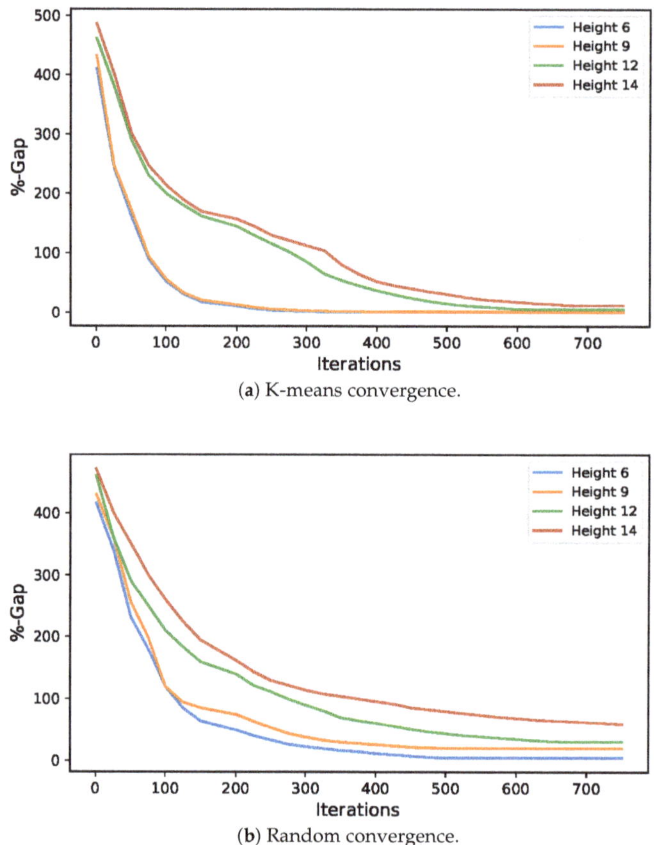

(**a**) K-means convergence.

(**b**) Random convergence.

Figure 10. K-means and Random convergence chart for cost optimization.

5.3. Comparisons

To evaluate our algorithm in a different scenario than that of a random operator, in this Section we compare the results obtained by the algorithm that uses discretization by k-means, with the results published in [34,67]. In these works, the harmony search algorithm was used to optimize the buttress retaining wall. To evaluate the comparison, the best value will be analyzed for each of the different heights of the wall, in addition to comparing the distribution of the total solutions obtained for the different heights through violin plots. To determine that the difference is statistically significant, we will use the non-parametric Wilcoxon signed-rank test.

In Figures 11 and 12, the best values obtained for the k-means and HS algorithms are compared. In the comparison, all the parameters were kept fixed except for the height, which like the previous experiments, varied from 6 to 14 m. When analyzing the heights of 6 to 8 m in Figure 11, it is observed that the cost of the solutions obtained by both algorithms is similar. As the wall grows in height, the curves have a greater separation, with the height of 14 (m) obtaining the greatest difference, this being 4.87% in favor of k-means. In the case of the comparison of the algorithms when optimizing the emissions of CO_2, the curves behave similarly to that of the cost optimization case. For small values of wall height, very similar values are obtained. As the height of the wall increases, the quality of the k-means solutions improves compared to the HS. this is seen in Figure 12.

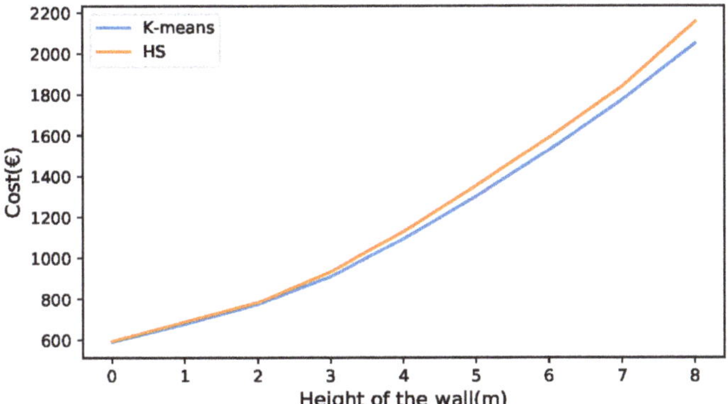

Figure 11. Comparison between the best solutions obtained by the k-means and HS algorithms in cost optimization.

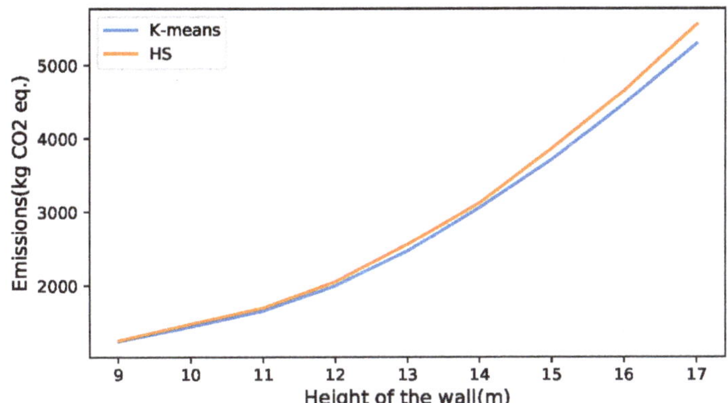

Figure 12. Comparison between the best solutions obtained by the k-means and HS algorithms in emission optimization.

Figures 13 and 14 show the comparison of the total solutions obtained by both algorithms. In the case of the cost optimization shown in Figure 13, we see that the interquartile range of the solutions obtained by k-means has better performance than those of HS. Up to height 12, the dispersion of the solutions is kept quite small in both algorithms. From height 12 onwards, this dispersion increases considerably in both k-means and HS. The case of the optimization of emissions presents a similar behavior, increasing its dispersion considerably from height 13. The above is observed in Figure 14.

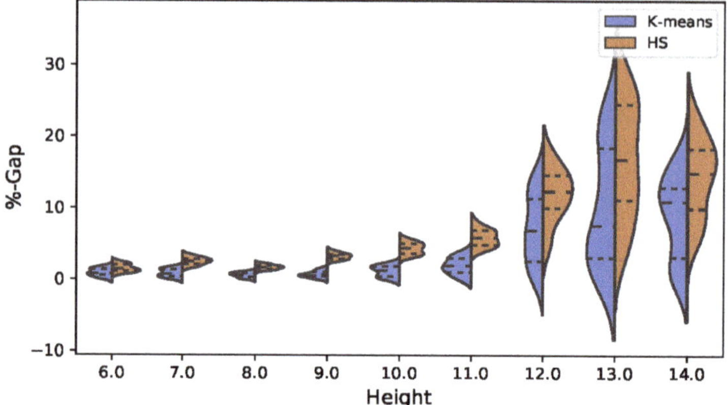

Figure 13. Violin plot comparison between k-means and HS results for cost in euros.

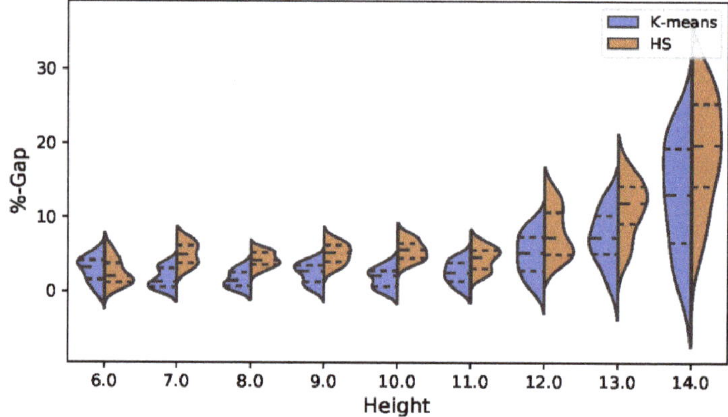

Figure 14. Violin plot comparison between k-means and HS results for CO_2 emissions.

6. Conclusions

In this article, a hybrid algorithm was proposed which uses the unsupervised k-means learning technique to construct discrete versions of optimization algorithms that work in continuous spaces. The optimization problem of a counterfort retaining wall was addressed, considering the cost and emission of CO_2 as objective functions. The cuckoo search optimization algorithm was used to be discretized. Additionally, a random operator was constructed to determine the contribution of the k-mean operator in the optimization process. It was concluded that k-means produces better results than the random operator and in many cases this does it systematically, thus reducing the dispersion of the solutions. In addition, when we compare k-means with HS, we observe that as we increase the height, where the optimization problem becomes more difficult because it is more difficult to obtain stability of the wall with respect to overturning and sliding, k-means is more robust than HS reaching the height of 14 (m) at a difference of 4.76% in favor of k-means in optimizing emissions and 4.87% in minimizing costs. On the other hand, when we analyze the dispersion of the set of solutions, we see that k-means once again perform better than HS, especially for heights greater than 12 (m).

There are several possible directions for further extensions and improvements of the present work. The first line arises from observing the configuration parameters presented in Table 4. The configuration procedure can be simplified and improved by incorporating adaptive mechanisms that allow the parameters to be modified in accordance with the feedback obtained from the candidate solutions.

A second line considers the incorporation of an intelligent agent that uses value gradient policies or action methods frequently used in reinforcement learning, in order to have information on the performance of the optimization algorithms with which we could modify the parameters dynamically. Finally, another possible line of research is to explore the population management of solutions dynamically. Through analyzing the history of exploration and exploitation of the search space, one can identify regions where it is necessary to increase the population and others where it is appropriate to decrease it.

Author Contributions: Conceptualization, V.Y., J.V.M. and J.G.; methodology, V.Y., J.V.M. and J.G.; software, J.V.M. and J.G.; validation, V.Y., J.V.M. and J.G.; formal analysis, J.G.; investigation, J.V.M. and J.G.; ; data curation, J.V.M.; writing—original draft preparation, J.G.; writing—review and editing, V.Y., J.V.M. and J.G.; funding acquisition, V.Y. and J.G. All authors have read and agreed to the published version of the manuscript.

Funding: The first author was supported by the Grant CONICYT/FONDECYT/INICIACION/11180056, the other two authors were supported by the Spanish Ministry of Economy and Competitiveness, along with FEDER funding (Project: BIA2017-85098-R)

Conflicts of Interest: The authors declare no conflict of interest.

References

1. García, J.; Altimiras, F.; Peña, A.; Astorga, G.; Peredo, O. A binary cuckoo search big data algorithm applied to large-scale crew scheduling problems. *Complexity* **2018**, *2018*. [CrossRef]
2. García, J.; Moraga, P.; Valenzuela, M.; Crawford, B.; Soto, R.; Pinto, H.; Peña, A.; Altimiras, F.; Astorga, G. A Db-Scan Binarization Algorithm Applied to Matrix Covering Problems. *Comput. Intell. Neurosci.* **2019**, *2019*. [CrossRef] [PubMed]
3. Al-Madi, N.; Faris, H.; Mirjalili, S. Binary multi-verse optimization algorithm for global optimization and discrete problems. *Int. J. Mach. Learn. Cybern.* **2019**, *10*, 3445–3465. [CrossRef]
4. Kim, M.; Chae, J. Monarch Butterfly Optimization for Facility Layout Design Based on a Single Loop Material Handling Path. *Mathematics* **2019**, *7*, 154. [CrossRef]
5. García, J.; Crawford, B.; Soto, R.; Astorga, G. A clustering algorithm applied to the binarization of Swarm intelligence continuous metaheuristics. *Swarm Evol. Comput.* **2019**, *44*, 646–664. [CrossRef]
6. García, J.; Lalla-Ruiz, E.; Voß, S.; Droguett, E.L. Enhancing a machine learning binarization framework by perturbation operators: Analysis on the multidimensional knapsack problem. *Int. J. Mach. Learn. Cybern.* **2020**. [CrossRef]
7. García, J.; Moraga, P.; Valenzuela, M.; Pinto, H. A db-Scan Hybrid Algorithm: An Application to the Multidimensional Knapsack Problem. *Mathematics* **2020**, *8*, 507. [CrossRef]
8. Saeheaw, T.; Charoenchai, N. A comparative study among different parallel hybrid artificial intelligent approaches to solve the capacitated vehicle routing problem. *Int. J. Bio-Inspir. Comput.* **2018**, *11*, 171–191. [CrossRef]
9. Crawford, B.; Soto, R.; Astorga, G.; García, J. Constructive metaheuristics for the set covering problem. In *International Conference on Bioinspired Methods and Their Applications*; Springer: Berlin, Germany, 2018; pp. 88–99.
10. Valdez, F.; Castillo, O.; Jain, A.; Jana, D.K. Nature-inspired optimization algorithms for neuro-fuzzy models in real-world control and robotics applications. *Comput. Intell. Neurosci.* **2019**, *2019*, 9128451. [CrossRef]
11. Penadés-Plà, V.; García-Segura, T.; Yepes, V. Robust Design Optimization for Low-Cost Concrete Box-Girder Bridge. *Mathematics* **2020**, *8*, 398. [CrossRef]
12. García-Segura, T.; Yepes, V.; Frangopol, D.M.; Yang, D.Y. Lifetime reliability-based optimization of post-tensioned box-girder bridges. *Eng. Struct.* **2017**, *145*, 381–391. [CrossRef]
13. Yepes, V.; Martí, J.V.; García, J. Black Hole Algorithm for Sustainable Design of Counterfort Retaining Walls. *Sustainability* **2020**, *12*, 2767. [CrossRef]
14. Marti-Vargas, J.R.; Ferri, F.J.; Yepes, V. Prediction of the transfer length of prestressing strands with neural networks. *Comput. Concr.* **2013**, *12*, 187–209. [CrossRef]
15. Fu, W.; Tan, J.; Zhang, X.; Chen, T.; Wang, K. Blind parameter identification of MAR model and mutation hybrid GWO-SCA optimized SVM for fault diagnosis of rotating machinery. *Complexity* **2019**, *2019*, 3264969. [CrossRef]

16. Sierra, L.A.; Yepes, V.; García-Segura, T.; Pellicer, E. Bayesian network method for decision-making about the social sustainability of infrastructure projects. *J. Clean. Prod.* **2018**, *176*, 521–534. [CrossRef]
17. Crawford, B.; Soto, R.; Astorga, G.; García, J.; Castro, C.; Paredes, F. Putting continuous metaheuristics to work in binary search spaces. *Complexity* **2017**, *2017*, 8404231. [CrossRef]
18. Shi, Y. Particle swarm optimization: Developments, applications and resources. In Proceedings of the 2001 Congress on Evolutionary Computation, Seoul, Korea, 27–30 May 2001; Volume 1, pp. 81–86.
19. Hatamlou, A. Black hole: A new heuristic optimization approach for data clustering. *Inf. Sci.* **2013**, *222*, 175–184. [CrossRef]
20. Yang, X.S.; Deb, S. Cuckoo search via Lévy flights. In Proceedings of the 2009 World Congress on Nature & Biologically Inspired Computing (NaBIC), Coimbatore, India, 9–11 December 2009; pp. 210–214.
21. Yang, X.S. A new metaheuristic bat-inspired algorithm. In *Nature Inspired Cooperative Strategies for Optimization (NICSO 2010)*; Springer: Berlin, Germany, 2010; pp. 65–74.
22. Yang, X.S. Firefly algorithms for multimodal optimization. In *International Symposium on Stochastic Algorithms*; Springer: Berlin, Germany, 2009; pp. 169–178.
23. Pan, W.T. A new fruit fly optimization algorithm: Taking the financial distress model as an example. *Knowl.-Based Syst.* **2012**, *26*, 69–74. [CrossRef]
24. Li, X.L.; Shao, Z.J.; Qian, J.X. An optimizing method based on autonomous animats: Fish-swarm algorithm. *Syst. Eng. Theory Pract.* **2002**, *22*, 32–38.
25. Rashedi, E.; Nezamabadi-Pour, H.; Saryazdi, S. GSA: A gravitational search algorithm. *Inf. Sci.* **2009**, *179*, 2232–2248. [CrossRef]
26. Calvet, L.; de Armas, J.; Masip, D.; Juan, A.A. Learnheuristics: Hybridizing metaheuristics with machine learning for optimization with dynamic inputs. *Open Math.* **2017**, *15*, 261–280. [CrossRef]
27. Caserta, M.; Voß, S. Matheuristics: Hybridizing Metaheuristics and Mathematical Programming. In *Metaheuristics: Intelligent Problem Solving*; Springer: Berlin, Germany, 2009; pp. 1–38.
28. Talbi, E.G. Combining metaheuristics with mathematical programming, constraint programming and machine learning. *Ann. Oper. Res.* **2016**, *240*, 171–215. [CrossRef]
29. Juan, A.A.; Faulin, J.; Grasman, S.E.; Rabe, M.; Figueira, G. A review of simheuristics: Extending metaheuristics to deal with stochastic combinatorial optimization problems. *Oper. Res. Perspect.* **2015**, *2*, 62–72. [CrossRef]
30. Chou, J.S.; Nguyen, T.K. Forward Forecast of Stock Price Using Sliding-Window Metaheuristic-Optimized Machine-Learning Regression. *IEEE Trans. Ind. Inform.* **2018**, *14*, 3132–3142. [CrossRef]
31. Sayed, G.I.; Tharwat, A.; Hassanien, A.E. Chaotic dragonfly algorithm: An improved metaheuristic algorithm for feature selection. *Appl. Intell.* **2019**, *49*, 188–205. [CrossRef]
32. De León, A.D.; Lalla-Ruiz, E.; Melián-Batista, B.; Moreno-Vega, J.M. A Machine Learning-based system for berth scheduling at bulk terminals. *Expert Syst. Appl.* **2017**, *87*, 170–182. [CrossRef]
33. García, J.; Crawford, B.; Soto, R.; Castro, C.; Paredes, F. A k-means binarization framework applied to multidimensional knapsack problem. *Appl. Intell.* **2018**, *48*, 357–380. [CrossRef]
34. Molina-Moreno, F.; Martí, J.V.; Yepes, V. Carbon embodied optimization for buttressed earth-retaining walls: Implications for low-carbon conceptual designs. *J. Clean. Prod.* **2017**, *164*, 872–884. [CrossRef]
35. Voß, S. Meta-heuristics: The state of the art. In *Workshop on Local Search for Planning and Scheduling*; Springer: Berlin, Germany, 2000; pp. 1–23.
36. Bishop, C.M. *Pattern Recognition and Machine Learning*; Springer: Berlin, Germany, 2006.
37. Asta, S.; Özcan, E.; Curtois, T. A tensor based hyper-heuristic for nurse rostering. *Knowl.-Based Syst.* **2016**, *98*, 185–199. [CrossRef]
38. Martin, S.; Ouelhadj, D.; Beullens, P.; Ozcan, E.; Juan, A.A.; Burke, E.K. A multi-agent based cooperative approach to scheduling and routing. *Eur. J. Oper. Res.* **2016**, *254*, 169–178. [CrossRef]
39. García, J.; Crawford, B.; Soto, R.; Astorga, G. A percentile transition ranking algorithm applied to binarization of continuous swarm intelligence metaheuristics. In *International Conference on Soft Computing and Data Mining*; Springer: Johor, Malaysia, 2018; pp. 3–13. [CrossRef]
40. Vecek, N.; Mernik, M.; Filipic, B.; Xrepinsek, M. Parameter tuning with Chess Rating System (CRS-Tuning) for meta-heuristic algorithms. *Inf. Sci.* **2016**, *372*, 446–469. [CrossRef]
41. Ries, J.; Beullens, P. A semi-automated design of instance-based fuzzy parameter tuning for metaheuristics based on decision tree induction. *J. Oper. Res. Soc.* **2015**, *66*, 782–793. [CrossRef]

42. Li, Z.Q.; Zhang, H.L.; Zheng, J.H.; Dong, M.J.; Xie, Y.F.; Tian, Z.J. Heuristic evolutionary approach for weighted circles layout. In *International Symposium on Information and Automation*; Springer: Berlin, Germany, 2010; pp. 324–331.
43. Yalcinoz, T.; Altun, H. Power economic dispatch using a hybrid genetic algorithm. *IEEE Power Eng. Rev.* **2001**, *21*, 59–60. [CrossRef]
44. Kaur, H.; Virmani, J.; Thakur, S. A genetic algorithm-based metaheuristic approach to customize a computer-aided classification system for enhanced screen film mammograms. In *U-Healthcare Monitoring Systems*; Advances in Ubiquitous Sensing Applications for Healthcare; Dey, N., Ashour, A.S., Fong, S.J., Borra, S., Eds.; Academic Press: Cambridge, MA, USA, 2019; pp. 217–259. [CrossRef]
45. Faris, H.; Hassonah, M.A.; Ala'M, A.Z.; Mirjalili, S.; Aljarah, I. A multi-verse optimizer approach for feature selection and optimizing SVM parameters based on a robust system architecture. *Neural Comput. Appl.* **2018**, *30*, 2355–2369. [CrossRef]
46. Faris, H.; Aljarah, I.; Mirjalili, S. Improved monarch butterfly optimization for unconstrained global search and neural network training. *Appl. Intell.* **2018**, *48*, 445–464. [CrossRef]
47. Chou, J.S.; Thedja, J.P.P. Metaheuristic optimization within machine learning-based classification system for early warnings related to geotechnical problems. *Autom. Constr.* **2016**, *68*, 65–80. [CrossRef]
48. Pham, A.D.; Hoang, N.D.; Nguyen, Q.T. Predicting compressive strength of high-performance concrete using metaheuristic-optimized least squares support vector regression. *J. Comput. Civ. Eng.* **2015**, *30*, 06015002. [CrossRef]
49. Göçken, M.; Özçalıcı, M.; Boru, A.; Dosdoğru, A.T. Integrating metaheuristics and artificial neural networks for improved stock price prediction. *Expert Syst. Appl.* **2016**, *44*, 320–331. [CrossRef]
50. Chou, J.S.; Pham, A.D. Nature-inspired metaheuristic optimization in least squares support vector regression for obtaining bridge scour information. *Inf. Sci.* **2017**, *399*, 64–80. [CrossRef]
51. Kuo, R.; Lin, T.; Zulvia, F.; Tsai, C. A hybrid metaheuristic and kernel intuitionistic fuzzy c-means algorithm for cluster analysis. *Appl. Soft Comput.* **2018**, *67*, 299–308. [CrossRef]
52. Mann, P.S.; Singh, S. Energy efficient clustering protocol based on improved metaheuristic in wireless sensor networks. *J. Netw. Comput. Appl.* **2017**, *83*, 40–52. [CrossRef]
53. De Alvarenga Rosa, R.; Machado, A.M.; Ribeiro, G.M.; Mauri, G.R. A mathematical model and a Clustering Search metaheuristic for planning the helicopter transportation of employees to the production platforms of oil and gas. *Comput. Ind. Eng.* **2016**, *101*, 303–312. [CrossRef]
54. Faris, H.; Mirjalili, S.; Aljarah, I. Automatic selection of hidden neurons and weights in neural networks using grey wolf optimizer based on a hybrid encoding scheme. *Int. J. Mach. Learn. Cybern.* **2019**, *10*, 2901–2920. [CrossRef]
55. De Rosa, G.H.; Papa, J.P.; Yang, X.S. Handling dropout probability estimation in convolution neural networks using meta-heuristics. *Soft Comput.* **2018**, *22*, 6147–6156. [CrossRef]
56. Tuba, M.; Alihodzic, A.; Bacanin, N. Cuckoo search and bat algorithm applied to training feed-forward neural networks. In *Recent Advances in Swarm Intelligence and Evolutionary Computation*; Springer: Berlin, Germany, 2015; pp. 139–162.
57. Rere, L.; Fanany, M.I.; Arymurthy, A.M. Metaheuristic algorithms for convolution neural network. *Comput. Intell. Neurosci.* **2016**, *2016*, 1537325. [CrossRef] [PubMed]
58. Rashid, T.A.; Hassan, M.K.; Mohammadi, M.; Fraser, K. Improvement of variant adaptable LSTM trained with metaheuristic algorithms for healthcare analysis. In *Advanced Classification Techniques for Healthcare Analysis*; IGI Global: Hershey, PA, USA, 2019; pp. 111–131.
59. Jothi, R.; Mohanty, S.K.; Ojha, A. DK-means: A deterministic k-means clustering algorithm for gene expression analysis. *Pattern Anal. Appl.* **2019**, *22*, 649–667. [CrossRef]
60. García, J.; Pope, C.; Altimiras, F. A Distributed-Means Segmentation Algorithm Applied to Lobesia botrana Recognition. *Complexity* **2017**, *2017*, 5137317. [CrossRef]
61. Arunkumar, N.; Mohammed, M.A.; Ghani, M.K.A.; Ibrahim, D.A.; Abdulhay, E.; Ramirez-Gonzalez, G.; de Albuquerque, V.H.C. K-means clustering and neural network for object detecting and identifying abnormality of brain tumor. *Soft Comput.* **2019**, *23*, 9083–9096. [CrossRef]
62. Abdel-Basset, M.; Wang, G.G.; Sangaiah, A.K.; Rushdy, E. Krill herd algorithm based on cuckoo search for solving engineering optimization problems. *Multimed. Tools Appl.* **2019**, *78*, 3861–3884. [CrossRef]

63. Chi, R.; Su, Y.X.; Zhang, D.H.; Chi, X.X.; Zhang, H.J. A hybridization of cuckoo search and particle swarm optimization for solving optimization problems. *Neural Comput. Appl.* **2019**, *31*, 653–670. [CrossRef]
64. Li, J.; Xiao, D.D.; Lei, H.; Zhang, T.; Tian, T. Using Cuckoo Search Algorithm with Q-Learning and Genetic Operation to Solve the Problem of Logistics Distribution Center Location. *Mathematics* **2020**, *8*, 149. [CrossRef]
65. Pan, J.S.; Song, P.C.; Chu, S.C.; Peng, Y.J. Improved Compact Cuckoo Search Algorithm Applied to Location of Drone Logistics Hub. *Mathematics* **2020**, *8*, 333. [CrossRef]
66. Yepes, V.; Alcala, J.; Perea, C.; González-Vidosa, F. A parametric study of optimum earth-retaining walls by simulated annealing. *Eng. Struct.* **2008**, *30*, 821–830. [CrossRef]
67. Molina-Moreno, F.; García-Segura, T.; Martí, J.V.; Yepes, V. Optimization of buttressed earth-retaining walls using hybrid harmony search algorithms. *Eng. Struct.* **2017**, *134*, 205–216. [CrossRef]
68. Ministerio de Fomento. *EHE: Code of Structural Concrete*; Ministerio de Fomento: Madrid, Spain, 2008.
69. Ministerio de Fomento. *CTE. DB-SE. Structural Safety: Foundations*; Ministerio de Fomento: Madrid, Spain, 2008. (In Spanish)
70. Huntington, W.C. *Earth Pressures and Retaining Walls*; Literary Licensing, LLC: Whitefish, MT, USA, 1957.
71. Calavera, J. *Muros de Contención y Muros de Sótano*; INTEMAC: Madrid, Spain, 2001. (In Spanish)
72. CEB-FIB. *Model Code. Design Code*; Thomas Telford Services Ltd.: London, UK, 2008.
73. Hays, W.L.; Winkler, R.L. *Statistics: Probability, Inference, and Decision*; Holt, Rinehart, and Winston: New York, NY, USA, 1971.
74. Wilcoxon, F. Individual comparisons by ranking methods. In *Breakthroughs in Statistics*; Springer: Berlin, Germany, 1992; pp. 196–202.

Article

A Mathematical Pre-Disaster Model with Uncertainty and Multiple Criteria for Facility Location and Network Fortification

Julia Monzón [1], Federico Liberatore [2,3,*] and Begoña Vitoriano [4]

1 Department of Statistics and Operational Research, Complutense University of Madrid, 28040 Madrid, Spain; jmonzon@ucm.es
2 School of Computer Science & Informatics, Cardiff University, Cardiff CF24 3AA, UK
3 UC3M-Santander Big Data Institute (IBiDat), Charles III University of Madrid, 28903 Getafe, Spain
4 Department of Statistics and Operational Research and Institute of Interdisciplinary Mathematics, Complutense University of Madrid, 28040 Madrid, Spain; bvitoriano@mat.ucm.es
* Correspondence: liberatoref@cardiff.ac.uk

Received: 28 February 2020; Accepted: 22 March 2020; Published: 3 April 2020

Abstract: Disasters have catastrophic effects on the affected population, especially in developing and underdeveloped countries. Humanitarian Logistics models can help decision-makers to efficiently and effectively warehouse and distribute emergency goods to the affected population, to reduce casualties and suffering. However, poor planning and structural damage to the transportation infrastructure could hamper these efforts and, eventually, make it impossible to reach all the affected demand centers. In this paper, a pre-disaster Humanitarian Logistics model is presented that jointly optimizes the prepositioning of aid distribution centers and the strengthening of road sections to ensure that as much affected population as possible can efficiently get help. The model is stochastic in nature and considers that the demand in the centers affected by the disaster and the state of the transportation network are random. Uncertainty is represented through scenarios representing possible disasters. The methodology is applied to a real-world case study based on the 2018 storm system that hit the Nampula Province in Mozambique.

Keywords: stochastic programming; decision making; inventory prepositioning; network fortification; pre-disaster phase; humanitarian logistics; emergency management

1. Introduction

After a disaster strikes a territory it is imperative to deliver relief items to the affected population to reduce human casualties and suffering. This is especially true in underdeveloped and developing countries that suffer from a lack of resources and reliable infrastructures. Additionally, poor planning and structural damage to the transportation network could hinder the relief effort or, in some cases, completely frustrate them. Damaged roads, resulting from the effect of the disaster, could completely disconnect affected populations from the rest of the transportation network, thus, making impossible to undertake any relief operation. This was the case in Haiti, after the 2010 earthquake, where, despite the large volume of emergency goods available, the victims could not receive support for a long time due to the damage suffered by the road network [1–4]. The 2011 Japan earthquake and tsunami interdicted about three-fourths of the highways in the region, hampering the emergency response operations [5]. More recently in Puerto Rico, after Hurricane Maria hit the island in 2017, shortages in vehicle drivers and disruptions to the road network made it impossible to deliver emergency supplies, including food, water, and medicine, resulting in at least 10,000 containers idly sitting at the port [6]. Locating emergency inventories and fortifying vulnerable elements of the road infrastructure prior to an emergency helps to mitigate the impact of a disaster, facilitates the relief operations, and increases the effectiveness of the disaster response.

In this paper, a pre-disaster model that tackles these issues is proposed. The model jointly optimizes the fortification of elements of the distribution network, the location of emergency inventories, and the definition of their capacity. The goal is to support the distribution operations of relief goods in the best possible way. Due to the uncertainty that characterizes disasters, the model is stochastic in nature. Multiple objectives are considered. However, the objective of minimizing the expected unsatisfied demand has the highest priority, as not providing relief to populations affected by a disaster might result in human casualties. This objective has multiple optima, therefore, the optimal solutions are further evaluated according to two time measures: the expected maximum arrival time at a demand node and the expected total service time. The computational experiments show little trade-off between these time criteria. For this reason, the solutions are evaluated using a pay-off matrix, rather than by applying specific multicriteria methods. More generally, the methodology has been designed to be applicable in situations with limited data on the availability of resources and the status of the road infrastructure, which is especially relevant to underdeveloped and developing countries. In fact, the model has been tailored to a real case study on the 2018 tropical depression that hit the Nampula Province (Mozambique) and considers only the data that could be obtained from official sources. For this reason, and given its focus on pre-disaster operations, the model relies on a number of simplifying assumptions concerning the distribution operations. For instance, these are represented by an underlying flow model. As a consequence, most details of a real supply chain are disregarded, such as, multiple commodities, multiple transportation modes, vehicles availability and capacity, and distribution queues. However, this is consistent with the application context and the scope of the model.

1.1. Literature Review

According to Anysia and Kopczak [7], Humanitarian Logistics is "the process of planning, implementing and controlling the efficient, cost-effective flow and storage of goods and materials, as well as related information, from the point of origin to the point of consumption for the purpose of alleviating the suffering of vulnerable people." The seminal work by Aghani and Oh [8] presents a large-scale multi-commodity multi-modal network flow model with time windows in the context of disaster relief operations, which the authors solve using heuristics. In recent years, due to the relevance of the subject, the investigation in Humanitarian Logistics models has experienced a golden age and a wealth of research contributions have been published. Due to the large number of works in the field, the review presented in this paper focuses on relief distribution models that combine location operations with structural operations on the transportation network. However, pointers to exhaustive literature reviews are provided in the following. The interested reader is referred to the excellent literature review on operations management in the context of Operations Research (OR) and Management Sciences (MS) by Altay and Green [9], which Ortuño et al. [10] and Galindo and Batta [11] expand and update. Liberatore et al. [12] provide an annotated bibliography specialized on uncertainty in humanitarian logistics. Çelik [4] reviews the literature on network restoration and recovery in humanitarian operations. Aslan and Çelik [13] give in their research paper an overview of recent studies on emergency inventory propositioning involving uncertainty. Finally, the review paper by Sabbaghtorkan, Batta, and He is specialized on prepositioning of assets and supplies in disaster operations management [14].

To the best of the authors' knowledge, the first model to consider the coordination of the activities of relief transportation, road recovery, and inventory prepositioning is by Wisetjindawat et al. [15]. The authors consider ongoing road repairs as constraints on the availability of roads, rather than as part of the decision making process, i.e., all links in the transportation network must be available after 24 h. Therefore, the problem is formulated as a location-routing problem that explicitly consider routes that satisfy the availability time constraint, at a given confidence level. The approach proposed by the authors is an interesting combination of simulation, vehicle routing, and location analysis. However, it does not explicitly incorporate neither the stochastic elements of the problem nor the recovery operations in the formulation.

Later, Iloglu and Albert [16] study the relationship between network restoration and emergency response operations in the early stages of a disaster. A multi-period p-median model determines the location of emergency responders that attend emergency calls. The formulation includes scheduling constraints that plan the repair team operations of network recovery. The problem should be used right after a disaster strikes, therefore no stochasticity is considered. However, the formulation allows repair crews to access arcs that are completely disconnected from the rest of the network, which is unrealistic.

This initial model is later expanded by the authors in [17], where the underlying multi-period p-median model is substituted by a multi-period maximal multiple coverage model. As the emergency responders can be relocated at every period, this more recent formulation limits the choices for relocation to those within a certain radius from the previous one. Also, the multiple coverage model considers allows for backup emergency service during large volume of emergency service requests after disasters. Obviously, the new formulation improves the realism and applicability of the model. However, it still allows repair teams to access arcs disconnected from the rest of the network. Therefore, despite the update, the major flaw of the model is still present.

In another recent study, Aslan and Çelik [13] propose a two-stage stochastic programming model to design a multi-echelon humanitarian response network. In the first stage, the model defines the location of the inventories and their capacity for each commodity. In the second stage, relief transportation decisions and road repair operations are made jointly. A limitation of this contribution is that the model assumes that there are sufficient restoration resources and these resources can start repairing any damaged arc immediately after the disaster. The authors recognize the limited realism of this approach, however, they claim that the assumptions help to simplify the solution of the problem and understand its structure. More recently, Sanci and Daskin [18] introduce an additional level of complexity in the model. Their article expands on the previous contributions by integrating facility location and network restoration models to locate both emergency response facilities and restoration equipment before a disaster. Also, it allows to allocate more than one restoration resource to a damaged arc to reduce the recovery time. Finally, the model ensures that damaged arcs can be repaired only if they are accessible from the initial locations of the repair resources, therefore, addressing the major flaw in [16,17]. Interestingly, the model can find a balance between unmet demand and transportation costs by changing the value of a penalization coefficient. This approach clearly contradicts the rationale behind Humanitarian Logistics, that gives absolute priority to the minimization of human casualties and suffering. The authors suggest setting the penalty for unmet demand to a value that ensures that all the demand is attended. However, this clearly shows that the authors recognize that the formulation is indeed flawed, and that a lexicographical approach should have adopted instead of using a penalization coefficient.

All the studies reviewed consider post-disaster network recovery operations. The model proposed in this article, however, optimizes pre-disaster network fortification operations, rather than recovery operations. Therefore, to the best of the authors' knowledge, this is the first fully pre-disaster model to combine relief transportation, road recovery, and inventory prepositioning. Also, the formulation presented in this study considers a simpler supply chain than those of the previous contributions: one commodity, three types of nodes, and unlimited capacities. This is a desired feature. Differently from the papers presented in the review, our model has been designed to being applicable in contexts with limited information availability, such as underdeveloped and developing countries.

1.2. Contributions

The contribution of this research is twofold. In particular, this work extends the literature by introducing:

1. The first pre-disaster stochastic model in the literature that jointly optimizes the location of emergency inventories and the fortification of transportation network elements, to support humanitarian logistics operations in the response phase. The model is specially tailored to underdeveloped and developing countries, where data availability is an issue. In particular, its design is based on a real case study.
2. A novel public dataset on the January 2018 tropical depression that affected the Nampula Province (Mozambique). The dataset has been assembled using real data obtained from local agencies, including the Mozambique Institute for Disaster Management (INGC), and is comprised of a real deterministic scenario and stochastic scenarios. Due to the low requirements in terms of data, the methodology presented to generate the scenarios can be easily extended to other case studies.

The rest of the paper is organized as follows. Section 2 presents the optimization model and the formulation. The dataset on the Nampula Province case study is detailed in Section 3. Section 4 is devoted to the computational experiments and provides a discussion of the results. Finally, the paper is concluded with some insights and guidelines for future research (Section 5).

2. Model

The model proposed in this research is a two-stage stochastic model. The first stage involves all the decisions taken before the disaster strikes, such as determining the inventories' locations and choosing the road sections to fortify. The problem assumes that a fortified road cannot be interdicted. The second stage takes place after the disaster strikes, which determines the number of people affected in the population centers and the roads that have been interdicted and cannot be used. In this stage, the emergency goods distribution takes place and is modeled as a flow problem with an unlimited supply. The objective is maximizing the number of affected people that receive relief. However, this objective presents multiple optima. Therefore, a second optimization step chooses among these optima the solution that minimizes the distribution time. Two different time measures are considered. Finally, the model also provides information regarding the desired inventory capacities: each inventory should be able to relieve as many persons as its maximum supply across all the scenarios. The optimization model is presented in the following.

2.1. Sets

Let $G(V, A)$ be a directed graph, being V the set of vertices, indexed by i and j, and A the set of arcs (i, j). The vertices represent population centers and road crossings. On the other hand, the arcs represent the road sections; specifically, each road segment is modeled by a pair of arcs, i.e., (i, j) and (j, i). Finally, Ω is the set of stochastic scenarios and is indexed by ω.

2.2. Parameters

Two attributes, P and Q, specify the number of inventories to locate and the number of road sections to fortify, respectively. Vertices are characterized by a demand, $demand_i^{\omega}$, which represent the population affected by the disaster in a community and depends on the scenario. Arcs are characterized by a length, $length_{ij}$, which represents the traversal time. Also, each scenario specifies which arcs are not interdicted. This information is represented by the parameter $safe_{ij}^{\omega}$, that takes value 1 if the arc (i, j) can be traversed in scenario ω, and 0 if it is interdicted. Finally, the scenarios have an assigned probability distribution, p_{ω} $\forall \omega \in \Omega$, which verifies $\sum_{\omega \in \Omega} p_{\omega} = 1$.

2.3. Variables

First-Stage Variables:

- $X_{ij} = \begin{cases} 1 & \text{if arc } (i,j) \text{ is fortified,} \\ 0 & \text{otherwise.} \end{cases}$
- $Y_i = \begin{cases} 1 & \text{if an inventory is located in vertex } i, \\ 0 & \text{otherwise.} \end{cases}$

Second-Stage Variables:

- $flow_{ij}^{\omega} \geq 0$, units of flow on arc (i,j) in scenario ω.
- $supply_i^{\omega} \geq 0$, supply available at vertex i in scenario ω.
- $time_i^{\omega} \geq 0$, arrival time at vertex i in scenario ω.
- $T^{\omega} \geq 0$, maximum arrival time across all the demand vertices in scenario ω.
- $cross_{ij}^{\omega} = \begin{cases} 1 & \text{if arc } (i,j) \text{ is used for distribution in scenario } \omega, \\ 0 & \text{otherwise.} \end{cases}$
- $reach_i^{\omega} = \begin{cases} 1 & \text{if vertex } i \text{ can be reached from an inventory vertex in scenario } \omega, \\ 0 & \text{otherwise.} \end{cases}$

2.4. Constraints and Objective Functions

Arc Constraints:

$$\sum_{(i,j)\in A} X_{ij} = 2 \cdot Q \tag{1}$$

$$X_{ij} = X_{ji} \qquad \forall (i,j) \in A \tag{2}$$

$$cross_{ij}^{\omega} \leq safe_{ij}^{\omega} + X_{ij} \qquad \forall (i,j) \in A, \forall \omega \in \Omega \tag{3}$$

$$flow_{ij}^{\omega} \leq K^{\omega} \cdot cross_{ij}^{\omega} \qquad \forall (i,j) \in A, \forall \omega \in \Omega \tag{4}$$

$$X_{ij} \in \{0,1\} \qquad \forall (i,j) \in A \tag{5}$$

$$cross_{ij}^{\omega} \in \{0,1\} \qquad \forall (i,j) \in A, \forall \omega \in \Omega \tag{6}$$

$$flow_{ij}^{\omega} \geq 0 \qquad \forall (i,j) \in A, \forall \omega \in \Omega \tag{7}$$

K^{ω} is a constant that takes a value that is greater than or equal to the largest possible flow traversing an arc in a scenario ω. A trivial upper bound to K^{ω} is given by:

$$K^{\omega} = \sum_{i\in V} demand_i^{\omega} \qquad \forall \omega \in \Omega \tag{8}$$

Constraint (1) enforces that the number of fortified arcs is exactly $2Q$, as each road segment is represented by a pair of arcs. In fact, constraints (2) specify that if arc (i,j) is fortified, then also arc (j,i) must be fortified, and vice-versa. Next, constraints (3) impose that, in a specific scenario ω, an arc can be used in the flow model only if it is not interdicted in the scenario or if it is fortified in the first stage. Constraints (4) relate $flow_{ij}^{\omega}$ to $cross_{ij}^{\omega}$ by enforcing that an arc (i,j) can have a positive flow in a given scenario ω only if $cross_{ij}^{\omega} = 1$. Finally, constraints (5)–(7) present the condition of existence for variables X_{ij}, $cross_{ij}^{\omega}$, and $flow_{ij}^{\omega}$, respectively.

Vertex and Flow Constraints:

$$\sum_{i \in V} Y_i = P \tag{9}$$

$$supply_i^{\omega} \leq K^{\omega} \cdot Y_i \qquad \forall i \in V, \forall \omega \in \Omega \tag{10}$$

$$\sum_{j:(j,i) \in A} flow_{ji}^{\omega} + supply_i^{\omega} = \sum_{j:(i,j) \in A} flow_{ij}^{\omega} + demand_i^{\omega} \cdot reach_i^{\omega} \qquad \forall i \in V, \forall \omega \in \Omega \tag{11}$$

$$Y_i \in \{0,1\} \qquad \forall i \in V \tag{12}$$

$$supply_i^{\omega} \geq 0 \qquad \forall i \in V, \forall \omega \in \Omega \tag{13}$$

$$reach_i^{\omega} \in \{0,1\} \qquad \forall i \in V, \forall \omega \in \Omega \tag{14}$$

Constraint (9) imposes that the number of inventories is exactly P. Only inventories can have a positive supply capacity, as enforced by constraints (10) which relate $supply_i^{\omega}$ to Y_i. Constraints (11) define the flow on the arcs, the supply capacity at each vertex, and the vertices that can be reached, in each scenario. Finally, constraints (12)–(14) establish the condition of existence for variables Y_i, $supply_i^{\omega}$, and $reach_i^{\omega}$, respectively.

Time Constraints:

$$time_j^{\omega} \geq time_i^{\omega} + length_{ij} - K' \cdot (1 - cross_{ij}^{\omega}) \qquad \forall (i,j) \in A, \forall \omega \in \Omega \tag{15}$$

$$T^{\omega} \geq time_i^{\omega} \qquad \forall i \in V, \forall \omega \in \Omega : demand_i^{\omega} > 0 \tag{16}$$

$$time_i^{\omega} \geq 0 \qquad \forall i \in V, \forall \omega \in \Omega \tag{17}$$

$$T^{\omega} \geq 0 \qquad \forall \omega \in \Omega \tag{18}$$

where K' is a constant that takes a value that is greater than or equal to the largest possible arrival time at a vertex. A trivial upper bound is given by:

$$K' = \sum_{(i,j) \in A} length_{ij} \tag{19}$$

Constraints (15) assign consistent arrival times to all the vertices for all the scenarios, while constraints (16) compute the maximum arrival time to a demand vertex in each scenario. Constraints (17) and (18) define the condition of existence for variables $time_i^{\omega}$ and T^{ω}, respectively.

Objective Functions:

Due to the disruptions in the infrastructure network caused by the disaster, some of the demand vertices might be disconnected from all the inventories and, therefore, unreachable. Leaving an affected community to its own devices during an emergency leads to human suffering and might result in casualties. Thus, minimizing the population that is not reached by the distribution operations takes precedence over any other objective.

$$\min \sum_{\omega \in \Omega} p_{\omega} \cdot \sum_{i \in V} demand_i^{\omega} \cdot (1 - reach_i^{\omega}) \tag{20}$$

Due to the stochastic nature of the model, objective function (20) minimizes the expected unsatisfied demand. This objective presents multiple optima, therefore, to discriminate among them, the solutions are evaluated in terms of the time required by the distribution operations. Two different time measures are proposed, presented in the following:

$$\min \sum_{\omega \in \Omega} p_\omega \cdot T^\omega \tag{21}$$

$$\min \sum_{\omega \in \Omega} p_\omega \cdot \sum_{(i,j) \in A} length_{ij} \cdot flow_{ij}^\omega \tag{22}$$

The first objective (21) minimizes the expected maximum arrival time to a demand vertex across all the scenarios, while the second (22) is based on the classical min-cost flow objective function and minimizes the expected total distribution time across all scenarios.

Overall, the model includes $|A| + |V| + |\Omega| \cdot (1 + 2 \cdot |A| + 3 \cdot |V|)$ variables. Of these, $|A| + |V| + |\Omega| \cdot (|A| + |V|)$ are binary. The number of constraints in the model is: $2 + |A| + 3 \cdot |\Omega| \cdot (|A| + |V|)$.

2.5. Solution

A solution to the model provides an answer to the following questions:

- Q: What road sections should be fortified?
  A: $\{(i,j) \in A : X_{ij} = 1\}$.
- Q: Where to locate emergency inventories?
  A: $\{i \in V : Y_i = 1\}$.
- Q: What capacity should the inventories have?
  A: $capacity_i = \max_{\omega \in \Omega}\{supply_i^\omega\}, \forall i \in V : Y_i = 1$.

3. Case Study

3.1. Background

Mozambique is a coastal country in Southern Africa, which borders on the south with South Africa, on the southwest with Eswatini, to the west with Zimbabwe, Zambia, and Malawi, and to the north with Tanzania. On the east side stands the Mozambique Canal, with Madagascar and the Comoros Islands as overseas neighbors. According to the 2019 Human Development Index (HDI) [19], Mozambique is among the 10 least developed countries, ranking 180 out of 189. The population of the country is of 28.9 million inhabitants, approximately (Mozambique National Institute of Statistics) and the most populous province is Nampula, representing almost 20% of the total. Figure 1 shows the location of the Nampula Province withing the Mozambican territory. The Nampula Province is prone to frequent floods due to tropical storms. The frequency and intensity of these phenomena has been growing in the last years. In fact, since 2015, the Nampula Province has been hit by the Tropical Storm Chedza (January 2015), the Tropical Cyclone Dineo (February 2017), a tropical depression (January 2018), the Tropical Cyclone Idai (March 2019), and the Tropical Cyclone Kenneth (April 2019).

This case study focuses on the 2018 low-pressure system that formed in the Mozambique Channel on 13 January 2018 and evolved into the tropical depression stage on 16 January. The tropical depression penetrated the territory of Mozambique from the Nampula Province, more specifically, from the Mossuril district. The storm system consisted of heavy rain, winds of 85 km/h, and affected the provinces of Nampula, Niassa and Cabo Delgado, accumulating 400 mm of rain in less than four days. At the same time, this disaster was exacerbated by the Congo air masses and Tropical Cyclone Berguitta. Figure 2 shows the evolution of the phenomenon through time.

Figure 1. Territory of Mozambique and of the Nampula Province, in red. (Source: Wikimedia Commons, license CC BY-SA 3.0)

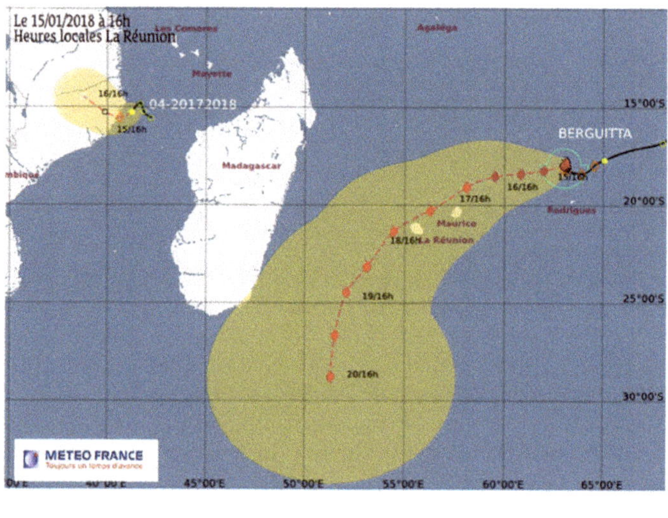

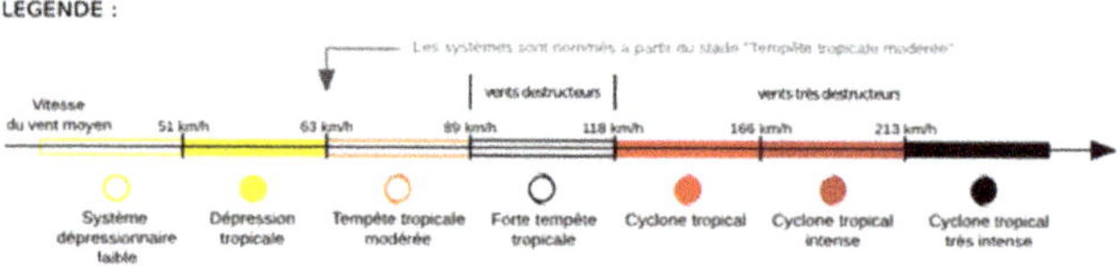

Figure 2. Evolution of the 2018 low-pressure system that formed in the Mozambique Channel on 13 January 2018. (Source: Meteo France, license [20])

In terms of damage, according to the INGC [21], the floods affected more than 80,000 people and killed 34, mainly in the Nampula Province, which accounted for 73,240 of the victims. Additionally, many roads were shut, which hampered the response and rescue operations.

3.2. Dataset

In the dataset, the set V is comprised of 38 vertices representing the 18 districts and the five municipalities in the Nampula Province, and 15 intersections. The set A consists of 106 arcs, corresponding to the 53 road sections that connect the vertices. The graph is illustrated in Figure 3. At the time of the emergency, the inventories were located in the cities of Nampula and Nacala, corresponding to vertices 1 and 19 in the figure. The traversal time of the arcs, $length_{ij}$, have been calculated using Google Maps and are expressed in minutes.

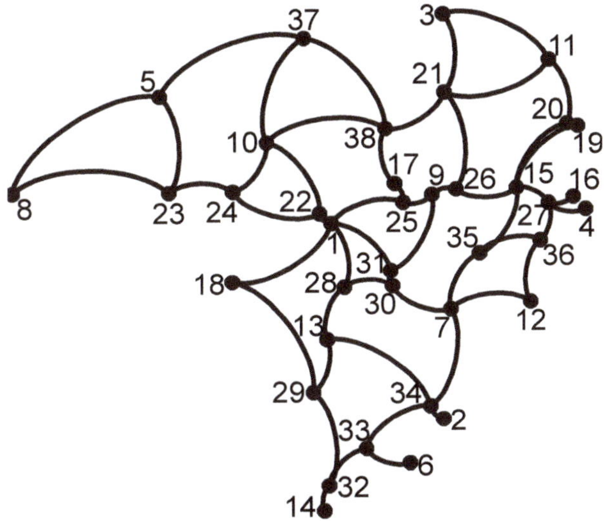

Figure 3. Graph representing the Nampula Province.

The report by the INGC [21] provides information relative to the impact of the tropical storm in the north of Mozambique, which includes the Nampula Province. The report illustrates the affected population in each municipality and the roads that have been shut and, therefore, it is possible to devise a single scenario that represents the real impact of the tropical storm in the area. From this point onward, we will refer to this scenario as the *deterministic scenario*.

The *stochastic scenarios* have been generated under the assumption that the tropical storm penetrated the territory of Mozambique from each of the coastal towns in the Nampula Province (i.e., Angoche, Ilha de Mozambique, Larde, Memba, Mogincual, Moma, Mossuril, Nacala, Nacala-a-Velha, and Lunga), including Mossuril that is the original entry point. Thus, 10 scenarios are included in the dataset. Each scenario $\omega \in \Omega$ is characterized by a probability, p_ω, the vertices demands, $demand_i^\omega$, and the availability of the road sections, $safe_{ij}^\omega$. Due to the lack of information, the scenarios are considered equiprobable, i.e., $p_\omega = \frac{1}{10}$. The demand at each vertex and the availability of the arcs have been estimated by means of regression models that made use of the information presented in the report by the INGC regarding the effects of the tropical storm [21], the 2017 Census [22], and data provided by the National Road Administration of Mozambique. More in detail, the demand of the vertices corresponding to intersections is set to zero. On the other hand, for all the vertices corresponding to population centers and for each scenario, the demand has been obtained from a linear regression model that expresses the logarithm of the ratio of the affected population in each community as a function of the following variables:

- Total population.
- Orthogonal polynomial of degree two of the difference in latitude between the population center and the entry point of the tropical storm in the scenario.

- The difference in longitude between the population center and the entry point of the tropical storm in the scenario.
- Orthogonal polynomial of degree two of the population center altitude.

All the independent variables are statistically significant and the coefficient of determination is $R^2 = 0.9221$, meaning that the model explains most of the variability in the data.

To determine the availability of the arcs, an exploratory analysis of logistic regression models has been done. Different types of independent variables have been considered: related to the road (e.g., length, materials, type, conditions), related to bridges in the road (e.g., number of bridges, total length of bridges, materials), and geographical (e.g., distance to the entry point, height). Unfortunately, no variable was statistically significant. This is probably due to the large number of missing values in the data, despite having it obtained from official sources. Therefore, the availability of all the pairs of arcs $\{(i,j),(j,i)\}$ in the graph is determined using Bernoulli trials with a fixed probability of 0.16981. This probability corresponds to the ratio of road sections interdicted by the storm system over the total number in the Nampula Province, i.e., $\frac{9}{53} = 0.16981$.

The dataset is publicly available and can be downloaded from [23].

4. Computational Experiments

In this section, the experiments and their results are presented and discussed. Two groups of experiments are considered. The first group concerns the deterministic scenario. This means that, in this setting, the set of scenarios Ω is comprised of a single scenario that represents the real impact that the tropical storm had on the Nampula Province in terms of affected population and road interdicted. The second group is run on the stochastic scenarios. In this context, the model optimizes the decision based on the average performance over the randomly generated scenarios.

Regarding the objective functions, the model is first solved with respect to the unsatisfied demand (Equation (20)). Then, to understand the relationship between the two time measures considered (Equations (21) and (22)), the payoff matrix is calculated. Calculating the payoff matrix requires solving the model four times. Therefore, a single run involves solving the problem five times. This is explained in Algorithm 1. In the first step, the model is solved while optimizing the unsatisfied demand (Equation (20)). The value obtained is fixed. Therefore, from this point on, the model will produce only solutions with that specific value of unsatisfied demand. Then, objective function (21) is optimized, giving the ideal value of the maximum arrival time ($\overline{\text{max time}}$). The maximum arrival time is set to $\overline{\text{max time}}$ and objective function (22) is optimized, obtaining the anti-ideal value of the total distribution time ($\underline{\text{total time}}$). Next, the maximum arrival time is unfixed and the total distribution time is optimized (Equation (22)) to calculate its ideal value, $\overline{\text{total time}}$. Finally, the total distribution time is fixed to $\overline{\text{total time}}$ and the maximum arrival time is optimized (Equation (21)), obtaining the anti-ideal value $\underline{\text{max time}}$.

Algorithm 1 Solution procedure.

1: unsatisfied demand ← solve model for optimal objective function (20)
2: **fix** unsatisfied demand
3: $\overline{\text{max time}}$ ← solve model for optimal objective function (21)
4: **fix** max time ← $\overline{\text{max time}}$
5: $\underline{\text{total time}}$ ← solve model for optimal objective function (22)
6: **unfix** max time
7: $\overline{\text{total time}}$ ← solve model for optimal objective function (22)
8: **fix** total time ← $\overline{\text{total time}}$
9: $\underline{\text{max time}}$ ← solve model for optimal objective function (21)

The optimization model has been programmed in GAMS (ver. 29.1.0) and solved using CPLEX (ver. 12.9.0.0) on a Dell Precision 5540 (sourced from Cardiff, UK), equipped with Intel® Core^{TM} i9-9880H CPU @ 2.30GHz $\times$ 16 and 16 GB RAM. The multithreading option of CPLEX has been used. A CPU time limit of 1800s has been set on all the optimization processes.

The following experiments are carried out:

- The model is solved only on the real scenario. Inventory locations are fixed and the model can determine which elements of the road network to fortify.
- The model is solved only on the real scenario and it can define both the inventory locations and the road fortifications.
- The model is solved on all the stochastic scenarios and can define both the inventory locations and the road fortifications.
- The solution of the model solved on all the stochastic scenarios is compared to a solution determined by independently running the model on each scenario and then choosing the most frequent inventory locations.

4.1. Deterministic Scenario: Fixed Inventories

In this experiment, only the deterministic scenario is considered and the inventories are located in Nampula and Nacala as in the real disaster. The model can determine the road sections to fortify. Table 1 presents the results.

Table 1. Results for the experiment on the deterministic scenario and with fixed inventories.

Q	Unsatisfied Demand (#)	$\overline{\text{Max Time}}$ (min)	$\underline{\text{Max Time}}$ (min)	$\overline{\text{Total Time}}$ (min)	$\underline{\text{Total Time}}$ (min)
0	0	246	246	5,424,026	5,424,026
1	0	233	233	5,406,686	5,406,686
2	0	227	227	5,398,586	5,398,586
3	0	227	227	5,398,586	5,398,586

The first column shows the number of fortified road sections. The second column presents the value of the unsatisfied demand (Equation (20)). The third and the fourth columns show the ideal (overlined) and anti-ideal (underlined) value for the maximum arrival time objective function (Equation (21)). The last two columns illustrate the ideal (overlined) and anti-ideal (underlined) value for the total time objective function (Equation (22)).

The first line in Table 1 can be used as a benchmark as it corresponds to the real setting: no road section had been fortified prior to the disaster and there are two emergency inventories located in Nampula and Nacala. The unsatisfied demand is equal to zero, so all the demand center could have been reached from at least one inventory. By increasing the number of fortified road sections (i.e., $Q \geq 0$), it can be observed that both the maximum time and the total time decrease. However, the maximum improvement is achieved for $Q = 2$, so it would not have been beneficial to fortify more than two road sections. Finally, we can observe that the ideal and the anti-ideal values of both time objectives are the same. Thus, no trade-off between the two time objectives is detected.

4.2. Deterministic Scenario: Model-Defined Inventory Locations

In this experiment, only the deterministic scenario is considered. However, differently from the previous one, the model can decide both the inventory locations and the road sections to be fortified. Table 2 illustrates the results of the experiment.

Table 2. Results for the experiment on the deterministic scenario with model-defined inventory locations.

P	Q	Unsatisfied Demand (#)	$\overline{\text{Max Time (min)}}$	$\underline{\text{Max Time (min)}}$	$\overline{\text{Total Time (min)}}$	$\underline{\text{Total Time (min)}}$
2^{fix}	0	0	246	246	5,424,026	5,424,026
1	0	0	279	348	3,117,624	4,994,886
1	1	0	233	296	3,060,684	7,900,516
1	2	0	227	289	2,992,794	7,892,416
1	3	0	227	289	2,992,284	7,892,416
1	4	0	227	289	2,992,284	7,892,416
2	0	0	233	292	1,799,560	2,527,329
2	1	0	191	286	1,793,830	7,474,306
2	2	0	191	233	1,722,910	5,569,478
2	3	0	191	233	1,719,340	5,554,178
2	4	0	191	233	1,719,340	5,554,178
3	0	0	181	292	1,418,585	2,298,989
3	1	0	135	286	1,412,015	3,296,969
3	2	0	135	233	1,341,935	3,296,969
3	3	0	135	233	1,338,365	3,296,969
3	4	0	135	233	1,338,365	3,296,969
4	0	0	135	233	1,127,135	3,085,939
4	1	0	123	233	1,127,135	2,276,549
4	2	0	123	222	1,112,480	2,276,549
4	3	0	123	222	1,108,910	2,276,549
4	4	0	123	222	1,108,910	2,276,549

The first two columns shows the number of inventories and fortified road sections, respectively. The third column presents the value of the unsatisfied demand (Equation (20)). The fourth and the fifth columns show the ideal (overlined) and anti-ideal (underlined) value for the maximum arrival time objective function (Equation (21)). The last two columns illustrate the ideal (overlined) and anti-ideal (underlined) value for the total time objective function (Equation (22)). The first line shows the benchmark from the previous experiment, that is, the solution obtained without fortifying any road segment and fixing the inventories to the real locations.

As expected, the unsatisfied demand is equal to zero as all the demand centers are connected to at least one inventory. Again it is possible to identify a maximum value for Q. According to the results, there is no point in fortifying more than three road sections, as the solution values do not improve. Differently from the previous case, a trade-off between the time measures can be detected. Also, it can be observed that the solution with one inventory and no fortified roads has a lower total distribution time than the benchmark. This result emphasizes the importance of an adequate pre-location of the emergency inventories, as it can lead to faster distribution with fewer resources. Finally, the decrease in the solution times becomes less prominent for the solutions with more than two inventories and one fortified road.

4.3. Stochastic Scenarios

In the following experiments the set Ω is comprised of the 10 scenarios generated as explained in Section 3.2. Table 3 presents the solution values for different configurations of the parameters P and Q.

Table 3. Solution values for the experiments on the stochastic scenarios.

P	Q	Unsatisfied Demand (#)	$\overline{\text{Max Time}}$ (min)	$\underline{\text{Max Time}}$ (min)	$\overline{\text{Total Time}}$ (min)	$\underline{\text{Total Time}}$ (min)
2^{fix}	0	8,758.3	351.1*	351.1	14,004,413.3	14,004,413.3
1	0	10,140.3	386.8*	386.8	11,831,402.1	11,831,402.1
1	2	2,826.2	393.2*	393.2	14,177,321.2	14,177,321.2
1	4	1,250.5	378.1*	378.1	13,648,683.8	13,648,683.8
1	6	353.7	378.1*	378.1	13,836,765.8	13,836,765.8
1	8	284.4	379.1*	379.1	12,989,075.8	12,989,075.8
1	10	278.5	385*	385	12,991,352.9	12,991,352.9
2	0	2,632.5	399.2	399.2	7,885,249.6	7,885,249.6
2	2	547.9	399.2	399.2	8,385,472.7	8,385,472.7
2	4	75.2	445.9	445.9	15,308,719.9	15,308,719.9
2	6	5.9	443.1	443.1	14,462,577	14,462,577
2	8	0	447.2	447.2	14,465,288.2	14,465,288.2
2	10	0	362.4*	362.4*	12,502,890.4	12,502,890.4
3	0	1,250.5	365.1	365.1	6,011,233.4	6,011,233.4
3	2	269.4	346.3*	346.3	5,244,866.3	5,244,866.3
3	4	29.3	346.3*	346.3	5,265,656	5,265,656
3	6	0	443.6	443.6	14,299,408.8	14,299,408.8
3	8	0	346.5*	359.3	4,991,146.1	5,052,085.1*
3	10	0	321.7*	322.1	4,603,674.2	4,606,679.2
4	0	547.9	335	335	5,502,793.5	5,502,793.5
4	2	75.2	334.6*	346.3	4,654,130.6	4,920,867.2*
4	4	4.6	301.9*	302	5,106,823.6	5,111,715.4
4	6	0	255*	327	4,464,166.7	8,788,346.8*
4	8	0	261.3*	274.5	3,061,713.6	3,131,276.8*
4	10	0	260.6*	271.2	2,973,307.4	3,013,568*

The first two columns shows the number of inventories and fortified road sections, respectively. The third column presents the expected unsatisfied demand (Equation (20)). The fourth and the fifth columns show the ideal (overlined) and anti-ideal (underlined) value for the maximum arrival time objective function (Equation (21)). The last two columns illustrate the ideal (overlined) and anti-ideal (underlined) value for the total time objective function (Equation (22)). Solution values with an asterisk (*) are sub-optimal as the execution was halted due to the time limit. The first line presents the benchmark, obtained without fortifying any road segment and fixing the inventories to the real locations.

Considering the humanitarian context of the problem, it is imperative to satisfy all the demand. Therefore, according to the results, the configurations that should be considered for implementation are:

- $(P, Q) = (2, 8)$, i.e., locating two inventories and fortifying eight road sections, or
- $(P, Q) = (3, 6)$, i.e., locating three inventories and fortifying six road sections.

The choice between them should be driven by the costs of fortifying a road and opening an emergency inventory. In terms of the time objectives, it is possible to detect a trade-off. However, all the instances that present different ideal and anti-ideal values could not identify at least one of the optimal solutions within the time limit. Therefore, these results are not conclusive. Finally, the solution of the instance $(P, Q) = (2, 0)$ can be compared with the benchmark. When using the proposed model to define the location of the inventories, the expected unsatisfied demand improves by 69.94%, corresponding to 6,125.8 people rescued on average. The distribution times are not comparable since the solution of the model reaches more demand centers than the benchmark, which implies a larger distribution operation in terms of demand centers relieved and emergency goods delivered. However, despite that, the expected total delivery time still improves by 43.69%.

4.4. Comparing Deterministic and Stochastic Solutions

This subsection compares the solutions obtained by the stochastic model with those of a deterministic model that considers one scenario at a time, to understand the usefulness of the presented approach. Table 4 presents the solutions obtained by the two models. The parameters $(P, Q) = (4, 2)$ and only the solutions corresponding to the ideal total time and the anti-ideal max time have been chosen for illustrative purposes. However, similar results have been obtained with all the configurations.

Table 4. Solution comparison between the stochastic and the deterministic model, for $(P, Q) = (4, 2)$.

Scenario	Inventory Locations	Fortified Arcs
stochastic	{3, 14, 16, 19}	{(2, 34), (4, 27)}
Angoche	{7, 9, 15, 20}	{(7, 34), (17, 38)}
Ilha de Mozambique	{3, 9, 14, 15}	{(3, 11), (15, 27)}
Larde	{7, 9, 15, 20}	{(11, 20), (23, 24)}
Memba	{11, 15, 19, 25}	{(8, 23), (13, 28)}
Mongicual	{2, 6, 9, 15}	{(21, 26), (22, 24)}
Moma	{7, 15, 20, 25}	{(6, 33), (17, 25)}
Mossuril	{3, 9, 14, 15}	{(4, 27), (15, 20)}
Nacala	{3, 9, 19, 20}	{(1, 25), (2, 34)}
Nacala-a-Velha	{1, 3, 15, 16}	{(2, 34), (3, 21)}
Lunga	{1, 3, 14, 26}	{(14, 32), (15, 19)}

The table illustrates in the first row the solution of the stochastic model and in the following 10 rows the individual solutions of the deterministic model, one for each scenario. The first column identifies the scenario considered. The second and the third column present the set of inventory locations and the set of fortified edges, respectively.

From the table, it is possible to observe that addressing each scenario separately would not allow the identification of the best global solution, as each deterministic solution is *ad hoc*. In fact, there is little overlap between the deterministic inventory locations and the stochastic one. This is clearly illustrated in Figure 4, which shows the bar plot of the frequencies of the inventory locations in the deterministic solution.

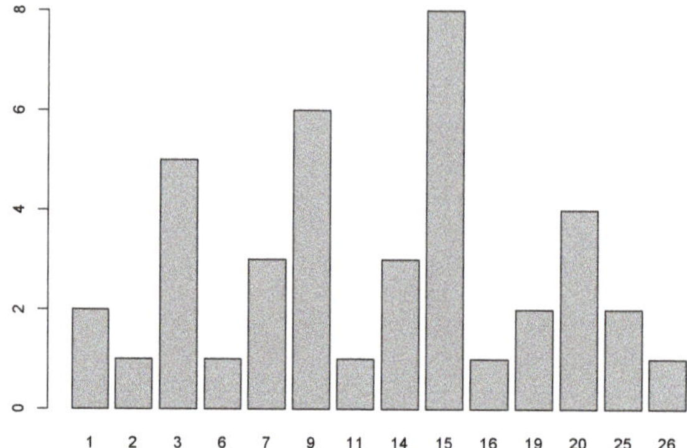

Figure 4. Bar plot of the frequencies of the inventory locations in the deterministic solutions. The x-axis represent the locations and the y-axis their frequencies in the deterministic solutions.

According to the plot, if a decision-maker were to use the most frequent locations to select the best configuration of inventory locations, the set {3, 9, 15, 20} would be chosen. However, the performance of this solution is strongly sub-optimal, as shown in Table 5.

Table 5. Objective function values comparison between the stochastic and the deterministic solutions.

Solution	Inventory Locations	Unsatisfied Demand	Max Time	Total Time
stochastic	{3, 14, 16, 19}	75.2	346.3	4,654,130.6
deterministic	{3, 9, 15, 20}	1,250.5	354.3	9,881,185.1

The table presents the objective function values of the stochastic solution, {3, 14, 16, 19}, and of the deterministic solution, {3, 9, 15, 20}. The first column shows the model considered. The second column illustrates the inventory locations identified by the model. The remaining columns present the corresponding objective function values. The solutions are compared on the stochastic model, that is, considering all the scenarios. Generally speaking, the deterministic solution is expected to perform worse than the stochastic, as the former is a heuristic solution obtained by solving each scenario independently and then choosing the most frequent locations across all the solutions (Figure 4), while the latter is the optimal solution to the stochastic model. From the table it can be seen that the deterministic solution is strongly sub-optimal. In fact, in terms of unsatisfied demand, the stochastic solution allows to relieve 1,175.3 more affected persons, on average, than the deterministic, corresponding to an improvement of 93.99%. Despite attending more demand, the stochastic solution is also more efficient in terms of distribution time, improving by 2.26% the max time and by 52.90% the total time of the deterministic solution.

Although this analysis only considered the inventory locations, similar conclusions can be drawn on the road fortifications.

5. Conclusions

This paper presents the first pre-disaster stochastic model that jointly optimizes the location of emergency inventories and the fortification of transportation network elements in Humanitarian Logistics. Another important feature of the model is that it is parsimonious in terms of data requirements. This is fundamental to be able to realistically apply the model to underdeveloped and developing countries, that usually have limited information on resources available and infrastructure status. The problem has been modeled as a two-level lexicographic problem, where the first level minimizes the expected unsatisfied demand, while the second level considers two time measures: expected maximum arrival time at the demand vertices, and expected total distribution time. As a weak trade-off is detected between the time measures, their relationship is assessed using the pay-off matrix, rather than relying on more complex multicriteria methods. The model is tested on a novel dataset based on a case study on the storm system that hit the Nampula Province (Mozambique) in January 2018. The model is used to evaluate the real response to the emergency and identify what the best course of action should have been. The experiments show that the solution obtained by the model is better than the current policy and improves on the expected unsatisfied demand by 69.94% (corresponding to 6,125.8 more people relieved, on average) and on the expected total delivery time by 43.69%. Also, the model provides two solutions that allow to service all the demand under all the scenarios. The choice between these two solutions depends on the costs of opening an emergency inventory and fortifying a road section, and it is left to the decision-maker. Finally, we illustrated empirically how useful it is to integrate uncertainty in the optimization.

Differently from other investigators, our future research plans do not involve making the model more complicated by factoring in additional operations that, in turn, require more information, as this would be against the rationale behind this contribution. Our objective is to devise models that help decision-makers in emergency management as much as possible while keeping the data required to a realistic level. For this same reason, we are currently considering to represent the uncertainty in the disaster outcome by applying robust optimization rather than stochastic optimization. Moving from an expected cost objective to a minimax formulation has two major advantages. First, it does not need the probability distributions of the uncertain parameters. Second, it results in more conservative solutions that cater for worst-case outcomes [24], which is a desirable feature in the context of a disaster. A different approach which would be worth exploring is formulating the problem to consider Recoverable Robustness [25]. A solution is recovery robust if it can be recovered by limited means in all likely scenarios. This provides a middle ground between classical Robust Optimization and Stochastic Programming. Another interesting extension to the model would be introducing a temporal dimension to consider multiple emergencies over time. This is motivated by the application context, as many disasters (e.g., floods and hurricanes) are seasonal. Therefore, the decision-maker could plan the fortification of the road network over multiple periods, while prepositioning the inventories before every emergency. Finally, an alternative line of research concerns the generation of stochastic scenarios. Namely, we are interested in looking for data-efficient ways of building realistic scenarios which take into account correlations between vertices and arcs disruptions.

Author Contributions: Conceptualization, F.L., J.M. and B.V.; methodology, F.L. and J.M.; software, J.M., F.L. and B.V.; validation, B.V., J.M. and F.L.; formal analysis, F.L., B.V. and J.M.; investigation, B.V. and F.L.; resources, B.V.; data curation, J.M.; writing—original draft preparation, F.L.; writing—review and editing, F.L. and B.V.; visualization, J.M. and B.V.; supervision, B.V.; project administration, F.L. and B.V.; funding acquisition, B.V. All authors have read and agreed to the published version of the manuscript.

Funding: The research was partially funded by the European Commission's Horizon 2020 research and innovation programme under the Marie Sklodowska-Curie, grant number MSCA-RISE 691161 (GEO-SAFE); the Government of Spain, grant MTM2015-65803-R; the Complutense University of Madrid, scholarship *Ayudas para la realización de estancias en proyectos de cooperación para el desarrollo sostenible.* All the financial support is gratefully acknowledged.

Acknowledgments: The authors would like to thank the following entities, agencies and organizations for all the help and support provided: National Institute of Statistics (Mozambique), National Road Administration (Mozambique), National Institute of Disaster Management (Mozambique), Red Cross (Spain), Maputo International Airport, and the NGO ONGAWA.

Conflicts of Interest: The authors declare no conflict of interest.

Abbreviations

The following abbreviations are used in this manuscript:

OR	Operations Research
MS	Management Sciences
HDI	Human Development Index
INGC	Mozambique Institute for Disaster Management
NGO	Non-Governmental Organization

References

1. Vitoriano, B.; Ortuño, M.T.; Tirado, G.; Montero, J. A multi-criteria optimization model for humanitarian aid distribution. *J. Glob. Optim.* **2011**, *51*, 189–208. [CrossRef]
2. Ortuño, M.T.; Tirado, G.; Vitoriano, B. A lexicographical goal programming based decision support system for logistics of Humanitarian Aid. *Top* **2011**, *19*, 464–479. [CrossRef]
3. Liberatore, F.; Ortuño, M.T.; Tirado, G.; Vitoriano, B.; Scaparra, M.P. A hierarchical compromise model for the joint optimization of recovery operations and distribution of emergency goods in Humanitarian Logistics. *Comput. Oper. Res.* **2014**, *42*, 3–13. [CrossRef]
4. Çelik, M. Network restoration and recovery in humanitarian operations: Framework, literature review, and research directions. *Surv. Oper. Res. Manag. Sci.* **2016**, *21*, 47–61. [CrossRef]
5. Asaly, A.N.; Salman, F.S. Arc selection and routing for restoration of network connectivity after a disaster. In *Global Logistics Management*; Taylor & Francis Group: Abingdon, UK, 2014; p. 165.
6. Gillespie, P.; Romo, R.; Santana, M. Puerto Rico aid is trapped in thousands of shipping containers. *CNN*. Available online: https://edition.cnn.com/2017/09/27/us/puerto-rico-aid-problem/index.html (accessed on 24 March 2020).
7. Thomas, A.S.; Kopczak, L.R. From logistics to supply chain management: The path forward in the humanitarian sector. *Fritz Inst.* **2005**, *15*, 1–15.

8. Haghani, A.; Oh, S.C. Formulation and solution of a multi-commodity, multi-modal network flow model for disaster relief operations. *Trans. Res. Part A Policy Pract.* **1996**, *30*, 231–250. [CrossRef]
9. Altay, N.; Green, W.G., III. OR/MS research in disaster operations management. *Eur. J. Oper. Res.* **2006**, *175*, 475–493. [CrossRef]
10. Ortuño, M.; Cristóbal, P.; Ferrer, J.; Martín-Campo, F.; Muñoz, S.; Tirado, G.; Vitoriano, B. Decision aid models and systems for humanitarian logistics. A survey. In *Decision Aid Models for Disaster Management and Emergencies*; Springer: Berlin/Heidelberg, Germany, 2013; pp. 17–44.
11. Galindo, G.; Batta, R. Review of recent developments in OR/MS research in disaster operations management. *Eur. J. Oper. Res.* **2013**, *230*, 201–211. [CrossRef]
12. Liberatore, F.; Pizarro, C.; de Blas, C.S.; Ortuño, M.; Vitoriano, B. Uncertainty in humanitarian logistics for disaster management. A review. In *Decision Aid Models for Disaster Management and Emergencies*; Springer: Berlin/Heidelberg, Germany, 2013; pp. 45–74.
13. Aslan, E.; Çelik, M. Pre-positioning of relief items under road/facility vulnerability with concurrent restoration and relief transportation. *IISE Trans.* **2019**, *51*, 847–868. [CrossRef]
14. Sabbaghtorkan, M.; Batta, R.; He, Q. Prepositioning of assets and supplies in disaster operations management: Review and research gap identification. *Eur. J. Oper. Res.* **2019**, *284*, 1–19. [CrossRef]
15. Wisetjindawat, W.; Ito, H.; Fujita, M. Integrating stochastic failure of road network and road recovery strategy into planning of goods distribution after a large-scale earthquake. *Transp. Res. Rec.* **2015**, *2532*, 56–63. [CrossRef]
16. Iloglu, S.; Albert, L.A. An integrated network design and scheduling problem for network recovery and emergency response. *Oper. Res. Perspect.* **2018**, *5*, 218–231. [CrossRef]
17. Iloglu, S.; Albert, L.A. A maximal multiple coverage and network restoration problem for disaster recovery. *Oper. Res. Perspect.* **2020**, *7*, 100132. [CrossRef]
18. Sanci, E.; Daskin, M.S. Integrating location and network restoration decisions in relief networks under uncertainty. *Eur. J. Oper. Res.* **2019**, *279*, 335–350. [CrossRef]
19. United Nations Development Programme. 2019 Human Development Index Ranking. Available online: http://hdr.undp.org/en/content/2019-human-development-index-ranking (accessed on 24 March 2020).
20. *Météo France*. Droits de Reproduction (Reproduction Rights). Available online: http://www.meteofrance.com/droits-de-reproduction (accessed on 24 March 2020).
21. INGC. *Ponto de Situação do NíVel de Resposta na Zona Norte (State of Response Level in the North Zone)*; Technical Report; Instituto Nacional de Gestão de Calamidades (Mozambique Institute for Disaster Management): Maputo, Mozambique, 2018.
22. Instituto Nacional de Estatistica (National Statistics Institute). IV Censo 2017 (IV Census 2017). Available online: http://www.ine.gov.mz/iv-censo-2017 (accessed on 24 March 2020).
23. Monzón, J.; Liberatore, F.; Vitoriano, B. 2018 Nampula Province Dataset. Available online: http://blogs.mat.ucm.es/humlog/wp-content/uploads/sites/6/2020/02/MonzonLiberatoreVitoriano2020.zip (accessed on 24 March 2020).
24. Snyder, L.V. Facility location under uncertainty: A review. *IIE Trans.* **2006**, *38*, 547–564. [CrossRef]
25. Liebchen, C.; Lübbecke, M.; Möhring, R.H.; Stiller, S. *Recoverable Robustness*; Technical Report ARRIVAL-TR-0066; Technische Universität Berlin: Berlin, Germany, 2007.

Article

A New Ant Colony-Based Methodology for Disaster Relief

José M. Ferrer [1,*], M. Teresa Ortuño [2] and Gregorio Tirado [1]

[1] Department of Financial and Actuarial Economics & Statistics, Interdisciplinary Mathematics Institute, Complutense University of Madrid, HUMLOG Research Group, 28040 Madrid, Spain; gregoriotd@ucm.es

[2] Department of Statistics and Operational Research, Interdisciplinary Mathematics Institute, Complutense University of Madrid, HUMLOG Research Group, 28040 Madrid, Spain; mteresa@ucm.es

* Correspondence: jmferrer@ucm.es

Received: 2 March 2020; Accepted: 27 March 2020; Published: 3 April 2020

Abstract: Humanitarian logistics in response to large scale disasters entails decisions that must be taken urgently and under high uncertainty. In addition, the scarcity of available resources sometimes causes the involved organizations to suffer assaults while transporting the humanitarian aid. This paper addresses the last mile distribution problem that arises in such an insecure environment, in which vehicles are often forced to travel together forming convoys for security reasons. We develop an elaborated methodology based on Ant Colony Optimization that is applied to two case studies built from real disasters, namely the 2010 Haiti earthquake and the 2005 Niger famine. There are very few works in the literature dealing with problems in this context, and that is the research gap this paper tries to fill. Furthermore, the consideration of multiple criteria such as cost, time, equity, reliability, security or priority, is also an important contribution to the literature, in addition to the use of specialized ants and effective pheromones that are novel elements of the algorithm which could be exported to other similar problems. Computational results illustrate the efficiency of the new methodology, confirming it could be a good basis for a decision support tool for real operations.

Keywords: Ant Colony Optimization; humanitarian logistics; last mile distribution; disaster relief

1. Introduction

Disaster management remains a crucial challenge for the international community. National and supranational institutions and non-governmental organizations have to deal with high impact disasters such as the 2010 Haiti earthquake, the 2013 Typhoon Haiyan or the 2018 Sulawesi earthquake and tsunami. This global concern is reflected in the scientific literature and specifically in the Operations Research community, where the increasing interest on humanitarian logistics and disaster management entails the publication of surveys [? ?], books [? ?], special issues [? ?] and journals (such as the Journal of Humanitarian Logistics and Supply Chain Management).

Within the disaster management cycle, the response phase includes the actions to be performed once the disaster has occurred. Research on the response phase may focus on specific types of disasters, such as earthquakes [? ?] or floods [?], or can address particular problems for generic disasters, such as location-allocation [?], evacuation [?], or debris management [?]. Among the operations to be carried out after a disaster strikes, the distribution of aid to the affected people plays an important role. Along with the usual objectives in commercial logistics, mainly the operation cost, in humanitarian logistics other attributes have to be considered [? ?], such as the response time, the equity of the distribution, etc. Therefore, multi-criteria models are necessary to face many realistic problems in this context. ?] provide a review on multi-criteria decision making in humanitarian aid classifying the existing models as deterministic or stochastic and pre-disaster or post-disaster.

Many times the distribution operations must be performed under unsecure conditions [?]. Assaults to relief vehicles are sadly frequent and this complicates the work of organizations that execute the aid missions. Avoiding as much as possible the most dangerous roads and forcing the vehicles to travel together forming convoys are two ways of improving the security of the operations.

In this paper, we develop an algorithm based on Ant Colony Optimization (ACO) metaheuristic to solve the multi-criteria last mile distribution problem in an unsecure environment stated in [?]. The complexity of this problem makes it necessary to use a heuristic approach to solve realistic cases. Besides, due to the strong links between the elements of a solution, it is difficult to develop heuristics that perform progressive alterations of a solution while ensuring feasibility, as it is the case of evolutionary algorithms or local search heuristics. Instead, algorithms based on construction of solutions, such as Greedy Randomized Adaptive Search Algorithm (GRASP) or Ant Colony Optimization metaheuristics, fit much better.

The ACO methodology that we propose in this work presents some novelties that make it suitable for routing and distribution problems in which the vehicles must travel in convoys. The use of convoys is appropriate not only in humanitarian operations, but in any logistics operation performed in an environment of insecurity, in which attacks on delivery vehicles to steal the commodities are possible. The algorithm considers two types of ants—vehicles and aid kits—four types of standard pheromones, and especially the use of effective pheromones, updatable in the course of each solution building process. This new concept allows a better balance between the elements of a solution and a higher diversification in the set of solutions that can be built. Effective pheromones may be applied to a wide variety of problems, especially vehicle routing problems involving a large number of vehicles. The results obtained show that the new ACO Algorithm proposed in this paper is a useful decision aid tool. The ACO Algorithm provides good quality solutions in a short time in realistic test cases, frequently improving the solutions obtained with the GRASP Algorithm developed in [?].

The rest of this paper is organized as follows. Section **??** reviews the relevant literature on multi-criteria optimization for the last mile distribution problem, as well as some applications of Ant Colony Optimization in disaster management. Section **??** presents the problem of transportation for last mile distribution, including the data notation and the description of the six attributes considered. Section **??** introduces the heuristic method proposed to solve the problem. An Ant Colony Optimization algorithm is developed together with the subordinate procedure that guides the construction of feasible solutions. Section **??** is devoted to showing and analyzing the results obtained when applying the proposed metaheuristic to solve two test cases based on real disasters. Finally, Section **??** draws some conclusions derived from this work.

2. Literature Review

Some of the most significant papers that apply multi-criteria techniques in humanitarian last mile distribution are mentioned below and summarized in Table **??**. ?] develop a multicommodity mixed integer programming model in order to maximize the total population covered and minimize the total travel time for earthquake response. ?] present a multicommodity, multimodal and multiperiod last mile distribution problem and establish a scalar objective function that aggregates minimization of transportation costs and maximization of demand covered. ?] consider three attributes (reliability of the transporting tours, coverage and total travel time) in a distribution problem where infrastructure may be damaged in a post-disaster situation. ?] use the scalarization approach to deal with a three-objective (total unsatisfied demand, total travel time and equity of the distribution) integer programming model that allows split deliveries. Two heuristic methods are proposed to solve the model. ?] propose a robust stochastic compromise programming model for disaster relief logistics, minimizing the sum of the expected value and the variance of the total cost and minimizing the sum of the maximum unsatisfied demands. ?] consider four attributes (amount of aid distributed, time of the operation, equity and cost) in a dynamic flow model that is solved via lexicographical goal programming. ?] develop a quadratic multi-objective programming model to face an allocation and distribution problem. Three attributes are considered: lifesaving utility, delay cost and equity of the distribution. This model allows vehicles to visit a node more than once and provides the individual itineraries of the vehicles. ?] aggregate the distribution time, the penalty cost of unsatisfied demand and the costs of opening distribution centers into a single objective function for a location-routing problem considering network failure, random travel times and multiple uses of vehicles. ?] apply a multi-objective evolutionary algorithm to solve a multiperiod and multicommodity distribution model which aims to minimize the unsatisfied demand and to minimize the risk of choosing damaged roads. ?] present a multimodal model considering dynamic and stochastic demands in order to minimize the travel time and the unmet demand. ?] develop a biobjective multiperiod stochastic location-transportation model that allows multiple trips to the vehicles not only in different periods, but also in the same period. The authors propose different heuristic techniques to solve the model. Outside the scope of aid distribution, ?] address a post-disaster waste management problem through a possibilistic programming model that minimizes both the cost of waste processing and the carbon emissions and maximizes the job opportunities.

Despite the frequency with which aid distribution is to be carried out in unsafe conditions, only a few papers consider the security as an attribute. ?] integrate seven attributes (quantity distributed, time of response, cost, equity, priority, reliability and security) into a linear lexicographical goal programming flow model. This model is multimodal, multidepot, allows split deliveries and vehicles travel in convoys so that they can be escorted. An alternative version of this model is established in ?], where the authors set a fixed amount to distribute and combine the six remaining attributes into a non-lexicographical model. Nevertheless, these models do not provide the individual route of each vehicle, so it is not possible to easily implement plans for the relief distribution. ?] propose a GRASP metaheuristic to solve a nonlinear multi-objective model that allows vehicles to visit a node more than once and provides the individual itineraries of the vehicles. The mathematical formulation of this model is established in ?], where an extensive comparison of the characteristics of the models is made with other studies that deal with the last mile distribution problem.

Table 1. The literature related to multi-criteria techniques in humanitarian last mile distribution.

Reference	# Criteria	Security	Multi-Criteria Approach	Heuristic Approach
?]	3		Scalarization	Variable Neighborhood Search
?]	2		Scalarization	-
?]	2		Compromise Programming	-
?]	6	✓	Compromise Programming	GRASP
?]	3		Interactive Method	-
?]	3		Scalarization	-
?]	3		Scalarization	Genetic Algorithm, ad-hoc heuristic
?]	2		Scalarization	-
?]	2		Hierarchical Optimization, Scalarization	Fix and Optimize, ad-hoc heuristic
?]	3		Pareto Optimization	Memetic Algorithm, ad-hoc heuristic
?]	7	✓	Goal Programming	-
?]	3		Goal Programming	-
?]	2		Pareto Optimization	-
?]	6	✓	Goal Programming	-
?]	2		Pareto Optimization	Evolutionary Algorithm
This study	6	✓	Compromise Programming	Ant Colony Optimization

In this study, we propose an Ant Colony Optimization methodology to solve the model developed in [?] and [?]. Even though ACO has been applied frequently to different routing and distribution problems [? ? ? ?], it has barely been used in the context of disaster management. The most remarkable work is ?], where the authors apply this metaheuristic to solve a distribution and evacuation problem. More recently, ?] combine different techniques, including ACO, to face a location-allocation problem of relief centers after an earthquake. Therefore, this paper presents the first application of Ant Colony Optimization to a multi-criteria last mile distribution problem. Furthermore, the proposed ACO methodology introduces novelties such as specialized ants, representing both vehicles and aid kits, and effective pheromones, that conveniently balance the elements of the built solutions.

3. Problem Description

The last mile distribution problem addressed in this paper consists in deciding how to deal out a predetermined amount of humanitarian aid from distribution centers to populations in need by moving a set of vehicles through a transportation network. We consider that the distribution is performed in an insecure and uncertain environment and, as a consequence, the convoys transporting the aid may be attacked while traveling and some roads may be unavailable due to damages produced by the disaster.

Table **??** lists the notation used to represent the main elements of the problem. The transportation network is formed by nodes and arcs. Each node is of one of three types: supply, demand or connection. A supply node represents a distribution center where a certain known amount of aid is available, while the demand nodes represent populations in need or help centers. Each demand node is associated with a certain amount of aid that is desirable to receive and a priority level. A connection node represents an intermediate point, where there is neither available nor demanded aid, but may be used by the vehicles for aid trans-shipment.

Table 2. Problem notation.

Sets and indexes	
I, J, K :	Sets of nodes ($i \in I$), vehicles ($j \in J$), and arcs ($k \in K$), respectively
$I_D \subset I$:	Subset of demand nodes
$I_S \subset I$:	Subset of supply nodes
C :	Set of classes of vehicles ($c \in C$), considering that one vehicle class is determined by the vehicle type and the depot at which the vehicles are parked initially
G :	Set of objectives ($g \in G$)
Parameters	
$qglobal$:	Amount of humanitarian aid that is planned to be distributed in the operation
$dist_k$:	Length of arc k
$cback_{j,i}$:	Return cost of vehicle j from i to its original depot by using the least expensive path
dem_i :	Amount of humanitarian aid demanded by node i
qav_i :	Amount of humanitarian aid available at node i
r_k :	Reliability of arc k
$vela_k$:	Maximum speed through arc k
$velv_j$:	Maximum speed of vehicle j
car_c :	Number of vehicles of class c

The arcs represent links (roads, pathways, streets, etc.) connecting the nodes. The links that can be traversed in both directions have two associated arcs, one for each direction. Each arc has associated a length, a speed, an assault probability and a probability of being available (the arc can be used).

The vehicles are initially located at some nodes, which are not necessarily supply nodes. We will call depots to the nodes where there are vehicles at the beginning of the operation. There are different types of vehicles to carry out the distribution. Vehicles of the same type have the same capacity, velocity and costs. For technical reasons, we classify the vehicles in classes: vehicles belonging to a same class are of the same type and are located also in the same depot.

The humanitarian aid is packed in kits composed of water, food, medicines, blankets, etc. The operation to be performed consists of delivering $qglobal$ of such kits.

In addition, the design of the routes must be carried out under the following assumptions:

– Vehicles can visit a node more than once. Likewise, they can pick up or deliver humanitarian aid more than once, even in the same node.

- Transhipment of aid between vehicles can be made at any node.
- Split pick up and delivery are allowed in all supply and demand nodes, respectively.
- For security reasons, all vehicles traveling through an arc must do it together forming a convoy. In this way, the convoys can be escorted. As a result, each vehicle can pass through an arc just once.
- The probability of assault to a convoy decreases with the size of the convoy.
- Vehicles must return to their respective depots at the end of the operation. The return of each vehicle will be made through the least expensive path, being $cback_{j,i}$ the return cost of vehicle j to node i. Assaults on return itineraries are not considered.
- In the last trip of its route (before it returns), each vehicle must transport some aid. This is imposed to prevent trips of empty vehicles with the only purpose of increasing convoy safety.

As usual in humanitarian logistics, this problem is intrinsically multi-criteria. In this case, we consider six different attributes that will be incorporated into the model as objectives to be minimized. In what follows, a description of the attribute measures used in the model is provided; however, since they are not necessary for the understanding of the paper, we decided to omit the mathematical formulation. The interested reader can see [?] for further details.

Time. The distribution is done in an emergency situation, so we consider the time of the operation as an attribute. The operation finishes when the last kit of aid is delivered to a demand point, the return times are not considered.

Cost. The total cost of the operation, including fixed and variable costs and the return costs, is also an attribute. In many cases, there may be a budget that can not be exceeded.

Equity. Since the available aid or the amount of aid planned to distribute are usually lower than the total demanded aid, very frequently it is not possible to fully meet the demand. Thus, it is necessary to consider the equity of the distribution as an attribute. In a perfectly fair deal all demand nodes must receive aid in the same proportion, that will be called *ideal demand proportion*, as established in (**??**).

$$idp = \frac{qglobal}{\sum_{i \in I_D} dem_i} \tag{1}$$

The equity measure is defined as the square root of the quadratic deviation of the proportions of satisfied demand from the ideal demand proportion over all demand nodes.

Priority. The priority attribute is measured by the sum of relative unsatisfied demands over all demand nodes weighted by their priority levels. The priority levels are considered for two reasons. On the one hand, a higher level of priority may indicate greater urgency in the demand node. On the other hand, there can be nodes that, due to having difficult access or belonging to insecure or damaged areas, are not visited at all. Then, a high priority level is required for such nodes.

Security. Dealing with an insecure environment is a main feature of the problem we are considering. In addition to the probabilities of assault to the convoys, how severe an assault is must also be taken into account. For this purpose, we use an elaborated measure that evaluates the expected gravity of suffering attacks due to the amount of aid that would be lost.

Reliability. As in the security attribute, not only the availability of the arcs must be considered, but also the consequences of finding not passable arcs. Thus, the reliability measure evaluates the expected aid loss over all arcs.

The objectives can be minimized separately or in an aggregated function as in (??), where weights $w_g, g \in G$ represent the preferences of the decision maker and f_g is the objective function associated with the attribute g.

$$f = \sum_{g \in G} w_g f_g \tag{2}$$

For obtaining solutions that take all attributes into account in a balanced way, we apply compromise programming (CP), introduced in ?]. CP is a multi-criteria decision making technique which consists in minimizing the distance to the ideal point. The components of the ideal point are the optimal values of the attributes when optimized independently. To normalize the attributes, the difference between the ideal and the anti-ideal is used, where the components of the anti-ideal point are the worst values for the attributes on the set of non-dominated solutions.

The CP objective function when using the p-norm is denoted as $\tilde{f}_p$ and defined as follows:

$$\tilde{f}_p = \left[\sum_{g \in G} w_g \left(\frac{f_g - i_g^+}{i_g^- - i_g^+} \right)^p \right]^{\frac{1}{p}} \tag{3}$$

Again, $w_g, g \in G$ represent the preferences of the decision maker, while i_g^+ and i_g^- are the ideal and the anti-ideal components associated with attribute g, respectively.

4. Heuristic Approach

In this section we present an algorithm based on Ant Colony Optimization, that was introduced in ?]. The ACO metaheuristic emulates the way in which the ants of a colony manage to find the shortest path from the anthill to a source of food through the pheromone trails deposited by the ants when they move. Extensive information about ACO metaheuristic and its multiple variants can be found in [? ? ?]. The metaheuristic algorithm that we propose in this paper takes some elements of the Ant Colony System (ACS) developed in ?] to establish a multiple-ant colony system variant. Section ?? is devoted to the main program of the metaheuristic (ACO Algorithm), and Section ?? to the subordinate procedure that performs the construction of solutions (Pheromone Trail Constructive Algorithm or PTC Algorithm).

Table ?? shows the main parameters and variables that are used in the metaheuristic. According to the ACS, two types of pheromone updates are made. The global update is performed every time m consecutive solutions are obtained and is intended to improve the construction of future solutions. The local update is performed after each single solution is obtained, and its purpose is to diversify the construction. Parameters ρ_g and ρ_l are the corresponding evaporation rates. Our algorithm uses two types of ants, representing vehicles and humanitarian aid kits, and four types of pheromones, that are deposited by the vehicles or by the aid kits in different situations. First type (respectively second type) pheromones are deposited by the vehicles (aid kits) when moving through the arcs, third type pheromones are associated with the amounts of aid leaving supply nodes, and fourth type pheromones correspond to the amounts of aid remaining at the nodes at the end of the operation. $\tau^1_{c,k}$, τ^2_k, τ^3_i and τ^4_i are the variables that indicate the quantities of pheromone of the four types that have been deposited at an arc k or at a node i.

The construction of solutions takes into account both pheromones and visibility. Pheromones indicate how often each element candidate to be added to the current partial solution was selected by previous ants, while visibility, established as greedy functions, indicates how good an element could be according to each of the objectives considered.

The PTC Algorithm builds new solutions by adding elements iteratively. As a result, the values of the pheromones associated with the added elements should be decreased to prevent an excessive concentration of these elements in the solution. This is achieved through what we called *effective* pheromones, a new concept that refers to variables that at the beginning of the execution of the PTC

Algorithm take the same values as the corresponding pheromones variables, but which will be updated during the construction of the solution.

Table 3. Heuristic approach notation.

Parameters	
ρ_l :	Local evaporation rate
ρ_g :	Global evaporation rate
σ_1 :	Greedy functions weight at the beginning of the algorithm
σ_2 :	Growth rate of pheromones weight
m :	Number of solutions obtained between two consecutive global updates
q_0 :	Probability of choosing the element with the highest score in any of the decision situations
n :	Total number of solutions to be obtained
Variables	
λ :	Variable that controls pheromones weight versus greedy functions weight
$\tau^1_{c,k}$:	First type pheromone in arc k for the vehicles of class c
τ^2_k :	Second type pheromone in arc k
τ^3_i :	Third type pheromone in supply node $i \in IO$
τ^4_i :	Fourth type pheromone in node $i \in ID \cup IO$
ϱ^g_ζ :	Individual greedy function of objective g associated to element ζ
fs_i :	Final stock at node i in the current solution
lt_k :	Amount of aid transported by arc k in the current solution
$vt_{c,k}$:	Number of vehicles of class c traversing arc k in the current solution

4.1. Ant Colony Optimization Algorithm

In this subsection we introduce the ACO Algorithm, whose pseudocode is given in Algorithm **??**. Its main steps are explained below.

Algorithm 1 ACO algorithm

```
Perform preprocess.
Create a solution S* with the PTC Algorithm.
for it = 2, ··· , n do
    Create a new solution S with the PTC Algorithm.
    if S is better than S* then
        Do S* = S.
    end if
    Perform the local update.
    if it ≡ 0 (mod m) then
        Perform the global update.
    end if
    Increase λ.
end for
Output S*.
```

Preprocess. It comprises several tasks needed before starting the construction of solutions:

- Simplify the logistic network removing unnecessary arcs.
- For each vehicle, calculate the least expensive return paths from any node to its depot via Dijkstra's algorithm $\rightarrow cback_{j,i}$.
- Calculate bounds for the greedy functions $\rightarrow b^g$.
- Initialize λ and pheromones.

Pheromone update. The local update increases the pheromones of the elements that appear less frequently in the previous iteration in order to diversify the construction, as stated in Equations (??)–(??).

$$\tau^1_{c,k} = (1-\rho_l)\tau^1_{c,k} + \frac{\rho_l}{1+vt_{c,k}} \quad \forall c, \forall k \tag{4}$$

$$\tau^2_k = (1-\rho_l)\tau^2_k + \frac{\rho_l}{1+lt_k} \quad \forall k \tag{5}$$

$$\tau^3_i = (1-\rho_l)\tau^3_i + \frac{\rho_l}{1+qav_i - fs_i} \quad \forall i \in I_S \tag{6}$$

$$\tau^4_i = (1-\rho_l)\tau^4_i + \frac{\rho_l}{1+fs_i} \quad \forall i \in I_D \cup I_S \tag{7}$$

In the same way, the global update increases the pheromones of the elements that appear more in the current best known solution in order to guide the construction towards potentially better solutions, as stated in Equations (??)–(??). The auxiliary parameter cl_c in (??) denotes the number of vehicles of class c, while the asterisk in $vt^*_{c,k}$, lt^*_k and fs^*_i indicates that these auxiliary variables correspond to the current best known solution.

$$\tau^1_{c,k} = (1-\rho_g)\tau^1_{c,k} + \frac{vt^*_{c,k}}{cl_c} \quad \forall c, \forall k \tag{8}$$

$$\tau^2_k = (1-\rho_g)\tau^2_k + \frac{lt^*_k}{qglobal} \quad \forall k \tag{9}$$

$$\tau^3_i = (1-\rho_g)\tau^3_i + \frac{qav_i - fs^*_i}{qav_i} \quad \forall i \in I_S \tag{10}$$

$$\tau^4_i = \begin{cases} (1-\rho_g)\tau^4_i + \frac{fs^*_i}{qav_i} & \forall i \in I_S \\ (1-\rho_g)\tau^4_i + \frac{fs^*_i}{dem_i} & \forall i \in I_D \end{cases} \tag{11}$$

The elements that are taken into account in both updates are the following: for the first type pheromones, the class-arc pairs (??) and (??); for the second type pheromones, the amounts of transported aid in the arcs (??) and (??); for the third type pheromones, the amounts of aid that have come out from the supply nodes (??) and (??); and for the fourth type pheromones, the amounts of aid that remain in the supply and demand nodes at the end of the distribution (??) and (??).

Increase λ. Variable λ, which controls the weight of pheromones and visibility, grows over the course of the ACO Algorithm so that the importance of pheromones in the construction is increasing. The pheromones weight is updated according to (??), where *it* is the current iteration. The exponential expression makes the increase of λ to be progressively slower.

$$\lambda = 1 - \sigma_1 e^{-\sigma_2 \frac{it}{n}} \tag{12}$$

4.2. Pheromone Trail Constructive Algorithm

The pseudocode of PTC Algorithm is shown in Algorithm ??. Lines 1 and 2 are the constructive steps, where pheromone and greedy functions are introduced in order to guide the solution building process. Lines 3 to 12 coordinate the elements of the current solution and determine if it is feasible. It must be taken into account that, as all vehicles travel in convoys through the arcs, the solution must satisfy the precedence relation induced on the set of arcs. Lines 13 to 17 intend to improve the solution with the help of pheromone functions.

The solution is built iteratively by adding or removing elements, being the nature of the elements dependent on the algorithm step. An element can be, for example, a trip of a vehicle or a demand node where to drop an aid kit. If ζ is a candidate element to be added, ph_ζ and gr_ζ are, respectively, the effective pheromone and the greedy function associated to it. As a result, the score of element ζ is calculated as stated in (**??**).

$$sc_\zeta = \lambda ph_\zeta + (1 - \lambda) gr_\zeta \tag{13}$$

The reason why this expression is additive rather than multiplicative—as usual in ACO—is because, in this way, the gradual increase of the pheromone weight can be more accurately controlled.

Algorithm 2 PTC Algorithm

```
1:  Design itineraries, form convoys and calculate arriving times at nodes.
2:  Deliver aid through the network.
3:  if all planned aid is delivered then
4:      Determine arriving and departure events.
5:      Calculate stocks after events.
6:  else
7:      Go back to line 1.
8:  end if
9:  Try to eliminate all negative stocks at nodes.
10: if there exists any negative stock then
11:     Go back to line 1.
12: end if
13: Improve equity.
14: Adjust supply nodes.
15: Trim routes.
16: Eliminate aid crossings.
17: Exchange vehicles.
```

The greedy function of a candidate element ζ is stated in (**??**), where ϱ_ζ^g is the individual greedy function of objective g, w_g is the weight of objective g in the objective function and b^g is an upper bound for ϱ_ζ^g.

$$gr_\zeta = \sum_{g \in G} w_g \left(\frac{b^g - \varrho_\zeta^g}{b^g} \right) \tag{14}$$

The greedy functions are used only where it is possible to define them in a reasonable way. When there are no greedy functions, the score is simply $sc_\zeta = ph_\zeta$ if the decision regards which element should be added to the partial solution (it is convenient to select elements with more pheromone) or as stated in (**??**) if the decision regards which element should be removed from the partial solution (it is convenient to select elements with less pheromone).

$$sc_\zeta = \frac{1}{ph_\zeta} \tag{15}$$

In all steps of the PTC Algorithm, the elements are chosen according to the criterion established in ACS: if $q \leq q_0$, where q is a random variable uniformly distributed over $[0, 1]$, the candidate element with the highest score is chosen; otherwise, the election is made according to the probabilities stated in (**??**), which are proportional to the scores.

$$pr_\zeta = \frac{sc_\zeta}{\sum_{\tilde{\zeta}} sc_{\tilde{\zeta}}} \tag{16}$$

The main steps of the PTC Algorithm are described in detail in what follows.

Design itineraries. The routes to be followed by the vehicles are built iteratively, starting from an empty solution where all the vehicles are parked at their corresponding depots. At each iteration, the candidate elements to be added to the partial solution are the feasible trips of every single vehicle j through an available arc k. Therefore, each element is a pair $\zeta = (j,k)$. The effective pheromone associated to each element is defined in (**??**), where c is the vehicle class corresponding to vehicle j.

$$ph_\zeta = \left(1 - \frac{vt_{c,k}}{car_c}\right) \tau^1_{c,k} \tag{17}$$

On the one hand, the effective pheromone is proportional to the first type pheromone, $\tau^1_{c,k}$, which depends on the vehicles of class c traversing arc k in the previously obtained solutions. In particular, the more vehicles of class c transit through this arc in the current best known solution, the more effective pheromone candidate element (j,k) has. On the other hand, the effective pheromone depends on the number of vehicles of class c traversing arc k in the current partial solution, so the greater $vt_{c,k}$ is, the less effective pheromone candidate element (j,k) has. If this gradual adjustment of the effective pheromone is not performed, it could lead to an excess of vehicles of a class through an arc for those pairs (c,k) with higher first type pheromone. This eventual excess of elements in a solution justifies the use of the effective pheromone and it also applies to other steps of the PTC Algorithm.

All the greedy functions used in the PTC Algorithm are defined in the same way as in the Elite Set Constructive Algorithm (ESCA) described in [**?**]. For example, the individual greedy function of the time objective is the time that takes vehicle j to traverse arc k (**??**), while the individual greedy function of the reliability objective is the probability that arc k is not available (**??**).

$$\varrho^t_\zeta = \frac{dist_k}{\min\{vela_k, velv_j\}} \tag{18}$$

$$\varrho^f_\zeta = 1 - r_k \tag{19}$$

The score of each element is calculated as stated in (**??**). Once an element (j,k) is chosen, arc k is added to the route of vehicle j and the associated precedence relations are updated, in order to ensure that all vehicles travel together in convoys. The construction process is iterated until the routes cannot be continued any longer.

Form convoys and calculate arriving times. The convoys are formed by all vehicles traveling through each arc. The travel speed of a convoy is determined by the minimum between the maximum velocity of the slowest vehicle of the convoy and the maximum speed of the associated arc. The arriving times of the convoys at their destination nodes are calculated as the minimum times at which the convoys can finish their trips through the arcs taking into account the travel speeds and the precedence relation on the arcs.

Aid distribution. Once the itineraries of the vehicles are designed, the amount of aid transported by each convoy must be determined. For this purpose, an auxiliary transportation network is built, including all the nodes of the original network but only the arcs traversed by convoys, according to the itineraries determined previously. Several dummy arcs and two dummy nodes are also added: the source node (sc), linked to all supply nodes by dummy arcs, and the sink node (sn), linked to each demand node also by dummy arcs. The capacities of the dummy arcs leaving the source or arriving at the sink are equal to the aid availability or the demand of the corresponding supply or demand node, respectively. The capacity of each original arc is equal to the capacity of the convoy traversing it. The calculation of the flow of aid through this network is performed by applying a modified version of the Ford-Fulkerson algorithm [**?**].

The augmenting paths from the source to the sink are determined iteratively, node by node, as follows:

- If the current node is the source, the candidate elements are the supply nodes with available aid. The effective pheromone of each supply node, $\zeta = i$, is stated in (**??**), where $fl_{sc,i}$ is the current flow from the source to the supply node i, i.e., the amount of aid leaving the supply node i in the current partial flow solution. The effective pheromone depends positively on the third type pheromone (which is related to the amount of aid leaving supply node i in the previously obtained solutions) and negatively on the amount of aid leaving i in the current solution.

$$ph_\zeta = \left(1 - \frac{fl_{sc,i}}{qav_i}\right) \tau_i^3 \tag{20}$$

  No greedy function is defined in this case, so the score of each supply node i is equal to the corresponding effective pheromone.

- If the current node is a node i_0 different from the source, a candidate element is a node i linked to i_0 through an arc $k = (i_0, i)$ with positive residual capacity. The effective pheromone of each candidate node, $\zeta = i$, is calculated as stated in (**??**). It depends positively on the amount of aid transported through arc k in the previously obtained solutions and negatively on the amount of aid transported through arc k in the current solution.

$$ph_\zeta = \left(1 - \frac{fl_{i_0,i}}{qglobal}\right) \tau_k^2 \tag{21}$$

  The individual greedy function of the time objective is the arriving time of the convoy associated to arc k, the individual greedy function of the cost objective is the average cost of transporting an aid kit by arc k among the vehicles of the associated convoy, etc.

  If the current node i_0 is a demand node, in addition to the other candidates mentioned earlier, the sink sn must be considered as a candidate element as well. The selection of the sink would finish the augmenting path. Its effective pheromone (**??**) depends positively on the amount of aid remaining at demand node i_0 at the end of the operation in the previously obtained solutions and negatively on the remaining aid in the current solution.

$$ph_{sn} = \left(1 - \frac{fl_{i_0,sn}}{dem_{i_0}}\right) \tau_{i_0}^4 \tag{22}$$

  The individual greedy function of the equity objective associated to the sink measures how the deviation of the proportion of satisfied demand at node i_0 from the ideal demand proportion would vary if the sink is selected; the individual greedy function of the priority objective is obtained in a similar way; and the individual greedy functions of the other objectives take the value 0, because finishing the augmenting path does not worsen those objectives.

  Finally, the scores are established according to (**??**).

Each time a node is selected and labeled, the current node is updated. Once an augmenting path is determined, the flow is updated according to the Ford-Fulkerson algorithm. The process is iterated until all the planned amount of aid is distributed from the source to the sink or it is not possible to find an augmenting path, in which case the current solution would be discarded and the algorithm would be restarted at Step 1.

Determine events and calculate stocks. The events in each node correspond to the arrivals and departures of convoys. Once the timing of the events are determined, the stocks, i.e., the amounts of aid available after each event, are obtained.

Eliminate negative stocks. At this point, some negative stock may be found because a convoy departs from a node transporting some aid that has not arrived yet. In order to avoid this, additional precedence relations forcing the arrival of aid to occur before the corresponding departure are added when needed.

Improve equity. This step is executed if there is aid leaving a demand node in which the proportion of satisfied demand is less than the ideal demand proportion. A demand node that verifies this condition is called *improvable node*. In order to keep the maximum possible amount of aid at the improvable nodes, a variation of the maximum flow problem is solved, as in the Aid distribution step. The auxiliary transportation network includes all the nodes but only the used arcs. The source node is linked to all improvable nodes by dummy arcs, and the sink node is linked by dummy arcs arriving from the demand nodes in which the proportion of satisfied demand exceeds the ideal demand proportion. The capacity of each dummy arc leaving the source is the minimum between the amount of aid leaving that node and the residual amount of aid needed to reach the ideal demand proportion. The capacity of each dummy arc reaching the sink is the amount of aid exceeding the ideal demand proportion at the associated node. The capacity of each original arc is equal to the amount of aid transported through it.

The augmenting paths from the source to the sink are determined in a similar way as in the Aid distribution step:

- If the current node is the source, the candidate elements are the improvable nodes. The effective pheromone of each improvable node, $\zeta = i$, is stated in (**??**). The effective pheromone depends positively on the final stock at node i in the previously obtained solutions and negatively on the final stock at node i in the current solution.

$$ph_\zeta = \left(1 - \frac{fs_i}{dem_i}\right) \tau_i^4 \tag{23}$$

  The score of each improvable node i is equal to the corresponding effective pheromone.

- If the current node is a node i_0 different from the source, a candidate element is a node i linked to i_0 through an arc $k = (i_0, i)$ with positive residual capacity. The effective pheromone of each candidate node, $\zeta = i$, is calculated as stated in (**??**). It depends positively on the amount of aid transported through arc k in the previously obtained solutions and negatively on the amount of aid transported through arc k in the current solution.

$$ph_\zeta = \left(1 - \frac{lt_k - fl_{i_0,i}}{1 + qglobal}\right) \tau_k^2 \tag{24}$$

  Since we are interested in removing needless flow, the score of each improvable node i is inversely proportional to the effective pheromone, as established in (**??**). Note that the unit added in the denominator of the fraction in (**??**) prevents the effective pheromone from being 0.

  If the current node i_0 is a demand node linked to the sink, the sink must be considered as a candidate element as well. Its effective pheromone (**??**) depends positively on the final stock at demand node i_0 at the end of the operation in the previously obtained solutions and negatively on the final stock at i_0 in the current solution.

$$ph_{sn} = \left(1 - \frac{sf_{i_0}}{1 + dem_{i_0}}\right) \tau_{i_0}^4 \tag{25}$$

The score of the sink is established according to (**??**) as well.

Each time an augmenting path is obtained, the flow is updated according to the Ford-Fulkerson algorithm. The final stock at each demand node i, sf_i, and the amount of aid transported through each arc k, lt_k, are updated as well. The algorithm stops when the current flow cannot be increased further.

Adjust supply nodes. The purpose of this step is to eliminate as much incoming flow of aid as possible at supply nodes in which there is aid available at the end of the operation. Once again, a variation of the maximum flow problem is solved. The auxiliary transportation network includes all the nodes and the used arcs, which are now reversed because we try to return flow back from the supply nodes. All supply nodes are linked to the source and to the sink by dummy arcs. The capacity of each dummy arc leaving the source is the minimum between the amount of load arriving to the corresponding supply node and the final stock at it. The capacity of each dummy arc reaching the sink is the amount of aid available initially at the corresponding supply node. The capacity of each remaining arc is equal to the amount of aid transported through it (in opposite direction).

The calculation of the effective pheromones is as follows:

- If the current node is the source, the candidate elements are the supply nodes. The effective pheromone of each supply node, $\zeta = i$, is stated in (**??**). The effective pheromone depends positively on the final stock at node i in the previously obtained solutions and negatively on the final stock at node i in the current solution.

$$ph_\zeta = \left(1 - \frac{fs_i}{1 + qav_i}\right) \tau_i^4 \tag{26}$$

  Since we want part of the available (but unused) aid to leave the supply node and, therefore, the final stock decreases, the score of each supply node i is inversely proportional to the corresponding effective pheromone.

- If the current node is a node i_0 different from the source, the effective pheromone of each candidate node is calculated as stated in (**??**) and the score is obtained as in (**??**).

  If the current node i_0 is a supply node, the sink must be considered as a candidate element if the residual capacity of the corresponding dummy arc is positive and there are no more candidates. In that case, the sink would be selected directly, and the augmented path would be completed.

Trim routes. In the Design itineraries step, the routes of the vehicles are usually extended more than necessary. However, once the flow of aid is obtained, we can remove all the unnecessary trips iteratively. In each iteration, a candidate element is the final trip of a vehicle j if this trip is not necessary. Thus, each element is a pair $\zeta = (j, k)$ where k is the arc that vehicle j traverses in its final trip. The effective pheromone associated to each element is defined in (**??**), where c is the vehicle class corresponding to vehicle j.

$$ph_\zeta = \left(1 - \frac{vt_{c,k}}{1 + car_c}\right) \tau_{c,k}^1 \tag{27}$$

Since we want to trim the routes, the scores are inversely proportional to the corresponding effective pheromones. Once an element (j, k) is chosen, arc k is removed from the route of vehicle j and the variable $vt_{c,k}$ is updated.

Eliminate aid crossings. If there are vehicles transporting aid that traverse the same link in opposite directions, the solution is rearranged to avoid this crossing, provided that it leads to an improvement in the value of the objective function.

Exchange vehicles. Some vehicle exchanges are performed with a double purpose. On the one hand, to shorten the last parts of the itineraries. On the other hand, to decrease the total return cost of the vehicles.

5. Computational Study

5.1. Implementation Details and Calibration

The algorithms introduced in the previous sections have been implemented in Fortran 95. The results presented in this section have been obtained by executing the algorithms on a personal computer with an Intel Core i5-2450M processor with 2.5 Ghz and 8Gb RAM running Windows 7.

Before applying the ACO Algorithm to our test cases, it is necessary to tune several parameters, namely the number of solutions obtained between each couple of global updates (m), the probability of choosing the element with the highest score in any of the decision situations (q_0), the local and global evaporation rates (ρ_l and ρ_g) and the two parameters which determine how to update the weight of the pheromone and the greedy functions in the constructive process (σ_1 and σ_2). We randomly generated a set of five small test instances, with 10 to 17 nodes, 29 to 41 arcs and 20 to 30 vehicles, which allow us to do a large number of executions in order to obtain a set of parameters that makes the ACO Algorithm work well on realistic cases.

The calibration consists of two stages. In the first stage, the parameters were taken in three pairs, (m, q_0), (ρ_l, ρ_g) and (σ_1, σ_2), and for each one, the ACO Algorithm was executed on all test instances, varying the values of the two chosen parameters in a range of between ten and twenty values, while the other four parameters were fixed to arbitrary values. With the results obtained in this first stage we established a new assignment of values.

In the second stage of the calibration, we took the parameters one by one and executed the ACO Algorithm 100 times on each instance, to obtain 2000 solutions in each execution, varying the value of the chosen parameter and fixing the values of the other five parameters according to the assignment established after the first stage. We collected the average of the best objective function value over the one hundred executions and over the five test instances, that is represented in the vertical axis of the figures included in this subsection, while the values tested for the parameter are represented in the horizontal axis. Each execution required an average time of 1.5 s of computation.

To calibrate the number of solutions obtained between each couple of global updates, we varied $m = 0, 1, \ldots, 10$. In the literature, $m = 5$ or $m = 10$ have been commonly used and our results did not contradict it, so we decided to set $m = 5$, which requires less computation time.

In the same way, the probability of choosing the element with the highest score is usually set to a high value in the literature; however, we found that $q_0 = 0.3$ was the best value in our experiments, as shown in Figure **??** (note that the values higher than 0.5 have been rejected in the first stage of the calibration). This means that in 30% of the decision situations the candidate with the highest score will be selected directly, whereas in the remaining 70% the selection will be made at random, where the probability of each candidate is proportional to its score.

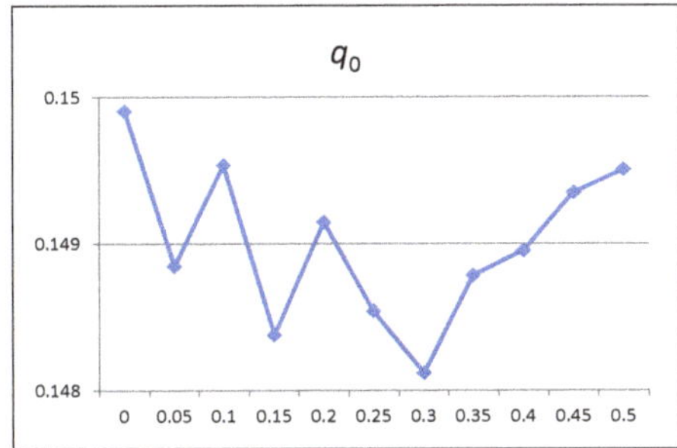

Figure 1. Calibration of the probability of choosing the element with the highest score.

Figure **??**a,b shows the results obtained for the evaporation rates. According to those, we set $\rho_l = 0.02$ and $\rho_g = 0.04$. Since both are small values, the pheromone trails will be very persistent, especially in the case of local updates, which are performed more frequently than the global ones.

Finally, we varied $\sigma_1 = 0, 0.05, \ldots, 0.5$ and $\sigma_2 = 0, 1, \ldots, 10$. The results led us to set $\sigma_1 = 0.2$, meaning that at the start of the metaheuristic the weight of the greedy functions is 0.2, and $\sigma_2 = 5$.

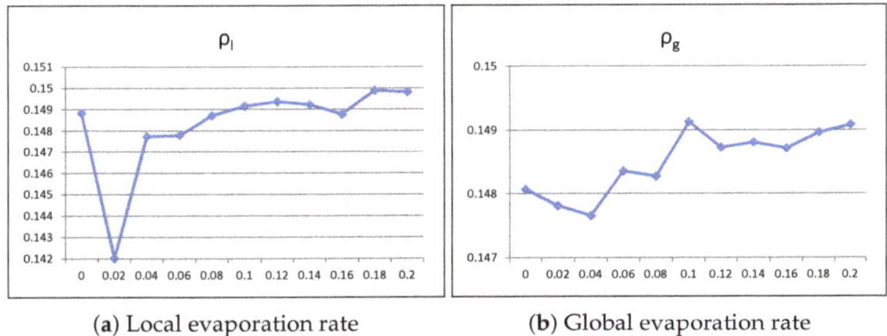

(**a**) Local evaporation rate (**b**) Global evaporation rate

Figure 2. Calibration of the evaporation rates.

5.2. Description of the Test Cases

The proposed ACO Algorithm has been tested on two realistic test cases with different characteristics: Haiti earthquake 2010 and Niger famine 2005.

The first test case is based on the earthquake that devastated Port-au-Prince, Haiti's capital, and its metropolitan area in 2010. It was introduced in [?], and it is very appropriate to illustrate the problem, since many of the characteristics that define it are present here. It develops mainly in an urban environment, but the effects of the disaster produced a great uncertainty about the transitability of many streets. Besides, aid operations were carried out in an environment of insecurity, with frequent attacks to convoys.

The logistic map, illustrated in Figure **??**, corresponds to the situation of the Port-au-Prince area fifteen days after the main earthquake. The network consists of 24 nodes and 84 arcs, corresponding to 42 two-way links. Nodes 1, 2 and 3 are supply nodes, with 60, 80 and 140 tons of humanitarian aid available, respectively. Nodes 10, 12, 13, 16, 17, 18, 20, 21 y 22 are demand nodes, with demands of 30, 40, 30, 30, 10, 30, 40, 20 y 20 tons of aid, respectively. Of these demand nodes, node 13 has special priority. The remaining 12 nodes are connection nodes and can be used to transship load.

The links have been classified by their quality (according to the thickness of the line: the thicker the line, the higher the speed) and by their reliability (according to the color scale shown in the figure). The arcs also have different ransack probabilities.

To perform the distribution operations there are 300 vehicles, classified in three different types according to their velocities, costs and capacities. There are 70 large vehicles, 95 medium size vehicles, and 135 small vehicles, which can carry 3, 2 and 1 tons of aid, respectively. The vehicles are located at the supply nodes and at the connection nodes 4, 6 and 18. The purpose of the operation is to deliver 150 tons of aid among the demand nodes.

The data for the Haiti test case come from various documents available in January 2010: a logistic map provided by the United Nations Office for the Coordination of Humanitarian Affairs (OCHA) [?], a map reporting the satellite-identified IDP concentrations, road and bridge obstacles in central Port-au-Prince, Haiti, created by the United Nations Institute for Training and Research (UNITAR), and a map focused only on the Port-au-Prince road conditions created by the World Food Programme (WFP) Emergency Preparedness and Response [?].

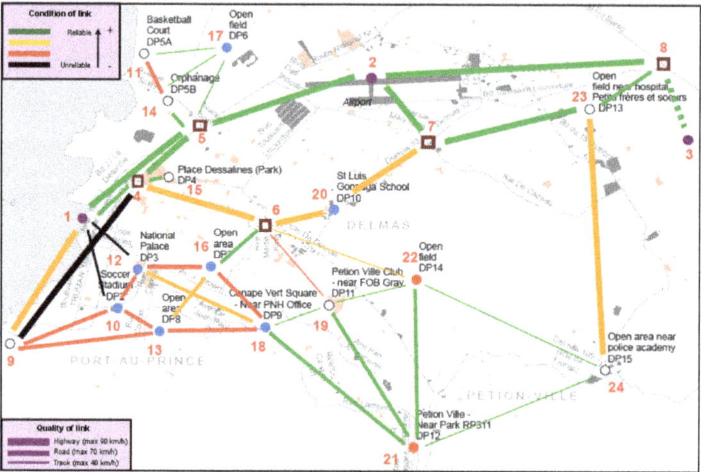

Figure 3. Transport network at Port-au-Prince.

The second test case, introduced in ?], is based on the 2005 Niger food crisis. The logistic network (see Figure **??**) consists of two supply nodes (Niamey and Gaya cities), two demand nodes (Agadez and Zinder), three connection nodes (Dosso, Tahoua and Maradi) and 11 two-way roads. There are a total of 376 vehicles available of three different types located at the supply and connection nodes to distribute 500 tons of humanitarian aid. The data for this test case come from the report of the real operation developed by the Spanish organization Fondo de Ayuda Humanitaria y de Emergencias de Farmamundi (FAHE) [?] and public websites such as Google Maps.

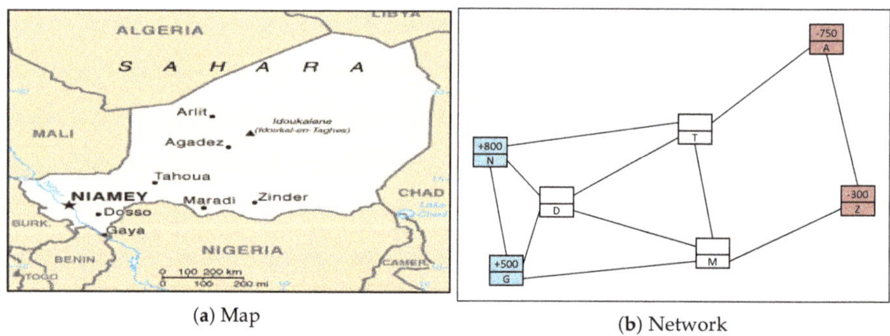

(**a**) Map (**b**) Network

Figure 4. Transport network at Niger.

Both test cases are explained in detail in [?] and their data are available on the website: www.mat.ucm.es/humlog.

5.3. Results

In this section we will compare and analyze the results obtained when applying the ACO Algorithm, proposed in this paper, and the GRASP Algorithm, introduced in [?], to the Haiti and Niger test cases. Furthermore, we compare as well the best solutions obtained by both algorithms with the optimal values for these test cases when they are available. These optimal values were obtained by solving to optimality relaxed flow models [? ?].

Firstly, we present the results obtained in the Haiti test case. Both algorithms—ACO and GRASP—have been executed ten times minimizing each objective independently and ten more times minimizing the aggregated function with equal weights for all attributes. The parameters of the algorithm were set according to the results of Section ??. A fixed computation time of 1500 s has been set for each of the 140 executions—two algorithms, seven objectives per algorithm and ten runs per objective.

Table ?? shows the computational results obtained with the ACO Algorithm and with the GRASP Algorithm in the Haiti test case. For each algorithm, the objective values of the best and the worst solutions among the ten executions are presented, as well as the average and the standard deviation of the objective values of the ten solutions. In this test case the operation time is given in minutes, the cost in dollars, and the security and the reliability in expected lost tons. We can see that the ACO Algorithm provides a better best solution when minimizing time, cost and security attributes, although for the cost the average is worse than with GRASP. For the reliability attribute, both algorithms provide the same best result, but this best is obtained in all GRASP executions—standard deviation is equal to zero—and only in two ACO executions. Both algorithms provide optimal solutions for the priority attribute in all executions and ACO performs worse than GRASP for the equity and for the aggregated function. The ACO algorithm provides around 10,000 feasible solutions on average while the GRASP algorithm provides 25,000. Nevertheless, this substancial difference does not translate into a proportional improvement in performance.

Table 4. Results obtained with ten executions per objective of GRASP and ACO algorithms in Haiti earthquake 2010.

	GRASP				ACO			
	Best	Average	Worst	Std. dev.	Best	Average	Worst	Std. dev.
Time	31.29	32.66	41.50	2.95	27.95	32.50	36.91	3.62
Cost	36,515.5	36,933.6	37,912.0	428.6	36,315.0	37,480.3	38,505.0	717.0
Equity	0.000	0.032	0.074	0.026	0.055	0.098	0.145	0.031
Priority	0.000	0.000	0.000	0.000	0.000	0.000	0.000	0.000
Security	37.14	38.31	42.50	1.73	37.13	38.07	42.38	1.58
Reliability	53.00	53.00	53.00	0.00	53.00	53.91	55.60	0.80
Aggregated	0.102	0.111	0.116	0.004	0.114	0.124	0.130	0.011

Table ?? shows the best values obtained among all tests performed during the experimentation—not only the ten executions of the previous table—in the Haiti test case, together with the optimal values for the attributes in which they are known. The optimal values for time and cost were obtained by using the static flow model described in [?], which although has some different assumptions, can serve as a reference for these two attributes that are measured in a very similar way. The ACO Algorithm provides the assumed optimal value for the time attribute and stays very close to the optimum for the cost, substantially improving the results obtained with the GRASP Algorithm. For the priority and the reliability attributes, both algorithms provide the same values, and almost the same for the security, whereas for the equity attribute, the ACO Algorithm performs worse. For the aggregated function, the GRASP Algorithm also works better, probably because of the higher influence of the equity attribute in the aggregated function, due to its tighter bound in comparison to the other attributes.

Figure **??** depicts the itineraries of the best solutions found for the time and the cost attributes. The numbers on the arcs represent the amounts of aid transported by the vehicles traversing them (a "0" on an arc indicates that an empty convoy is traversing the arc). In the best solution for the time (Figure **??**a), the three demand nodes farther from the supply nodes (18, 21 and 22) do not receive any aid and the distribution is mainly carried out by fast arcs, regardless of their dangerousness or reliability. The transportation is done by small and medium-sized vehicles, because they are faster than the large ones. When the cost attribute is optimized (Figure **??**b), only a few arcs are used and the convoys consist mainly of large vehicles, giving a better cost-capacity ratio. In this solution three demand nodes, including the priority node (13), are not visited.

Table 5. Best values obtained with several executions of GRASP and ACO algorithms in Haiti earthquake 2010.

	Optimum	GRASP	ACO
Time	27.95	(+6.7%) 29.83	(+0.0%) 27.95
Cost	35,835.0	(+1.6%) 36,408.0	(+0.3%) 35,945.0
Equity	0.000	0.000	0.05
Priority	0.000	0.000	0.000
Security	-	35.07	37.00
Reliability	-	53.00	53.00
Aggregated	-	0.098	0.110

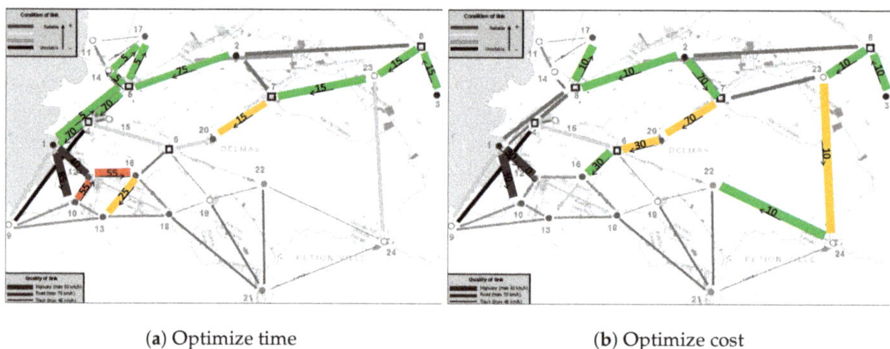

(**a**) Optimize time (**b**) Optimize cost

Figure 5. Best itineraries optimizing time and cost in Haiti earthquake 2010.

Figure **??** shows the solution obtained when minimizing $\tilde{f}_2$ with equal weights. In this case, the ACO algorithm has been executed until 25,000 feasible solutions have been generated. This compromise solution provides reasonably good values for all attributes, as it can be observed in Table **??**, all demand nodes receive some aid and, in particular, the priority node receives 26 out of 30 tons demanded.

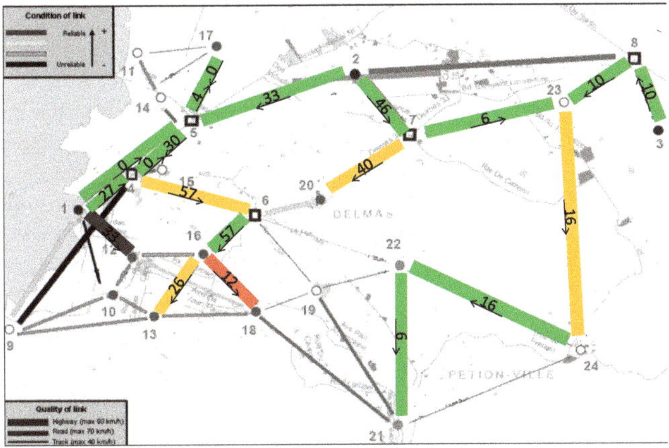

Figure 6. Best itineraries optimizing $\tilde{f}_2$ in Haiti earthquake 2010.

Table 6. Best objective values optimizing $\tilde{f}_2$ in Haiti earthquake 2010.

Time	Cost	Equity	Priority	Security	Reliability
41.41	49,767.00	0.30	0.13	87.59	88.45

Finally, we summarize the results obtained in the Niger test case. As in the Haiti test case, GRASP and ACO have been executed ten times minimizing each objective independently and ten times minimizing the aggregated function, running 1500 s per execution. Again, Table **??** shows the objective values of the best and the worst solutions among the ten executions for each attribute, together with the average and the standard deviation of the objective values of the ten solutions. The operation time is given in hours, the cost in euros, and the security and the reliability in expected lost tons. Both algorithms provide optimal solutions for the equity and priority objectives in the ten executions. The ACO Algorithm performs slightly better than GRASP for the security and significantly better for the remaining attributes—time, cost and reliability—and also for the aggregated function. The best value obtained with ACO for the cost is only 1.6% higher than the assumed optimal value of the cost attribute, which is EUR 63,240.0, obtained using the static flow model [**?**]. ACO is significantly better for the cost, the time, the reliability and the aggregated function. Despite the higher speed of GRASP generating feasible solutions in this test case—9000 on average per execution with GRASP versus 6400 with ACO—ACO performance is much better than that of GRASP.

Table 7. Results obtained with ten executions per objective of GRASP and ACO algorithms in Niger Famine 2005.

	GRASP				**ACO**			
	Best	Average	Worst	Std. dev.	Best	Average	Worst	Std. dev.
Time	40.96	45.72	46.25	1.59	22.21	34.59	47.00	10.32
Cost	64,860.0	65,196.9	65,497.5	238.3	64,250.0	64,389.2	64,852.5	164.7
Equity	0.000	0.000	0.000	0.000	0.000	0.000	0.000	0.000
Priority	0.367	0.367	0.367	0.000	0.367	0.367	0.367	0.000
Security	200.00	200.18	201.75	0.53	200.00	200.00	200.00	0.00
Reliability	789.10	809.96	830.30	11.87	763.10	764.54	765.60	0.60
Aggregated	0.167	0.172	0.175	0.002	0.154	0.159	0.172	0.007

It is remarkable that the number of feasible solutions generated using the same computation time is lower in the Niger test case than in the Haiti test case, even though the logistic network is much smaller in the former. This is because there are other factors that influence the computation time as, for example, the structure of the network, the number and the location of the vehicles, the relation between the amount to be distributed and the total capacity of the vehicles, etc.

The proposed compromise solution for the Niger test case is shown in Table **??**. All attribute values remain reasonably close to the best values obtained by optimizing each objective individually, except the value for the priority objective. This is partially due to the high confrontation between the equity and priority attributes. For optimizing the priority attribute, the node with the highest priority level (Zinder) should receive all the demanded aid, which would lead to an inequitable distribution.

Table 8. Best objective values optimizing $\tilde{f}_2$ in Niger famine 2005.

Time	Cost	Equity	Priority	Security	Reliability
65.00	6,802,000.0	0.066	0.711	259.85	956.20

In summary, the results obtained in both test cases show that the ACO Algorithm is able to obtain good solutions for all attributes when optimizing each of them individually, together with good compromise solutions as well. In particular, it provides the best known solutions for several attributes, such as the operation time in the Haiti test case or the reliability of the itineraries in the Niger test case. In addition, the computational effort required by the ACO algorithm is quite reasonable, making it suitable to solve realistic cases.

5.4. Managerial Insights

1. The security issues related to the distribution of humanitarian aid, mainly focused on avoiding possible assaults to the vehicles performing the distribution operations in dangerous areas, are a key element in certain contexts, as for example war zones or extremely poor places. In these situations, it is frequently required that the vehicles involved are grouped forming convoys, as a protection mechanism and to allow escort details monitoring. This condition introduces significant additional difficulties into the last mile distribution models that must be properly addressed, but it makes the models more realistic and useful. The proposed metaheuristic algorithm can be used to obtain the precise distribution plan in an efficient way.
2. In this work, two case studies, regarding an earthquake and a famine, a rapid and a slow onset disaster, respectively, have been considered. However, the proposed model and solution method can be applied to any kind of disaster, especially to the ones occurring in unsecure environments.

6. Concluding Remarks

This paper addresses a humanitarian last mile distribution problem in response to a disaster. The distribution must be done under unsecure conditions, so the vehicles must travel in convoys to try to prevent assaults. Furthermore, there is uncertainty about the state of the roads, that may have been damaged by the disaster. The proposed model is multimodal, multidepot, allows split deliveries and multiple trips and provides the individual itineraries of the vehicles. Besides, the model considers six performance criteria—time, cost, equity, priority, security and reliability—and is solved by a compromise programming approach.

The main contribution of this paper is the proposed methodology to solve the model. We develop an algorithm based on Ant Colony Optimization that considers several types of ants and pheromones. The methodology introduces the concept of effective pheromones to balance the elements of the provided solutions and to diversify the set of obtainable solutions. This methodology is especially appropriate to approach routing problems in which the vehicles are required to travel in convoys, but some of its most novel elements, as it is the case of the effective pheromones, could also be successfully applied to many other complex distribution and routing problems.

The algorithm is tested on two realistic case studies of public domain, and its performance is compared to that of the GRASP Algorithm, the only existing solution method in the literature. The ACO Algorithm performs well for all the objective functions tested in our experiments and provides, for both test cases considered, the best solutions obtained to date for several attributes. In comparison with the available solution method in the literature—the GRASP Algorithm—in the Haiti test case, both algorithms have their own strengths, but in the Niger test case, ACO clearly outperforms GRASP. The computation time required to obtain good solutions is reasonable in all cases, proving that the ACO Algorithm can be perfectly applied in an emergency context. Furthermore, GRASP and ACO algorithms actually complement each other and could even be used in combination in order to improve the quality of the final solutions provided to the decision makers.

One interesting future research line to continue this work comprises the development of a stochastic model in which the possibility of suffering assaults or finding blocked roads could be represented by random variables. Additionally, the design of simulation models to replicate real situations and test different solution strategies could also be worth considering.

Author Contributions: Conceptualization, J.M.F., M.T.O. and G.T.; Methodology, J.M.F., M.T.O. and G.T.; Software, J.M.F.; Validation, J.M.F., M.T.O. and G.T.; Formal analysis, J.M.F., M.T.O. and G.T.; Investigation, J.M.F., M.T.O. and G.T.; Data curation, J.M.F., M.T.O. and G.T.; Writing—original draft preparation, J.M.F.; Writing—review and editing, M.T.O. and G.T.; Visualization, J.M.F., M.T.O. and G.T.; Project administration, M.T.O. and G.T. All authors have read and agreed to the published version of the manuscript.

Funding: This research was funded by the European Commission grant H2020 MSCA-RISE 691161 (GEO-SAFE), the Government of Spain grant MTM2015-65803-R and the Government of Madrid, grant S2013/ICE-2845. The APC was funded by the European Commission grant H2020 MSCA-RISE 691161 (GEO-SAFE).

Conflicts of Interest: The authors declare no conflict of interest.

References

. Altay, N.; Green, W.G. OR/MS research in disaster operations management. *Eur. J. Oper. Res.* **2006**, *175*, 475–493. [CrossRef]
. Özdamar, L.; Ertem, M.A. Models, solutions and enabling technologies in humanitarian logistics. *Eur. J. Oper. Res.* **2015**, *244*, 55–65. [CrossRef]
. Tomasini, R.; van Wassenhove, L. *Humanitarian Logistics*; Palgrave Macmillan: London, UK, 2009.
. Vitoriano, B.; Montero, J.; Ruan, D. *Decision Aid Models for Disaster Management and Emergencies*; Atlantis Press: Paris, France, 2013.
. Pedraza-Martínez, A.J.; Van Wassenhove, L.N. Empirically grounded research in humanitarian operations management: The way forward. *J. Oper. Manag.* **2016**, *45*, 1–10. [CrossRef]
. Besiou, M.; Pedraza-Martínez, A.J.; Van Wassenhove, L.N. OR applied to humanitarian operations. *Eur. J. Oper. Res.* **2018**, *269*, 397–405. [CrossRef]
. Fiedrich, F.; Gehbauer, F.; Rickers, U. Optimized resource allocation for emergency response after earthquake disasters. *Saf. Sci.* **2000**, *35*, 41–57. [CrossRef]
. Viswanath, K.; Peeta, S. Multicommodity maximal covering network design problem for planning critical routes for earthquake response. *Transp. Res. Rec.* **2003**, *1857*, 1–10. [CrossRef]
. Kongsomsaksaku, S.; Yang, C.; Chen, A. Shelter location-allocation model for flood evacuation planning. *J. East. Asia Soc. Transp. Stud.* **2005**, *6*, 4237–4252.
. Barzinpour, F.; Saffarian, M.; Makoui, A.; Teimoury, E. Metaheuristic Algorithm for Solving Biobjective Possibility Planning Model of Location-Allocation in Disaster Relief Logistics. *J. Appl. Math.* **2014**, *2014*, 239868. [CrossRef]
. Yi, W.; Özdamar, L. A dynamic logistics coordination model for evacuation and support in disaster response activities. *Eur. J. Oper. Res.* **2007**, *179*, 1177–1193. [CrossRef]
. Habib, M.S.; Sarkar, B. An Integrated Location-Allocation Model for Temporary Disaster Debris Management under an Uncertain Environment. *Sustainability* **2017**, *9*, 716. [CrossRef]

- Huang, M.; Smilowitz, K.; Balcik, B. Models for relief routing: equity, efficiency and efficacy. *Transp. Res. Part E Logist. Transp. Rev.* **2012**, *48*, 2–18. [CrossRef]
- Holguín-Veras, J.; Pérez, N.; Jaller, M.; Van Wassenhove, L.; Aros-Vera, F. On the appropriate objective function for post-disaster humanitarian logistics models. *J. Oper. Manag.* **2013**, *31*, 262–280. [CrossRef]
- Gutjahr, W.J.; Nolz, P.C. Multicriteria optimization in humanitarian aid. *Eur. J. Oper. Res.* **2016**, *252*, 351–366. [CrossRef]
- Ferrer, J.M.; Ortuño, M.T.; Tirado, G. A GRASP metaheuristic for humanitarian aid distribution. *J. Heuristics* **2016**, *22*, 55–87. [CrossRef]
- Balcik, B.; Beamon, B.M.; Smilowitz, K. Last mile distribution in humanitarian relief. *J. Intell. Transp. Syst.* **2008**, *12*, 51–63. [CrossRef]
- Nolz, P.C.; Semet, F.; Doerner, K.F. Risk approaches for delivering disaster relief supplies. *OR Spectrum* **2011**, *33*, 543–569. [CrossRef]
- Lin, Y.H.; Batta, R.; Rogerson, P.A.; Blatt, A.; Flanigan, M. A logistics model for emergency supply of critical items in the aftermath of a disaster. *Socio-Econ. Plan. Sci.* **2012**, *45*, 132–145. [CrossRef]
- Bozorgi-Amiri, A.; Jabalameli, M.S.; Al-e Hashem, S.M.J.M. A multi-objective robust stochastic programming model for disaster relief logistics under uncertainty. *OR Spectrum* **2013**, *35*, 905–933. [CrossRef]
- Tirado, G.; Martín-Campo, F.J.; Vitoriano, B.; Ortuño, M.T. A lexicographical dynamic flow model for relief operations. *Int. J. Comput. Int. Syst.* **2014**, *7*, 45–57. [CrossRef]
- Huang, K.; Jiang, Y.; Yuan, Y.; Zhao, L. Modeling multiple humanitarian objectives in emergency response to large-scale disasters. *Transp. Res. Part E Logist. Transp. Rev.* **2015**, *75*, 1–17. [CrossRef]
- Ahmadi, M.; Seifi, A.; Tootooni, B. A humanitarian logistics model for disaster relief operation considering network failure and standard relief time: A case study on San Francisco district. *Transp. Res. Part E Logist. Transp. Rev.* **2015**, *75*, 145–163. [CrossRef]
- Zhou, Y.; Liu, J.; Zhang, Y.; Gan, X. A multi-objective evolutionary algorithm for multi-period dynamic emergency resource scheduling problems. *Transp. Res. Part E Logist. Transp. Rev.* **2017**, *99*, 77–95. [CrossRef]
- Maghfiroh, M.F.N.; Hanaoka, S. Last mile distribution in humanitarian logistics under stochastic and dynamic consideration. In Proceedings of the 2017 IEEE International Conference on Industrial Engineering and Engineering Management (IEEM), Singapore, 10–13 December 2017; pp. 1411–1415.
- Moreno, A.; Alem, D.; Ferreira, D.; Clark, A. An effective two-stage stochastic multi-trip location-transportation model with social concerns in relief supply chains. *Eur. J. Oper. Res.* **2018**, *269*, 1050–1071. [CrossRef]
- Habib, M.S.; Sarkar, B.; Tayyab, M.; Saleem, M.W.; Hussain, A.; Ullah, M.; Omair, M.; Iqbal, M.W. Large-scale disaster waste management under uncertain environment. *J. Clean. Prod.* **2019**, *212*, 200–222. [CrossRef]
- Ortuño, M.T.; Tirado, G.; Vitoriano, B. A lexicographical goal programming based decision support system for logistics of Humanitarian Aid. *TOP* **2011**, *19*, 464–479. [CrossRef]
- Vitoriano, B.; Ortuño, M.T.; Tirado, G.; Montero, J. A multi-criteria optimization model for humanitarian aid distribution. *J. Glob. Optim.* **2011**, *51*, 189–208. [CrossRef]
- Ferrer, J.M.; Martín-Campo, F.J.; Ortuño, M.T.; Pedraza-Martínez, A.J.; Tirado, G.; Vitoriano, B. Multi-criteria optimization for last mile distribution of disaster relief aid: Test cases and applications. *Eur. J. Oper. Res.* **2018**, *269*, 501–515. [CrossRef]
- Rizzoli, A.E.; Montemanni, R.; Lucibello, E.; Gambardella, L.M. Ant colony optimization for real-world vehicle routing problems. *Ann. Oper. Res.* **2007**, *1*, 135–151. [CrossRef]
- Yu, B.; Yang, Z.Z.; Xie, J.X. A parallel improved ant colony optimization for multi-depot vehicle routing problem. *J. Oper. Res. Soc.* **2011**, *62*, 183–188. [CrossRef]
- Zhang, L.Y.; Fei, T.; Zhang, X.; Li, Y.; Yi, J. Research about Immune Ant Colony Optimization in Emergency Logistics Transportation Route Choice. In *Communications and Information Processing*; Zhao, M., Sha, J., Eds.; Communications in Computer and Information Science; Springer: Berlin/Heidelberg, Germany, 2012; Volume 288, pp. 430–437.
- Hong, J.; Diabat, A.; Panicker, V.V.; Rajagopalan, S. A two-stage supply chain problem with fixed costs: An ant colony optimization approach. *Int. J. Prod. Econ.* **2018**, *204*, 214–226. [CrossRef]
- Yi, W.; Kumar, A. Ant colony optimization for disaster relief operations. *Transp. Res. Part E Logist. Transp. Rev.* **2007**, *43*, 660–672. [CrossRef]

. Saeidian, B.; Mesgari, M.S.; Pradhan, B.; Ghodousi, M. Optimized Location-Allocation of Earthquake Relief Centers Using PSO and ACO, Complemented by GIS, Clustering, and TOPSIS. *ISPRS Int. J. Geo-Inf.* **2018**, *7*, 1–25. [CrossRef]

. Cochrane, J.L.; Zeleny, M. *Multiple Criteria Decision Making*; University of South Carolina Press: Columbia, SC, USA, 1973.

. Dorigo, M.; Maniezzo, V.; Colorni, A. Ant system: optimization by a colony of cooperating agents. *IEEE Trans. Syst. Man Cybern. Part B Cybern.* **1996**, *26*, 29–41. [CrossRef] [PubMed]

. Dorigo, M.; Blum, C. Ant colony optimization theory: A survey. *Theor. Comput. Sci.* **2005**, *344*, 243–278. [CrossRef]

. Zhao, N.; Wu, Z.; Zhao, Y.; Quan, T. Ant colony optimization algorithm with mutation mechanism and its applications. *Expert Syst. Appl.* **2010**, *37*, 4805–4810. [CrossRef]

. Chandra Mohan, B.; Baskaran, R. A survey: Ant Colony Optimization based recent research and implementation on several engineering domain. *Expert Syst. Appl.* **2012**, *39*, 4618–4627. [CrossRef]

. Dorigo, M.; Gambardella, L.M. Ant colony system: a cooperative learning approach to the traveling salesman problem. *IEEE Trans. Evol. Comput.* **1997**, *1*, 53–66. [CrossRef]

. Ford, L.R.; Fulkerson, D.R. Maximal flow through a network. *Can. J. Math.* **1956**, *8*, 399–404. [CrossRef]

. United Nations Office for the Coordination of Humanitarian Affairs (OCHA). Available online: https://www.unocha.org/ (accessed on 17 March 2020).

. World Food Programme (WFP) Emergency Preparedness and Response. Available online: https://www.wfp.org/emergency-preparedness-and-response (accessed on 17 March 2020).

. Vitoriano, B.; Ortuño, M.T.; Tirado, G. HADS, a goal programming-based humanitarian aid distribution system. *J. Multi-Crit. Decis. Anal.* **2009**, *16*, 55–64. [CrossRef]

. Fondo de Ayuda Humanitaria y de Emergencias de Farmamundi (FAHE). Memoria. 2005. Available online: http://www.farmaceuticosmundi.org/farmamundi/descargas/pdf/memoria_2005_fahe.pdf (accessed on 17 March 2020).

Article

Robust Design Optimization for Low-Cost Concrete Box-Girder Bridge

Vicent Penadés-Plà [1], Tatiana García-Segura [2] and Víctor Yepes [1,*]

1 Institute of Concrete Science and Technology (ICITECH), Universitat Politècnica de València, 46022 Valencia, Spain; vipepl2@cam.upv.es

2 Department of Construction Engineering and Civil Engineering Projects, Universitat Politècnica de València, 46022 Valencia, Spain; tagarse@upv.es

* Correspondence: vyepesp@cst.upv.es

Received: 28 January 2020; Accepted: 8 March 2020; Published: 11 March 2020

Abstract: The design of a structure is generally carried out according to a deterministic approach. However, all structural problems have associated initial uncertain parameters that can differ from the design value. This becomes important when the goal is to reach optimized structures, as a small variation of these initial uncertain parameters can have a big influence on the structural behavior. The objective of robust design optimization is to obtain an optimum design with the lowest possible variation of the objective functions. For this purpose, a probabilistic optimization is necessary to obtain the statistical parameters that represent the mean value and variation of the objective function considered. However, one of the disadvantages of the optimal robust design is its high computational cost. In this paper, robust design optimization is applied to design a continuous prestressed concrete box-girder pedestrian bridge that is optimum in terms of its cost and robust in terms of structural stability. Furthermore, Latin hypercube sampling and the kriging metamodel are used to deal with the high computational cost. Results show that the main variables that control the structural behavior are the depth of the cross-section and compressive strength of the concrete and that a compromise solution between the optimal cost and the robustness of the design can be reached.

Keywords: robust design optimization; RDO; post-tensioned concrete; box-girder bridge; structural optimization; metamodel; kriging

1. Introduction

All structural designs involve variability and uncertainty [1,2]. The initial parameters, the structure dimensions, the mechanical characteristics of the materials, and the loads may differ from the design values [3,4]. Nevertheless, the design of a structure is made using the nominal value, which has a low probability of occurring (for example, the resistance of concrete is the resistance that has a 5% probability of failure). In addition, safety coefficients associated with a given probability of failure are assigned. However, a variation of these initial uncertain parameters can influence the variability of the structural behavior. Structural optimization usually uses a deterministic approach that does not consider the effects of the associated uncertainty [5–13]. This means that the structure has an optimum behavior only under the conditions initially defined, and the response can vary significantly when the values differ from the design values [14,15].

Unlike this approach, robust design has been studied to obtain designs in which the uncertainty of the initial parameters has the lowest possible influence on the objective response. This robust design is reached by a probabilistic optimization. Nowadays, there are two approaches to the optimal probabilistic design of a structure: Reliability-Based Design Optimization (RBDO) [16] and Robust Design Optimization (RDO) [17]. In RBDO, the probability of failure is studied from the variations of the initial parameters. RDO studies a design that is less sensitive to the variation of the initial parameters. The present paper focuses on the RDO. The concept of robust design was proposed by Taguchi in the 1940s and applied to optimization problems in 1980 [1]. This approach uses the mean and standard deviation to study the variability of the objective response.

The main limitation of RDO is the computational expense because assessing the sensitivity of the objective response of the problem requires a high number of runs [18]. Therefore, it is necessary to find methods that allow carrying out the optimization process more efficiently [4,19,20]. Metamodels allow the generation of a mathematical approximation of the objective response (an objective surface) from the assessment of points within the design space. Once the response surface has been generated, obtaining the value of the objective response given the inputs is much faster. These mathematical approximations or metamodels have already been used to solve RDO process problems [4]. The most widespread metamodels are polynomial regression, artificial neural networks (ANN) and kriging. ANN has been used in different works related to structural engineering [21,22]. However, the kriging model has been demonstrated to be useful to obtain great reliability in the assessment of the response due to its predictive accuracy in non-linear functions [23]. Penadés Plà et al. [24] compared standard heuristic optimization and heuristic optimization based on kriging models, demonstrating that the solutions obtained through optimization based on kriging models are close enough to the solutions obtained through standard heuristic optimization, but with high savings in computational costs.

In the present paper, the robust design methodology is applied to a continuous prestressed concrete box-girder footbridge to obtain a bridge that is optimal in terms of its cost objective function and robust in terms of structural stability. Its structural stability is measured by the variability of the vertical displacement in the middle of the bridge [19]. To this end, Latin hypercube sampling is used to obtain the initial sampling, the kriging model is used to obtain the mathematical approximation to the response, and then the simulated annealing optimization algorithm is used to obtain the robust optimum design. All this will be studied for different uncertain design parameters: the modulus of elasticity, the overload, and the prestressing force.

2. Robust Design Optimization

Robust design studies the variation of the objective response generated by the uncertain initial parameters. Therefore, robust design optimization (RDO) aims to reach the best objective response with the smallest deviation. It implies that the RDO problem is defined as a multi-objective optimization problem in which the objective response is the mean and the standard deviation (Equation (1)).

$$min\left\{\mu_{F(x,z)}(x_1, x_2, x_3, \ldots, x_n), \sigma_{F(x,z)}(x_1, x_2, x_3, \ldots, x_n)\right\} \tag{1}$$

where x_1, x_2, x_3, … … ,x_n are the deterministic values of the design variables or the probabilistic function of the uncertain initial parameters.

Commonly, the two objective functions to be minimized in an RDO problem are in conflict. This situation leads to a set of solutions that represent a Pareto frontier. Figure 1 shows an example of the difference between the optimal solution and the robust optimal solution in a design space of one design variable. Solution A is the optimal solution, point B is the most robust solution and point C is the robust optimal solution. It is possible to see that the same variation of the design variable (v) causes a higher variation in the objective function of the solution A (f_A) than it does in the solution C (f_C).

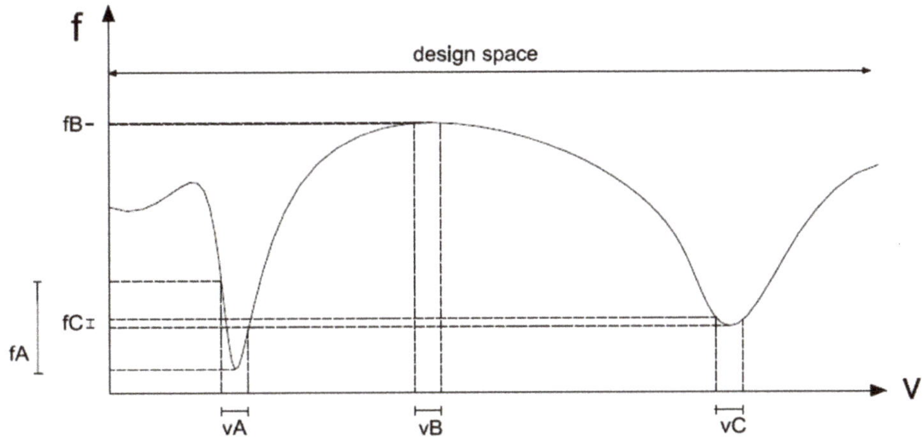

Figure 1. Robust design optimization.

3. Robust Design Optimization Using Metamodels

The main goal of metamodels is to obtain results more efficiently by creating a mathematical approximate model of a detailed simulation model (model of a model). This makes it possible to predict output data (objective response) from input data (variables or design parameters) of the design space. There are three main steps to create a metamodel: (1) obtaining the initial points of the input or sampling data set within the design space (size and position), (2) choosing the method to create the approximate mathematical model, and (3) choosing the fitted model. Each of these three steps can be performed using many different options [25]. In this study, Latin hypercube sampling is used to obtain the initial sampling, the kriging model is used to create the approximate mathematical model, and the search for the Best Linear Unbiased Predictor (BLUP) is used as the fitted model. Then, the mathematical approximation created is used to predict the objective functions according to the initial design variables and parameters. In this way, the optimization can be carried out more efficiently, saving a lot of computational costs, which is important in a probabilistic optimization. In addition, the simulated annealing algorithm is used to perform the optimization. Figure 2 shows a flowchart of the robust design optimization using these characteristics. A more detailed description of this approach can be seen in Penadés-Plà et al. [24].

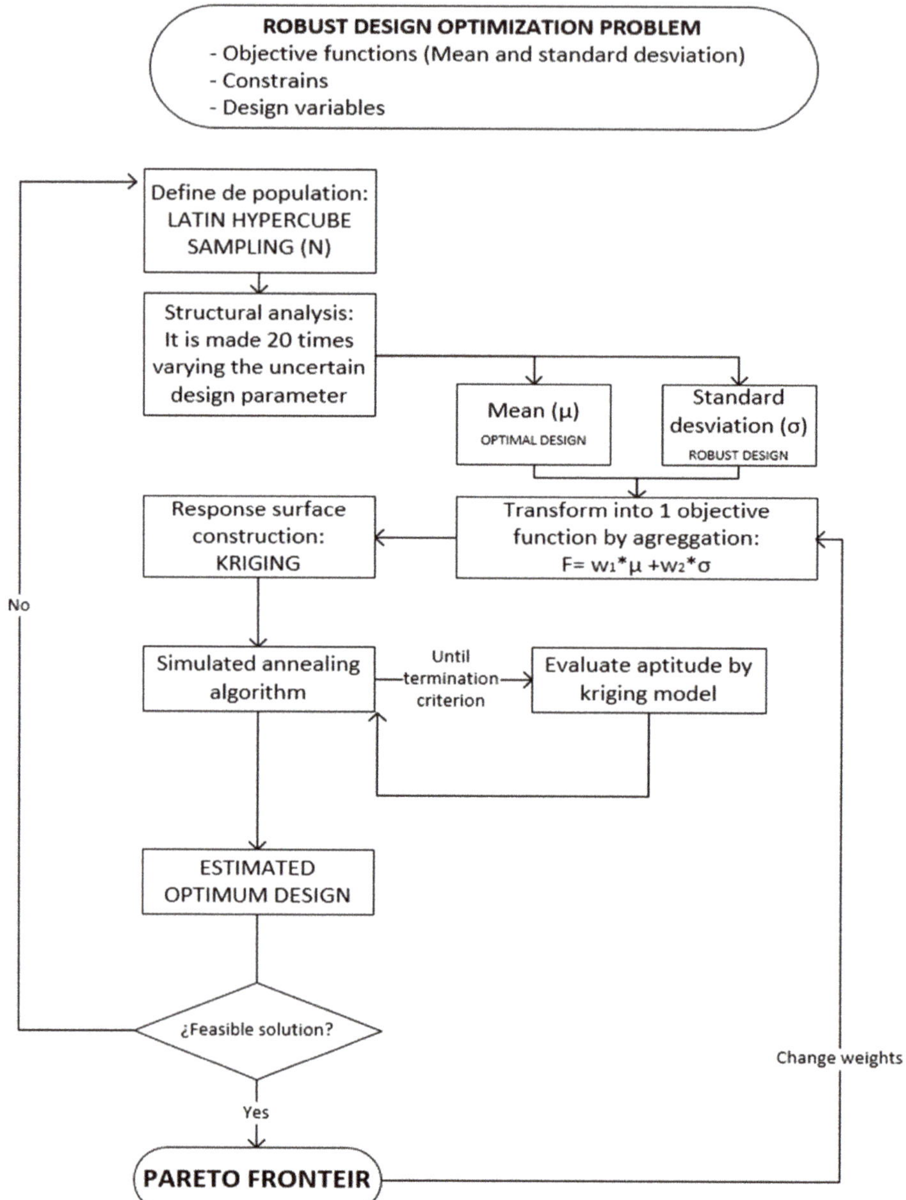

Figure 2. Flow diagram of robust design optimization.

3.1. Latin Hypercube Sampling

McKay et al. proposed in 1979 the Latin hypercube sampling (LHS) [26]. This method is a space-filling type of design of experiments. That means that this type of design of experiment trends to cover all of the design space by the positions of the initial sample points. In this way, local deformation of any area of the design space can be taken into account. For this purpose, a number

N of non-overlapping intervals must first be determined. These intervals divide each range of the design variables (v) into N sections, generating a mesh in the design space with Nv regions. Then, a combination of N random points is generated, so each point is placed in a combination of different intervals of each range of design variables. This guarantees that the initial sampling covers the entire range of each design variable. LHS has been used in several papers, showing its validity for the estimation of metamodel output data [20,27]. For this reason, in the present paper, a uniform distribution of the initial sample points by LHS is employed.

3.2. Kriging

The Kriging metamodel was originally created by Dannie Gerhardus Krige, later much research contributed to its development and finally, Matheron formalized the approach in 1963 [28]. The main idea of the kriging metamodel is that the deterministic response $y(x)$ can be described by

$$y(x) = (x) + Z(x) \tag{2}$$

where the approximation function is known is called $f(x)$, and $Z(x)$ is an execution of a stochastic process with a mean zero, variance σ^2 and a non-zero covariance. The first term of the equation, $f(x)$, offers a global approach to the design space that is similar to a regression model (Equation (3)). The second term, $Z(x)$, generates local deviations to interpolate the initial sample points using the kriging model (Equation (4)).

$$f(x) = \sum_{i=1}^{n} \beta_i \cdot f_i(x) \tag{3}$$

$$cov\left[Z(x_i), Z\left(x_j\right)\right] = \sigma^2 \cdot R\left(x_i, x_j\right) \tag{4}$$

where the process variance σ^2 scales the spatial correlation function $R(x_i,x_j)$ between two data points. The Gaussian correlation function (Equation (5)) is widely used in engineering design [29]. It can be defined with a single parameter (θ) that determines the area of influence of the adjacent points [30]. When the sample points have a high correlation, then θ is low, thus $Z(x)$ will be similar throughout the design space. As the θ grows, the closest points will have the greatest correlation, thus $Z(x)$ will vary according to the point in the design space:

$$R\left(x_i, x_j\right) = e^{-\sum_{k=1}^{m} \theta |x_k^i - x_k^j|^2} \tag{5}$$

3.3. The Fitted Model

The search for the Best Linear Unbiased Predictor (BLUP) is used by the formulation of kriging. Simpson et al. [31] have discussed fitting methods for most widely-used metamodels.

3.4. Mean and Variance

The mean (μ) and standard deviation (σ) of the responses of the objectives measure the robust optimum design. These statistical parameters have been obtained for four different levels of uncertainty (10%, 20%, 30%, and 40%). The value of the uncertain initial parameter has been calculated according to a uniform distribution depending on the level of uncertainty. In this way, the mean refers to the optimum design, and the standard variation refers to the robust design.

3.5. Optimization

Simulated annealing (SA) is the heuristic algorithm used to carry out the RDO. This algorithm has been used in a lot of research to solve optimization problems [32,33]. In the present paper, the method of Medina [34] is used to calibrate the initial temperature. This method suggests that the starting temperature is halved when the rate of acceptance is above 40% but doubled when it is below

20%. When a Markov chain ends, the temperature then drops in accordance with a cooling coefficient k based on the equation $T = k^*T$. In this study, the calibration showed that the length of the Markov chain of 1000 and a coefficient of cooling of 0.8 are suitable. The algorithm stops after three Markov chains without finding improvements.

4. Problem Design

In this section, the robust design optimization problems proposed are discussed. Section 4.1 describes the structure considered and Section 4.2 defines the characteristics of the problem. Section 4.2 includes the initial uncertain parameters considered and the objective functions studied.

4.1. Description of the Box-Girder Footbridge

The structure is a concrete pedestrian bridge with three continuous spans of 40–50–40 m long. The box-girder cross-section has a uniform width of 3 m, and seven variables define the remaining geometric dimensions of the cross-section (Figure 3): depth (h), web inclination width (d), bottom slab width (b), bottom slab thickness (ei), top slab thickness (es), external cantilever section thickness (ev), and webs slab thickness (ea). The variables are restricted to a range as shown in Table 1. The haunch (c), is determined from the other variables (Equation (6)) as recommended by Schlaich and Scheff [35]. Furthermore, the haunch must also allow space to enclose the ducts in the high and low points. This structure was used to compare the standard heuristic optimization and kriging-based heuristic optimization. In this work, the kriging-based heuristic optimization and RDO are applied to the same structure. More detailed information about this structure can be found in Penadés-Plà et al. [24].

$$t = max\left\{\frac{b - 2 \cdot ea}{5}, ei\right\} \tag{6}$$

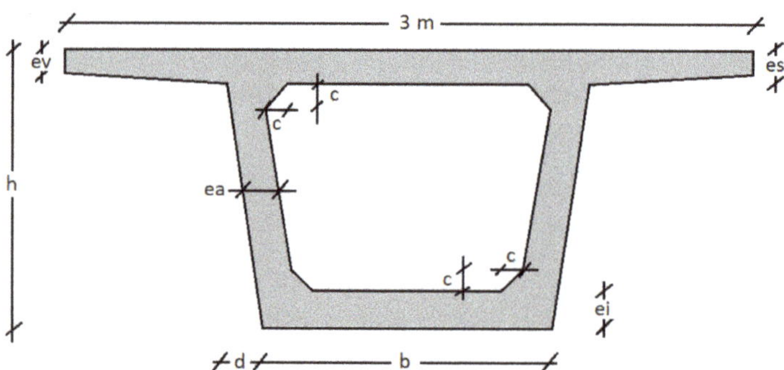

Figure 3. Box-girder cross-section.

Table 1. Main parameters of the analysis.

Design Variable	Min. Value (m)	Max. Value (m)	Precision (m)
Depth h	1.25	2.5	0.05
Width (b)	1.2	1.8	0.05
Inclination width (d)	0	0.4	0.05
Top slab thickness (es)	0.15	0.4	0.05
External cantilever section thickness (ev)	0.15	0.4	0.05
Bottom slab thickness (ei)	0.15	0.4	0.05
Webs slab thickness (ea)	0.3	0.6	0.05

Spanish regulations [36,37] and the Eurocodes [38,39] are used to carry out the structural verification defined by the ultimate and service limit states: bending, vertical shear, longitudinal shear, punching shear, compression and tension stress, torsion, torsion combined with bending and shear, cracking, and vibration. Moreover, compliance with constructability and geometrical criteria must are also be checked.

4.2. Description of the Robust Design Optimization Problem

The robust design optimization proposed in this paper is defined by two objective functions: the first one is the mean cost, and the second one is the structural stability represented by the vertical displacement in the middle of the bridge [19]. The statistical parameters for both objective functions are obtained varying the initial uncertain parameter (modulus of elasticity, overload, and prestressing force) according to a uniform distribution with three different levels of uncertainty (10%, 20%, and 30% for the modulus of elasticity and 10%, 20%, 30%, and 40% for the overload and prestressing force). These uncertain parameters were chosen after carrying out a sensitivity analysis of the vertical displacement and selecting the critical parameters.

In this way, the differences between the different Pareto frontiers obtained for each problem can be studied. Therefore, the goal is to obtain the design with the best cost that has the best structural stability for each RDO problem. Equations (7) and (8) correspond to these objective functions assessed.

$$\mu_{COST} = \sum_{i=1,n} e_i \times m_i(x_1, x_2, \ldots, x_n) \tag{7}$$

$$\sigma_{VERTICALDISPLACEMENT}(x_1, x_2, x_3, \ldots, x_n) \tag{8}$$

where $x_1, x_2, x_3, \ldots \ldots, x_n$ are the design variables.

The conventional objective function assesses the cost for the construction units taking into account the placement and material used. The BEDEC ITEC database provides unit costs [40]. The compressive strength grade determines the cost of the concrete. The unit costs of the problem are shown in Table 2. The measurements (m_i) relating to the construction units are calculated as defined by the design variables. The variation of the vertical displacement in the middle of the bridge has been obtained according to the standard deviation of 20 different cases varying the initial uncertain parameter. Each one of these vertical displacements has been calculated in accordance with Spanish regulations [36,37] along with the Eurocodes [38,39].

Table 2. Unit cost.

Unit Measurements	Cost (€)
m^3 of scaffolding	10.2
m^2 of formwork	33.81
m^3 of lighting	104.57
kg of steel (B-500-S)	1.16
kg of post-tensioned steel (Y1860-S7)	3.40
m^3 of concrete HP-35	104.57
m^3 of concrete HP-40	109.33
m^3 of concrete HP-45	114.10
m^3 of concrete HP-50	118.87
m^3 of concrete HP-55	123.64
m^3 of concrete HP-60	128.41
m^3 of concrete HP-70	137.95
m^3 of concrete HP-80	147.49
m^3 of concrete HP-90	157.02
m^3 of concrete HP-100	166.56

It is common that a multi-objective optimization problem is transformed into a mono-objective optimization where the objective function is an aggregation function [19] (Equation (9)).

$$Aggregation function = w_1 \cdot \mu_{COST} + w_2 \cdot \sigma_{VERTICALDISPLACEMENT} \quad (9)$$

Here, the mean and the standard deviation are the normalized values of the objective functions, and w_1 and w_2 are weights with values in the range [0,1] such that $w_1 + w_2 = 1$.

In this work, 200 different cases (N) are considered in such a way that w_1 runs from 0 to 1 with increasing $1/N$ and w_2 corresponds to $1\text{-}w_1$. In this way, 200 different optimizations are made and all the possible solutions of the Pareto frontier are covered.

5. Results

The results are subdivided into two parts according to the initial uncertain design parameter considered: modulus of elasticity and loads (overload and prestressing force). Each one of these sections provides an initial validation of the kriging surfaces generated, the Pareto frontiers obtained, and some comparisons. For this purpose, 200 points are created to verify the accuracy of the kriging surfaces (validation), and another 200 solutions are obtained from the robust design optimization problems carried out (Pareto frontier). After that, the results will be discussed.

5.1. Variation of Modulus of Elasticity

In this part, the uncertain design parameter studied is the modulus of elasticity. Three different RDO problems are studied. For this purpose, six kriging surfaces are generated depending on the objective function (μ_{cost} and $\sigma_{\text{vertical displacement}}$) and the variability considered of the modulus of elasticity (10%, 20%, and 30%). Table 3 shows the different validations of the different kriging surfaces obtained. The accuracy of the kriging surfaces that predict the mean costs are better than the kriging surfaces that predict the variability of the vertical displacement. The difference between the real and predicted mean value of the cost is lower than 2%, and the difference between the real and predicted standard deviation of the vertical displacement of the middle of the bridge is lower than 5% in all different uncertainties of the modulus of elasticity considered.

Table 3. Validation of the kriging surfaces while varying the modulus of elasticity.

Uncertainty of E (%)	10	20	30
μ Cost discrepancy	1.21%	1.28%	1.07%
σ Displacement discrepancy	4.63%	4.75%	4.03%

Figure 4 shows the Pareto frontiers for the different uncertainties of the modulus of elasticity considered. This figure represents the mean of the cost against the standard deviation of the vertical displacement of the middle of the bridge. It shows that an increment of the uncertainty of the modulus of elasticity causes a displacement of the Pareto frontier, moving away from the positive ideal point (lowest μ_{cost} and lowest $\sigma_{\text{vertical displacement}}$). This is because the design of the structure should resist all the possible values of the uncertain parameter. Therefore, a higher variation of the initial uncertain parameter imposes greater requirements on the design and an increment of the cost.

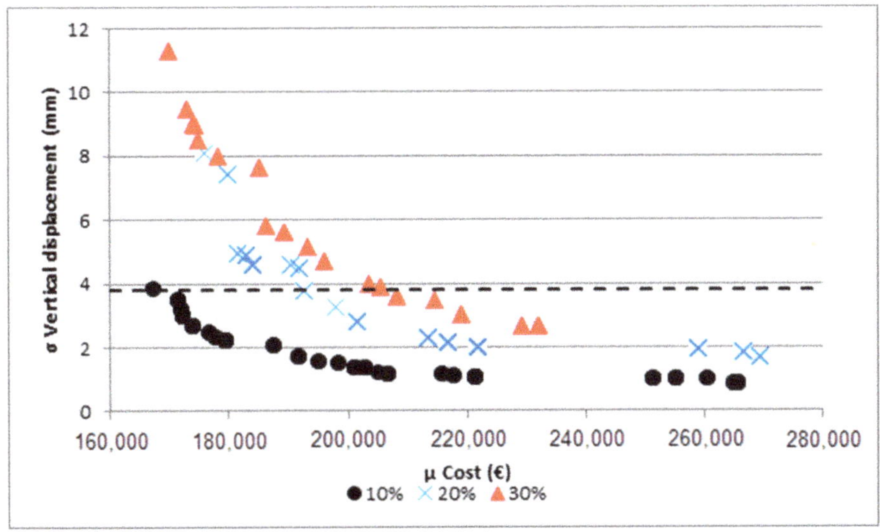

Figure 4. Pareto frontier of modulus of elasticity Robust Design Optimization (RDO) problems.

Table 4 shows a comparison between the designs of the different Pareto frontiers with the same structural behavior. In this case, the reference value taken into account is the standard deviation of the vertical displacement of the cheapest design of the Pareto frontier with the lowest variation of the modulus of elasticity studied. In this way, an imaginary horizontal line will intersect all the Pareto frontiers (dashed line of Figure 4). Solutions S10, S20, and S30 are selected, which correspond to a $\sigma_{\text{vertical displacement}}$ lower than 3.82 mm. It shows that to reach similar structural behavior, the price increases with an increment of the uncertainty of the modulus of elasticity and that the design variables that cause this increment of the price are the depth and f_{ck}. Both are higher for each increment of the variability of the modulus of elasticity.

Table 4. Comparison of design with the same structural behavior in modulus of elasticity RDO problems.

	b (mm)	h (mm)	d (mm)	e_v (mm)	e_s (mm)	e_a (mm)	e_i (mm)	f_{ck} (MPa)	c (mm)	μ_{cost} (€)	$\sigma_{v,displacement}$ (mm)
S10	1200	1450	0	150	150	350	225	45	225	167,370.9	3.811
S20	1200	1800	125	150	150	350	250	60	250	192,570.6	3.778
S30	1200	1950	0	150	150	350	225	80	225	208,111.9	3.548

Furthermore, if just one Pareto frontier is studied and three key designs are considered: (A) the optimum or lowest μ_{cost}, (B) the robust optimum or shortest to the positive ideal point, and (C) the most robust or lowest $\sigma_{\text{vertical displacement}}$, the same design variables are affected. For example, Table 5 shows these designs for the Pareto frontier with a 20% variability of the modulus of elasticity. As shown in Table 4, the values of depth and f_{ck} are higher when more robustness is required.

Table 5. Comparison of different designs of the Pareto Frontier with a 20% variation of the modulus of elasticity.

	b (mm)	h (mm)	d (mm)	e_v (mm)	e_s (mm)	e_a (mm)	e_i (mm)	f_{ck} (MPa)	c (mm)	μ_{cost} (€)	$\sigma_{v,displacement}$ (mm)
A	1200	1800	125	150	150	350	250	60	250	192,570.6	3.778
B	1200	1900	50	150	150	350	150	80	150	201,479.9	2.794
C	1800	2000	200	150	150	350	175	100	220	269,128.5	1.684

5.2. Variation of Loads: Overload and Prestressing Force

In this part, the uncertain design parameters studied are two loads. The first one is the overload due to its high uncertainty, and the second one is the prestressing force to know how the variability influences the behavior of the bridge. The overload is defined according to the IAP-11 [37], which corresponds to 5 kN/m^2. In this case, due to the higher uncertainty of these parameters, another increment of uncertainty in the loads is considered (40%). Therefore, four RDO problems are studied for each load. For this purpose, eight kriging surfaces are generated for each load depending on the objective function (μ_{cost} and $\sigma_{vertical\ displacement}$) and the variability considered of the modulus of elasticity (10%, 20%, 30%, and 40%). In these cases, the results discussed are the same as in the previous subsection. In this way, first, the validations of both loads are discussed (Tables 6 and 7) After that, the Pareto frontiers for each different uncertainty of the design parameter are shown (Figures 5 and 6), and finally some solutions are compared following the same rules as in the previous comparison: the overload (Tables 8 and 9), and the prestressing force (Tables 10 and 11).

Tables 6 and 7 show the different validations of the kriging surfaces obtained. As in the previous cases, the discrepancy of the mean value of the cost is lower than 2% in all cases. However, the discrepancy of the standard deviation of the vertical displacement of the middle of the bridge depends on the variability of the displacement, being higher when the vertical displacement variability is higher and lower when the vertical displacement variability is lower. The results show that when the variability of the overload is lower (10%), the kriging method cannot capture the variability of the displacement accurately. Thus, this uncertainty is not considered.

Table 6. Validation of the kriging surfaces varying the overload.

Uncertainty of Overload (%)	10	20	30	40
μ Cost discrepancy	1.32%	1.19%	1.17%	1.28%
σ Displacement discrepancy	38.61%	15.78%	11.53%	15.18%

Table 7. Validation of the kriging surfaces varying the prestressing force.

Uncertainty of P0 (%)	10	20	30	40
μ Cost discrepancy	1.34%	1.09%	1.06%	1.21%
σ Displacement discrepancy	13.5%	7.16%	3.47%	4%

Figures 5 and 6 represent the Pareto frontiers for the different variations of the loads. In both cases, the Pareto frontiers have the same behavior as before, moving away from the positive ideal point according to the increment of the uncertainty of the loads. In addition, the comparisons made (Tables 8–11) have similar behavior to the above.

Tables 8 and 10 show a comparison between different designs with the same structural behavior of the different Pareto frontiers. Table 8 corresponds to the RDO problems in which the overload is the uncertain parameter, and the $\sigma_{vertical\ displacement}$ of reference corresponds to 2.93 mm (dashed line of Figure 5). Table 9 corresponds to the RDO problems in which the prestressing force is the uncertain parameter, and the $\sigma_{vertical\ displacement}$ of reference corresponds to 11.06 mm (dashed line of Figure 6). In both cases, to reach a similar structural behavior the price increases with an increment of the uncertainty of the loads. As well as in the case of the RDO problems in which the modulus of elasticity is the uncertain parameter, the increment of the price is due to the increment of the depth and f_{ck}. The difference is that in the case (where the modulus of elasticity is the uncertain parameter) the depth and the value of f_{ck} increase in each increment of variability, and in the case where the uncertain parameter is the load, the increment of the depth and f_{ck} is not simultaneous. In these cases, a balance between these two design variables is achieved to reach a similar structural behavior. In addition, this increment of depth and f_{ck} is less significant in the case of the overload, due to the low differences among the different uncertainties. The same occurs when the comparison is made between the optimum or cheapest (A), the robust optimum or shortest to the positive ideal point (B), and the most robust or lowest variation of the vertical displacement (C) (Tables 9 and 11). As above, the key design variables to modify the structural behavior change are the depth and f_{ck}. These variables tend to be higher when higher robustness is required.

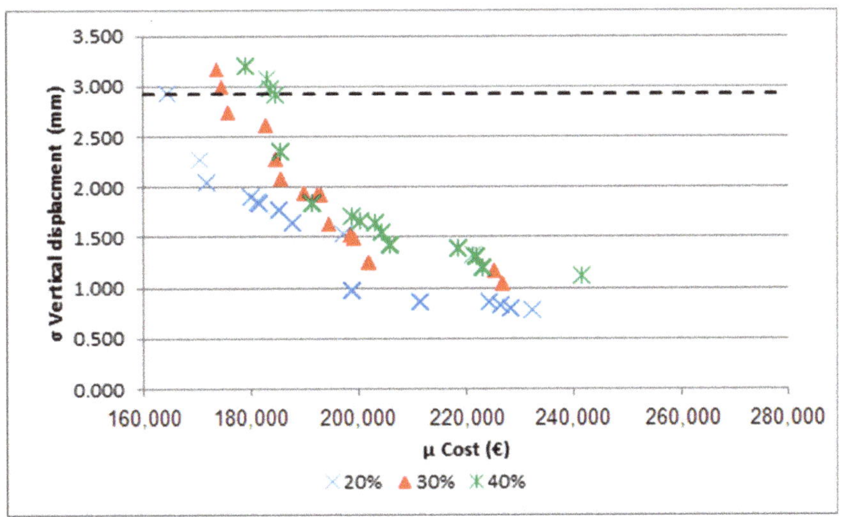

Figure 5. Pareto frontier of overload RDO problems.

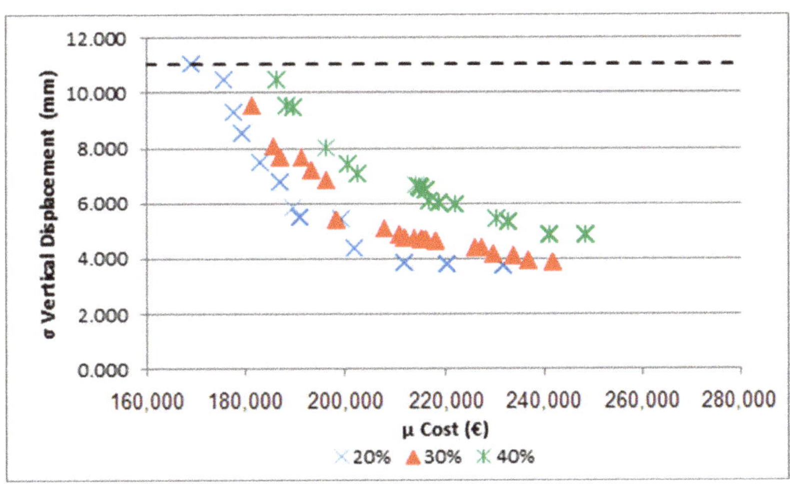

Figure 6. Pareto frontier of prestressing force RDO problems.

Table 8. Comparison of designs with the same structural behavior in overload RDO problems.

	b (mm)	h (mm)	d (mm)	e_v (mm)	e_s (mm)	e_a (mm)	e_i (mm)	f_{ck} (MPa)	c (mm)	μ_{cost} (€)	$\sigma_{v,displacement}$ (mm)
S20	1200	1250	0	150	150	350	200	60	200	164,594.2	2.924
S30	1200	1250	200	150	150	350	175	70	175	174,467.1	2.991
S40	1200	1700	25	175	175	350	250	50	250	184,821.6	2.917

Table 9. Comparison of different designs of the Pareto Frontier with a 20% variation of the overload.

	b (mm)	h (mm)	d (mm)	e_v (mm)	e_s (mm)	e_a (mm)	e_i (mm)	f_{ck} (MPa)	c (mm)	μ_{cost} (€)	$\sigma_{v,displacement}$ (mm)
A	1200	1350	100	150	150	350	175	80	175	180,240.5	1.913
B	1200	1850	200	175	175	350	225	60	225	198,687.3	0.971
C	1600	1800	150	275	275	350	225	70	225	238,573.8	0.753

Table 10. Comparison of designs with the same structural behavior in prestressing force RDO problems.

	b (mm)	h (mm)	d (mm)	e_v (mm)	e_s (mm)	e_a (mm)	e_i (mm)	f_{ck} (MPa)	c (mm)	μ_{cost} (€)	$\sigma_{v,displacement}$ (mm)
S20	1200	1350	0	150	150	350	200	60	200	168,833.9	11.058
S30	1200	1400	200	150	150	350	150	80	150	181,276.4	9.552
S40	1200	1750	125	150	150	350	200	55	200	186,380.7	10.497

Table 11. Comparison of different designs of the Pareto Frontier with a 20% variation of the prestressing force.

	b (mm)	h (mm)	d (mm)	e_v (mm)	e_s (mm)	e_a (mm)	e_i (mm)	f_{ck} (MPa)	c (mm)	μ_{cost} (€)	$\sigma_{v,displacement}$ (mm)
A	1200	1350	0	150	150	350	200	60	200	168,833.9	11.058
B	1200	1650	0	150	150	350	175	80	175	190,734.7	5.510
C	1300	2000	0	225	300	350	275	80	275	231,832.0	3.772

6. Conclusions

Currently, the design of structures is made according to a deterministic design. This approach has the result that when the design is optimized according to a conventional objective function, the behavior of the structure is really dependent on the initial values considered. This paper uses a probabilistic approach to consider the variation of the design parameters. In addition, to reduce the large computational cost of the probabilistic optimization, Latin hypercube sampling and kriging metamodels are used. Each point of the Latin hypercube sampling is calculated 20 times varying the initial uncertain parameters (modulus of elasticity, overload, and prestressing force) obtaining the mean of the cost and the standard deviation of the vertical displacement in the middle of the bridge. These values are used to create the kriging surface that predicts the objective response depending on the initial design variables. These surfaces have an error lower than 2% in the mean of the cost for all cases, lower than 5% in the standard deviation of the vertical displacement when the modulus of elasticity is the uncertain parameter, and an accuracy dependent on the value of the vertical displacement when the loads are the uncertain parameters. After that, 200 solutions have been calculated for each case to obtain the different Pareto frontiers.

The Pareto frontiers show that, for all RDO problems, an increment of the uncertainty causes a displacement of the Pareto frontier, moving away from the positive ideal point. That means that to obtain specific robustness when the uncertainty of the parameter is higher, the cost of the design will be higher. In addition, when just one Pareto frontier is taken into account, a more robust design implies an expensive design. In all cases, this increment of the price is due to an increment of two specific design variables: depth (h) and f_{ck}. Therefore, to obtain a robust design, it is necessary to increment the depth (h) and/or f_{ck}. However, these Pareto frontiers allow obtaining a compromise design between cost and robustness: the optimum robust design. This solution is the design closest to the positive ideal point.

This work shows that a probabilistic optimization can be carried out to obtain an optimum robust design. Nevertheless, the robust design optimization of complex problems requires a high computational cost. Therefore, the use of metamodels is necessary to carry out probabilistic optimization. In previous works, the computational cost saved and the validity of kriging metamodels were proven. This work shows that the kriging metamodel has an appropriate behavior to carry out the robust design optimization, and therefore can be used to carry out optimization where there is uncertain information.

Author Contributions: This paper represents a result of teamwork. The authors jointly designed the research. V.P.-P. drafted the manuscript. T.G.-S. and V.Y. edited and improved the manuscript until all authors are satisfied with the final version. All authors have read and agreed to the published version of the manuscript.

Funding: This research was funded by the Ministerio de Economía, Ciencia y Competitividad and FEDER funding grant number [BIA2017-85098-R].

Conflicts of Interest: The authors declare no conflict of interest.

References

1. Taguchi, G. *Introduction to Quality Engineering*; Asian Productivity Organisation: Tokyo, Japan, 1986.
2. Phadke, M.S.; Shridhar, M. *Quality Engineering Using Robust Design*; Prentice Hall: Upper Saddle River, NJ, USA, 1989.
3. Fowlkes, W.Y.; Creveling, C.M. *Engineering Method for Robust Product Design*; Addison-Wesley Publishing Company: Boston, MA, USA, 1995.
4. Lee, K.-H.; Kang, D.-H. A robust optimization using the statistics based on kriging metamodel. *J. Mech. Sci. Technol.* **2006**, *20*, 1169–1182. [CrossRef]
5. Carbonell, A.; González-Vidosa, F.; Yepes, V. Design of reinforced concrete road vaults by heuristic optimization. *Adv. Eng. Softw.* **2011**, *42*, 151–159. [CrossRef]
6. Ahsan, R.; Rana, S.; Nurul Ghani, S. Cost optimum design of posttensioned I-girder bridge using global optimization algorithm. *J. Struct. Eng.* **2012**, *138*, 272–283. [CrossRef]
7. García-Segura, T.; Yepes, V.; Martí, J.V.; Alcalá, J. Optimization of concrete I-beams using a new hybrid glowworm swarm algorithm. *Lat. Am. J. Solids Struct.* **2014**, *11*, 1190–1205. [CrossRef]
8. Pnevmatikos, N.G.; Thomos, G.T. Stochastic structural control under earthquake excitations. *Struct. Control Health Monit.* **2014**, *21*, 620–633. [CrossRef]
9. García-Segura, T.; Yepes, V. Multiobjective optimization of post-tensioned concrete box-girder road bridges considering cost, CO_2 emissions, and safety. *Eng. Struct.* **2016**, *125*, 325–336. [CrossRef]
10. Martí, J.V.; García-Segura, T.; Yepes, V. Structural design of precast-prestressed concrete U-beam road bridges based on embodied energy. *J. Clean. Prod.* **2016**, *120*, 231–240. [CrossRef]
11. Yepes, V.; Martí, J.V.; García-Segura, T.; González-Vidosa, F. Heuristics in optimal detailed design of precast road bridges. *Arch. Civ. Mech. Eng.* **2017**, *17*, 738–749. [CrossRef]
12. Sun, X.; Fu, H.; Zeng, J. Robust approximate optimality conditions for uncertain nonsmooth optimization with Infinite number of constraints. *Mathematics* **2018**, *7*, 12. [CrossRef]
13. Rodriguez-Gonzalez, P.T.; Rico-Ramirez, V.; Rico-Martinez, R.; Diwekar, U.M. A new approach to solving stochastic optimal control problems. *Mathematics* **2019**, *7*, 1207. [CrossRef]
14. Moayyeri, N.; Gharehbaghi, S.; Plevris, V. Cost-based optimum design of reinforced concrete retaining walls considering different methods of bearing capacity computation. *Mathematics* **2019**, *7*, 1232. [CrossRef]
15. Sierra, L.A.; Yepes, V.; García-Segura, T.; Pellicer, E. Bayesian network method for decision-making about the social sustainability of infrastructure projects. *J. Clean. Prod.* **2014**, *176*, 521–534. [CrossRef]
16. Valdebenito, M.A.; Schuëller, G.I. A survey on approaches for reliability-based optimization. *Struct. Multidiscip. Optim.* **2010**, *42*, 645–663. [CrossRef]
17. Doltsinis, I.; Kang, Z. Robust design of structures using optimization methods. *Comput. Methods Appl. Mech. Eng.* **2004**, *193*, 2221–2237. [CrossRef]
18. Simpson, T.W.; Booker, A.J.; Ghosh, D.; Giunta, A.A.; Koch, P.N.; Yang, R.-J. Approximation methods in multidisciplinary analysis and optimization: A panel discussion. *Struct. Multidiscip. Optim.* **2004**, *27*, 302–313. [CrossRef]
19. Martínez-Frutos, J.; Martí, P. Diseño óptimo robusto utilizando modelos Kriging: Aplicación al diseño óptimo robusto de estructuras articuladas. *Rev. Int. Métodos Numéricos para Cálculo y Diseño Ing.* **2014**, *30*, 97–105. [CrossRef]
20. Jin, R.; Chen, W.; Simpson, T.W. Comparative studies of metamodelling techniques under multiple modelling criteria. *Struct. Multidiscip. Optim.* **2001**, *23*, 1–13. [CrossRef]
21. Martí-Vargas, J.R.; Ferri, F.J.; Yepes, V. Prediction of the transfer length of prestressing strands with neural networks. *Comput. Concr.* **2013**, *12*, 187–209. [CrossRef]

22. Salehi, H.; Burgueno, R. Emerging artificial intelligence methods in structural engineering. *Eng. Struct.* **2018**, *171*, 170–189. [CrossRef]
23. Jin, R.; Du, X.; Chen, W. The use of metamodeling techniques for optimization under uncertainty. *Struct. Multidiscip. Optim.* **2003**, *25*, 99–116. [CrossRef]
24. Penadés-Plà, V.; García-Segura, T.; Yepes, V. Accelerated optimization method for low-embodied energy concrete box-girder bridge design. *Eng. Struct.* **2019**, *179*, 556–565. [CrossRef]
25. Barton, R.R.; Meckesheimer, M. Metamodel-based simulation optimization. *Handb. Oper. Res. Manag. Sci.* **2006**, *13*, 535–574.
26. McKay, M.D.; Beckman, R.J.; Conover, W.J. Comparison of three methods for selecting values of input variables in the analysis of output from a computer code. *Technometrics* **1979**, *21*, 239–245.
27. Chuang, C.H.; Yang, R.J.; Li, G.; Mallela, K.; Pothuraju, P. Multidisciplinary design optimization on vehicle tailor rolled blank design. *Struct. Multidiscip. Optim.* **2008**, *35*, 551–560. [CrossRef]
28. Matheron, G. Principles of geostatistics. *Econ. Geol.* **1963**, *58*, 1246–1266. [CrossRef]
29. Simpson, T.W.; Mauery, T.M.; Korte, J.; Mistree, F. Kriging models for global approximation in simulation-based multidisciplinary design optimization. *AIAA J.* **2001**, *39*, 2233–2241. [CrossRef]
30. Forrester, A.I.J.; Keane, A.J. Recent advances in surrogate-based optimization. *Prog. Aerosp. Sci.* **2009**, *45*, 50–79. [CrossRef]
31. Simpson, T.W.; Poplinski, J.D.; Koch, P.N.; Allen, J.K. Metamodels for computer-based engineering design: Survey and recommendations. *Eng. Comput.* **2001**, *17*, 129–150. [CrossRef]
32. Camp, C.V.; Huq, F. CO_2 and cost optimization of reinforced concrete frames using a big bang-big crunch algorithm. *Eng. Struct.* **2013**, *48*, 363–372. [CrossRef]
33. Martí, J.V.; González-Vidosa, F.; Yepes, V.; Alcalá, J. Design of prestressed concrete precast road bridges with hybrid simulated annealing. *Eng. Struct.* **2013**, *48*, 342–352. [CrossRef]
34. Medina, J.R. Estimation of incident and reflected waves using simulated annealing. *J. Watery Port Coast. Ocean Eng* **2001**, *127*, 213–221. [CrossRef]
35. Schlaich, J.; Scheef, H. *Concrete Box-Girder Bridges*; International Association for Bridge and Structural Engineering: Zürich, Switzerland, 1982.
36. Ministerio de Fomento. *EHE-08: Code on Structural Concrete*; Ministerio de Fomento: Madrid, Spain, 2008.
37. Ministerio de Fomento. *IAP-11: Code on the Actions for the Design of Road Bridges*; Ministerio de Fomento: Madrid, Spain, 2011.
38. European Committee for Standardization. *EN 1001-2: Eurocode 1: Actions on Structures-Part 2: Traffic Loads Bridges*; European Committee for Standardization: Brussels, Belgium, 2003.
39. European Committee for Standardisation. *EN1992-2: Eurocode 2: Design of Concrete Structures-Part 2: Concrete Bridge-Design and Detailing Rules*; European Committee for Standardisation: Brussels, Belgium, 2005.
40. Catalonia Institute of Construction Technology. *BEDEC PR/PCT ITEC Material Database*; Catalonia Institute of Construction Technology: Barcelona, Spain, 2016.

Article

A Variation of the ATC Work Shift Scheduling Problem to Deal with Incidents at Airport Control Centers

Antonio Jiménez-Martín *, Faustino Tello and Alfonso Mateos

Decision Analysis and Statistics Group, E.T.S.I. Informáticos, Universidad Politécnica de Madrid, Campus de Montegancedo S/N, 28660 Boadilla del Monte, Spain; faustino.tello@upm.es (F.T.); alfonso.mateos@upm.es (A.M.)

* Correspondence: antonio.jimenez@upm.es

Received: 19 January 2020; Accepted: 25 February 2020; Published: 2 March 2020

Abstract: This paper deals with a variation of the air traffic controller (ATC) work shift scheduling problem focusing on the tactical phase, in which the plan for the day of operations can be modified according to real-time traffic demand or other possible incidents (one or more ATCs become sick and/or there is an increase in unplanned air traffic), which may lead to a new sectorization and a lower number of available ATCs. To deal with these issues, we must reassign the available ATCs to the new sectorization established at the time the incident happens, but also taking into account the work done by the ATCs up to that point. We propose a new methodology consisting of two phases. The goal of the first phase is to build an initial possibly infeasible solution, taking into account the sectors that have been closed or opened in the new sectorization, together with the ATCs available after the incident. In the second phase, we use simulated annealing (SA) and variable neighborhood search (VNS) metaheuristics to derive a feasible solution in which the available ATCs are used and all the ATC labor conditions are met. A weighted additive objective function is used in this phase to account for the feasibility of the solution but also for the number of changes in the control center at the time the incident happens and the similarity of the derived solution with templates usually used by the network manager operations center, a center managing the air traffic flows of an entire network of control centers. The methodology is illustrated by means of seven real instances provided by the Air Traffic Management Research, Development and Innovation Reference Center (CRIDA) experts representing possible incidents that may arise. The solutions derived by SA outperform those reached by VNS in terms of both the number of violated constraints in all seven instances, and solution compactability in six out the seven instances, and both are very similar with regard to the number of control center changes at the time of the incident. Although computation times for VNS are clearly better than for SA, CRIDA experts were satisfied with SA computation times. The solutions reached by SA were preferred.

Keywords: air traffic management; tactical phase; work-shift scheduling problem; metaheuristics; performance analysis

1. Introduction

Air transport is growing exponentially, from the 3.8 billion air travelers in 2016 to 7.2 billion passengers expected to travel in 2035, according to the International Air Transport Association [1]. Therefore, air transport and the resulting shortage of laborers in the civil aviation industry has become a serious problem. Air traffic controller (ATC) requires long hours of training, with more than one year needed to train ATCs.

The role of the network manager (NM) is to establish a balance between air traffic demand and airspace/airport capacity in Europe. However, currently, this role is merely moderation between aircraft operators and capacity providers, since the NM has limited instruments to influence either capacity or demand side planning decisions [2,3]. The European Commission also recognizes that the lack of the NM's clear executive powers in practice means that the NM tends to decide by consensus, which often results in weak compromises [4]. The European Commission however stresses that an optimization of the network performance requires an extended operation scope of actions by the NM, a view also shared by Ryanair [5].

Although the NM initiates planning several months before the day of operations [2,3], most of demand-capacity imbalance situations are still resolved on the day of operations by means of demand management actions, predominantly by delaying flights. For instance, the total en-route air traffic flow management delay was 8.7 million minutes in Europe in 2016, corresponding to a traffic of more than 10 million flights [3,6].

More than 55% of total en-route delay is attributed to lack of capacity and staffing reasons, while approximately half of that delay occurred during peak summer months—June, July and August [3,6]. The Performance Review Commission notes that the capacity requirements are frequently not met by some area control centres, but also that maximum capacity is not delivered at the times when it is needed [3,6].

One of the underlying causes for capacity/demand mismatch is seasonal traffic variability. If traffic is highly variable and there is limited flexibility to adjust the capacity provision according to actual demand, the result may be poor service quality or an under utilization of resources [6]. If addressed proactively, traffic variability can be mitigated or resolved to a certain degree by utilizing previous experience, roster staffing levels to suit and to make more operations staff available by reducing ancillary tasks performed by ATCs during the peak period [6]. Meanwhile delay costs occur when there is no sufficient capacity provision for aircraft operators.

In Reference [7], air traffic control scenarios are classified using decision trees. The authors conclude that decision trees and classification rules perform well in prediction, stability and interpretability.

The activities of the network manager operations center (NMOC) are divided into four phases [8]—strategic, pre-tactical, tactical and post-operational. The strategic phase is related to capacity predictions at air traffic control centers by air navigation service provides, preparing a routing scheme with the help of NMOC seven days ahead of operations.

The pre-tactical phase is related to the definition of the initial network plan. The NMOC publishes the agreed plan for the day of operations, informing air traffic control units and aircraft operations about the air traffic flow and capacity management measures affecting European airspace from one to six days ahead of operations.

The tactical phase updates the plan for the day of operations according to real-time traffic demand where the NMOC monitors the situation and continuously optimizes capacity. Delays are minimized by providing aircraft affected by changes with alternative solutions on the day of operations.

The post-operational phase is related to operational process improvement by comparing planned and measured outcomes covering all air traffic flow and capacity management domains and units. Operational processes are measured in order to develop best practices and/or analyze lessons learned after the days of operations.

In this paper we focus on the tactical phase, that is, the day of operation. A specific sectorization has been established depending on the aircraft traffic for the considered day and the ATCs have been assigned to the open sectors. We assume that an incident arises, which involves a sectorization change and possibly less available ATCs, and that the available ATCs must be reassigned to the open sectors from that moment until the end of the shift taking into account the work done by each of them from the beginning of the shift to the instant of the incident.

The problem under consideration is similar but different to the shift and break assignment problems, which are referred to in the literature with different names, such as shift design [9,10], shift scheduling [11–14], break scheduling [15–17] and both process shift design and break assignment with a large number of breaks [16].

Such problems have been extensively investigated in Operations Research and have also recently been tackled with Artificial Intelligence techniques. Also, this is a timetabling and scheduling problem. Timetabling and scheduling problems are combinatorial problems, which, on the grounds of size and complexity, cannot be solved by exact methods within a reasonable computation time. For examples of other timetabling and scheduling problem-solving approaches, see References [18,19].

A general mathematical model and specific models for personnel scheduling problems are presented in Reference [20]. Complexity issues regarding personnel scheduling problems are addressed by identifying polynomial solvable and NP-hard special cases. More recently, a general mathematical model and specific models for personnel scheduling problems were proposed in Reference [21] enabling the implementation of various heuristic algorithms and their application to a wide range of problems.

Reference [22] conducted a literature review related to shiftwork management within air traffic management, addressing shiftwork impact on health and safety, productivity and efficiency and discussing issues concerning working time organization in accordance with EC Directive 93/104.

A study on shiftwork practices in both ATM and other fields, such as medicine, the police force and the airline industry, is presented in Reference [23]. It concludes that, although there are a range of software tools, in many cases involving ATM, they are costly and not completely suited to the needs. The strengths and weaknesses of automated scheduling tools have already been outlined [24].

Reference [25] used propositional satisfiability (SAT, [26]), MaxSAT, the pseudo-Boolean, satisfiability modulo theory, constraint satisfaction and integer linear problem solvers to address a number of different month- or year-long scheduling requirements. According to the results of applying three different optimization techniques jointly with the above problem solvers, SAT approaches appear to come out on top. Then Stojadinovic [27] combined SAT problem solving with the hill climbing method. The hill climbing method is applied to a feasible solution output by the SAT solver to solve more ATC requirements. The SAT solver is then applied again to further improve the solution. This cycle is iterated until the resulting solution is optimal.

A preliminary approach for modeling many of the features of the ATC scheduling problem was proposed in Reference [28]. The model divides time into 30-min slots, and an ATC should not work for more than 2 h, followed by at least a 30-min break.

A simplified version of the ATC work shift scheduling problem for Spanish airports was solved by minimizing the number of ATCs required to cover a given airspace sectoring in compliance with Spanish ATC working conditions [29] in the pre-tactical phase. This problem was mathematically modeled as a mixed integer problem. A simple sectorization for a whole day with 40 available ATCs involved 751,200 variables and more than 161,669 constraints. This makes it hard to reach good solutions in a reasonable time, leading to the use of a metaheuristic.

The search process employs regular expressions to check solution feasibility. Both search processes use regular expressions (regex) [30]. A regex is a sequence of characters that define a search pattern, used in search algorithms to find strings. The strings represent solutions. The patterns in our approach represent breaches of ATC working conditions.Thanks to the high testing speed and modularity of regex, the optimization model is simpler to implement and maintainable.

Both search processes use regular expressions (Regex) [30]. A Regex is a sequence of characters that define a search pattern, used in search algorithms to find strings. The strings represent solutions. The patterns in our approach represent breaches of ATC working conditions.Thanks to the high testing speed and modularity of regex, the optimization model is simpler to implement and maintainable.

A further optimization process is applied to the resulting optimal number of ATCs to balance ATC workloads. There are no constraints on ATC distribution across sectors in this simplified workshift scheduling problem, which accounts for a 24-h period in a core with just one sector.

The proposal reported in Reference [31] addresses cores with two (en-route and approach) sector types and ATCs with different credentials. Adopting a multi-objective approach, one shift is optimized for a sectorization specified during the pre-tactical phase according to a set number of ATCs considering ATC work and break periods, ATC positions and workload, control center changes and solution structure. The methodology is divided into three optimization phases using a rank-order centroid function to convert a multiple into a single optimization problem taking account ordinal information on objectives (i.e., objectives ranked by importance). First, a template-based method identified unfeasible solutions. Second, independent simulated annealing (SA) output feasible solutions applying regex to check for compliance with ATC working conditions. Third, further independent SA runs optimized the objective functions of these solutions again checking for ATC working condition compliance.

Tello et al. [32] replaced SA in the third phase of the above methodology with an adaptation of variable neighborhood search. They compared the performance of the two metaheuristics on four different sectorizations supplied by the Spanish ATM Research, Development and Innovation Reference Center (CRIDA). They also compared the run times between regex use and implementation in code to verify ATC working conditions (constraints).

Unlike the problems considered previously [31,32], in this paper we focus on the tactical rather than the pre-tactical phase, at the time an incident arises. As said before, a specific sectorization has been established depending on the aircraft traffic for the considered day and the ATCs have been assigned to the open sectors. We assume that an incident happens. Incidents can be of different types: ATCs who cannot continue their working hours because they are unwell or change the sectorization due to a significant increase in air traffic due to diverted air traffic from another airport that has closed for weather reasons.

The incident usually involves a sectorization change. From the moment the incidents occur until the end of the shift the ATCs must be reassigned to the new open sectors taking into account the work done by each of them from the beginning of the shift to the time of the incident and ATC working conditions.

It is important to note that response time is critical in the tactical phase for this variant of the problem. Thus, the use of metaheuristics is even more justified than in the pre-tactical phase.

To solve the problem we propose a methodology that utilizes the metaheuristics simulated annealing and variable neighborhood search, which consists of two phases. In the first phase, it derives a solution, which does not have to be feasible and even need a greater number of ATCs than available ones. This solution in the second phase, applying the metaheuristic, is transformed into a solution in which it need as many ATCs as the number of available ATCs and try to achieve feasibility.

The paper consists of three more sections. Section 2 shows the proposed methodology for a real time adaptation of the ATC work shift scheduling problem to deal with incidents in airports control centers. The adaptation of SA and VNS to the ATC work shift scheduling problem is explained in Section 2 Phase 1 and Phase 2, respectively. Section 3 illustrates the methodology in a real incident from the Barcelona control center. SA and VNS performance has been tested on a set of representative instances in Section 4. Finally, some conclusions are provided inn Section 5.

2. Problem Description

Airspace sectorization is the partitioning of the airspace into a given number of sectors. Sectors that are open at any one time should cover total airspace capacity. Additionally, sectors can be clustered together to constitute a *core*. Nevertheless, a sector may be part of more than one core. An air traffic control center may be in charge of managing one or more cores. Cores are managed separately, although ATCs can be assigned to two or more cores, if sectors they manage belong to more than one core. Sectors are divided into two types depending on their distance from the airport: approach (5 to 10 nautical miles from the airport) and en-route sectors (more than 10 nautical miles from the airport).

Each sector is assigned to a team of air traffic controllers, each of which can handle a limited amount of traffic. Sectors are operated by ATCs with different roles: planner ATCs and executive

ATCs. Planner ATCs foresee possible conflicts between aircraft that they report to executive ATCs who instruct pilots on how to avoid loss of minimum separation.

The sector configuration depends on the volume of air traffic. More smaller sectors are opened as the volumen of air traffic grows, increasing the demand for ATCs and vice versa. Therefore, sector configuration and number of ATCs changes as sectors are dynamically divided and merged according to air traffic variations.

Figure 1 shows an example of an airspace sectorization for the Barcelona eastern route. Each time slot has an associated configuration. The number stands for the number of open sectors, and the letter symbolizes the sector configuration. For example, sectorizations 5A and 5B have the same number of sectors with a different spatial distribution.

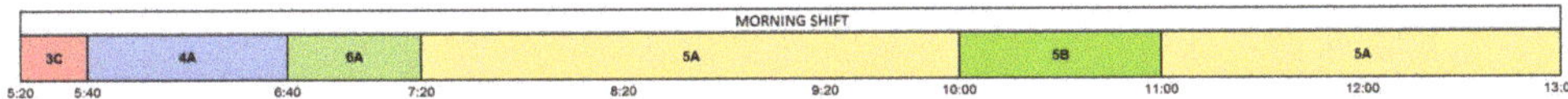

Figure 1. Barcelona eastern route airspace sectoring.

Besides, in Spain, working conditions are compiled and published in the Official State gazette, Royal Decree 1001/2010, and Law 9/2010. ATC working conditions are as follows:

1. (LC1) ATCs can only operate sectors belonging to the core for which they are qualified;
2. (LC2) CON ATCs can only operate en-route sectors;
3. (LC3) ATCs must rest for 25% of the work shift in day shifts (MS, LMS, AS and LAS). The night shift (NS) break periods account for 33% of working time;
4. (LC4) A sector opened during the whole night shift must be operated by four ATCs;
5. (LC5)ATCs cannot work for more than two hours without a break;
6. (LC6) ATCs can only operate during one shift;
7. (LC7) ATCs should rest at least half an hour every two hours;
8. (LC8) Sector changes are not allowed without resting unless there is an emergency. ATCs can change to an related sector without resting;
9. (LC9) The minimum work period length of a controller between two breaks is 15 min;
10. (LC10) Each break period should last at least 15 min;
11. (LC11) ATCs must remain in the same sector and position for at least 15 min (minimum time in a position);
12. (LC12) Each ATC can work in at most three different non-related sectors in the respective shift;
13. (LC13) Each work shift must have one ATC assigned. And each ATC must be responsible of one shift;
14. (LC14) Each ATC must work at least 15 min, can not rest all the shift.

More details about the problem description are available in Reference [31] .

3. Methodology

The tactical phase takes place on the day of operation, that is, during the work shift. A solution must be found immediately, if an unforeseen event arises, either due to the downgrading of an ATC or the increase in air traffic on any ground. To this end, a two-phase resolution methodology has been proposed.

The starting point for the first phase is the work shifts that were being applied until the unexpected incident occurs. Note that in the considered solutions, time is discretized in slots of five minutes. The slot in which the incident occurs will be called incident slot. This means that the contents of all slots before the incident slot cannot be modified.

We mark the incident slot with a vertical line, see Figure 2, which shows a solution matrix where each row is an ATC and each matrix element (i, j) represents the state of ATC i in time slot j. It is symbolized by letters. The value 111 represents a resting ATC, uppercase letters [A–Z] indicate that the ATC is working as an executive operator, whereas lowercase letters [a–z] are used for planner positions.

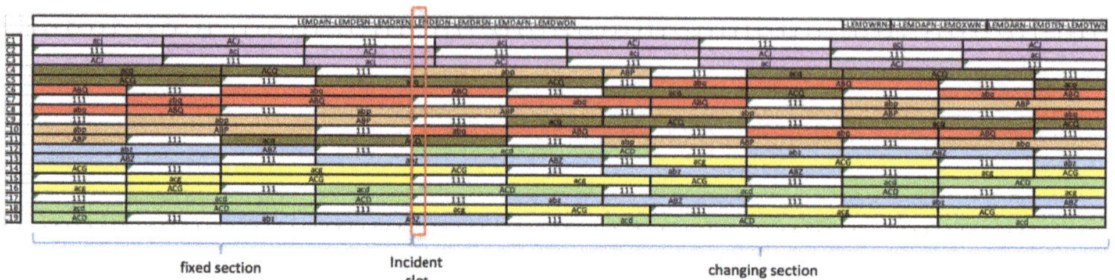

Figure 2. Solution matrix with an incident slot.

Depending on the unforeseen event, we will need to modify the sectorization, the solution matrix, and/or the list of available ATCs. For example, the absence of an ATC (by illness or urgency) does not always involve the closure of sectors. Sometimes, the others ATCs can cover the position that has been released until the replacement for a new ATC.

Another example is the closure of an airport because of adverse weather conditions. In that case, traffic will be diverted to nearby airports, which will lead to an increase in air traffic and, therefore, new sectors must be opened. As a result, there would be a change in sectorization and the solution matrix and, possibly, new ATCs must be incorporated to cover new sectors.

The necessary changes will be established by the network manager operations center (NMOC), and constitute an input to the solution algorithm together with the work shifts that are being used.

The aim of the first phase is to modify the work shifts and the list of ATCs in order to incorporate the work slots of the new open sectors in the solution matrix, eliminate the work slots corresponding to the closed sectors, introduce the new ATCs to the ATC list and the shift array, and indicate the downgrading of the ATCs that are not there. Moreover, the largest possible number of available ATCs should remain in the positions (and assigned sectors) that they were working in just after the incident so that not many changes in the control center occur simultaneously. In addition, the only possible solution uses more ATCs that are available in most cases.

The result of the first phase will be used as the initial solution for the second phase. The second phase uses an algorithm based on the metaheuristics simulated annealing or variable neighborhood search to derive a feasible solution that meets all the ATC working conditions (constraints) and reduces the number of shifts to the number of available ATCs.

Phase 1: Initialization Algorithm. The initialization algorithm aims at transforming the available solution before the unforeseen event (applied during part of the work shift) into a new solution containing the proposed changes to solve that incident. For this, we divide this phase into the following four steps:

Step 1: Closing Sectors. Many times a sector closes and at the same time two sectors related to the closed sector are opened. The objective of this step is to detect closed sectors and their related sectors, in order to replace closed sector working slots with working slots from the new related sectors. The algorithm searches for all closed sectors and checks for related sectors for each of them. In the event that no new opened sectors are found related to the closed sector, the working slots in that sector are replaced by breaks in the solution matrix. Otherwise, that is, there is a new sector to open and this is related to the closed sector, the closed sector working slots are replaced by open and related sector working slots.

Step 2: Opening Sectors. When the new sectors are not related to any of the closed sectors, or the closed sectors have already been associated with other related sectors, this step is carried out. The algorithm adds the working slots of the open sectors in work templates (see References [29,31,32]), and adds these templates to the solution matrix. There are several templates that can be used, but we have chosen to apply the template displayed in Figure 3. The reason for this choice is the ease with which changes can be made to this template without breaking many constraints due to its structure. That template uses three ATCs to cover an entire sector. In this step, more work shifts are added, so there may be more work shifts than ATCs available. However, introducing work slots between ATC breaks would generate shift arrays with a high number of unfulfilled and unstructured constraints or which would make it extremely difficult to the algorithm to work in Phase 2.

ATC 1	Working	Resting	Working	Working	Resting	Working	Working	...
ATC 2	Resting	Working	Working	Resting	Working	Working	Resting	...
ATC 3	Working	Working	Resting	Working	Working	Resting	Working	...

Figure 3. Template 1 × 3 (1 sector is controlled by 3 air traffic controllers (ATCs), two working, executive and planner, and one resting).

Step 3: Indisposed ATCs. Another possible incident is the indisposition of an ATC during his/her work shift. In this step, the work shift that the indisposed ATC has associated is modified, and the shift array indicates that this ATC is not available to take on any work from the time the downgrade is set. If his/her work can be assumed and added to the shift of another ATC, it is done, and otherwise, a new work shift will be generated and added to this shift.

Step 4: New ATCs. An ATC can be absent, and it is for this reason that there are guard ATCs to solve these kind of situations. The guard ATCs have a deadline to show up at the control center since the notice occurs. Therefore, we can count on new ATCs when it comes to spreading the workload. In this step, we add the new ATCs to the ATC' list and add a new shift for each new ATC, which supports work from the moment the ATC arrives at the control center. Finally, this shift is associated with the new ATC.

Once the four steps of the initialization algorithm have been performed, we have an initial solution to start with Phase 2. For this, the solution must comply with the following premises:

- All sectors must be covered in both work positions (executive and planner) all the time.
- All ATCs must have an associated shift in the shift list, although there can be work shifts not assigned to ATCs (assigned to imaginary ATCs).
- All ATCs must be listed in the ATC list.

Phase 2: Metaheuristic-based optimization process. Phase 2 consists of applying a metaheuristic: simulated annealing (SA) or variable neighborhood search (VNS). The goal of this phase is to reach a feasible solution that meets the modifications made to address the unforeseen event. For this, we simultaneously consider four objectives. First, an initial solution in Phase 1 may use more than the available number of ATCs. Therefore, they need to be reduced until the number of required ATCs matches the number of available ATCs.

Secondly, the solution must be feasible. All ATC's working conditions must be met and all open sectors in both positions (executive and planner) must be covered all time.

The third objective consists of minimizing the number of changes during the incident slot, that is, try to ensure that ATCs are kept in the same position just after the incident. At that point, several sectors may open or close and some ATCs change their job in a forced way, so we're going to try to minimize the number of changes to avoid that all ATCs get out of their position simultaneously as much as possible.

The last objective refers to a similar solution structure to previous template-based solution. Control center staff will find such a solution easier to understand, and this should facilitate any manual changes.

The target function in both SA and VNS is a weighted sum of the four objectives that are explained in detail below.

1. Reduce the number of shifts until the number of available ATCs: It is important to note that in the solutions we are analyzing, shifts are ordered from the least to the highest workload (in the first rows the lowest load and the last ones with the highest load).

   However, when sorting shifts, let's divide the array into two sections: shifts that have an assigned ATC and shifts that do not have assigned an ATC. In this way, shifts without an assigned ATC will occupy the first rows of the array sorted by workload, while shifts with an assigned ATC will be sorted between them subsequently. With this we aim to encourage the shifts that are going to be eliminated are the shifts without assigned ATCs. To do this, we use the following expression:

$$p = \sum_{k=1}^{nATC} \left(\frac{k \times h_k}{nATC^2} \right), \tag{1}$$

   where, $nATC$ is the number of shifts and h_k the number of shift work slots (in the k-th row of the solution coding).

   In Equation (1), the product of the numerator causes the p-value to be higher the higher are the shifts of the last rows (that is, those with the highest workload). Therefore, in the search process, the workload of the shifts from the first rows to those of the last rows will tend to be passed, causing the shifts of the first rows to run out of load and be able to be deleted. The denominator of the array is squared so that the solution is always improved by deleting a turn. Otherwise, it could happen that by removing a turn to worsen fitness.

   We must normalize Equation (1) to be used in the objective weight addition as follows:

$$f_1 = 1 - \frac{p_{max} - p}{p_{max} - p_{min}}, \tag{2}$$

   where p_{max} and p_{min} are maximum and minimum values of p, respectively.

   To describe the way p_{max} and p_{min} are computed, we must first introduce a new term, $nATC_{full}$. To calculate this value, the total workload is computed from the sectorization. This gives us a number of work slots that should be split across all shifts, denoted as C_{total} and to get $nATC_{full}$:

$$nATC_{full} = \left\lceil \frac{C_{total}}{nSlot} \right\rceil. \tag{3}$$

   Equation (3) refers to the number of ATCs needed that have to work the entire shift in order to cover the workload, C_{total}. Once this value is known, we can calculate the value of p_{max} as follows

$$p_{max} = \sum_{k=nATC-(nATC_{full}+1)}^{nATC} \left(\frac{k \times nSlot}{nATC^2} \right), \tag{4}$$

   where we assign the maximum workload to the shifts that are in the last rows, while the first ones are not assigned by the workload.

p_{min} is computed from the following expression:

$$p_{min} = \sum_{k=1}^{nATC_{full}+1} \left(\frac{k \times nSlot}{nATC^2} \right), \tag{5}$$

where we don't assign workload to shifts in the last rows, but shifts in the first rows are assigned the maximum workload.

2. Reduce the number of unfulfilled constraints: ATCs' shifts must comply with all working conditions. To make this possible, we have introduced the number of unfulfilled constraints within the target function, denoted by R_i.

   We transform the equation into a maximization form and normalize it, leading to the following expression:

$$f_2 = 1 - \frac{MaxR_i - R_i}{MaxR_i}, \tag{6}$$

   where $MaxR_i$ is the maximum value that can be achieved by breaking all constraints.

   For the calculation of $MaxR_i$ we must consider two cases:

   - Night shift. In this shift all constraints apply so $MaxR_i$ is calculated as follows: there is a total of 14 constraints. 8 constraints can be breached $nATC$ times, 2 constraints that can be breached $2nATC$ and 4 constraint can be breached $nATC$ multiplied every time it is not met in each slot of the shift by $1/nSlot$, that is, $nSlot \times nATC \times (1/nSlot)$. Then, we have:

$$MaxR_i = 8nATC + 2 \times 2nATC + 4(nATC + nSlot \times 1/nSlot \times nATC), \tag{7}$$

     that is, $MaxR_i = 20nATC$.
   - Day shifts (morning and afternoon). One constraint does not apply, so we'll use Equation (8):

$$MaxR_i = 8nATC + 1 \times nATC + 4(nATC + nSlot \times 1/nSlot \times nATC), \tag{8}$$

     that is, $MaxR_i = 18nATC$.

   As the incident management may lead to unfeasible solutions, weights representing the relative importance of constraint violation were introduced by experts from CRIDA (www.crida.es), a non-profit joint venture between ENAIRE, Spain's air navigation manager, the Universidad Politécnica de Madrid, and Ineco, a global infrastructure engineering and consultancy leader.

   First, a subset of the most important (priority) labor constraints was identified (LC1-LC7, LC9, LC12-LC14) (weight 5), followed by LC8 (weight 0.9), LC11 (weight 0.85), LC10 (weight 0.96) and LC9 (weight 0.5). Thus, we should attempt to meet the constraints in the most important set in preference to the other constraints.

3. Reduce the number of changes during the incident slot: To avoid a simultaneous change of a large number of ATCs just after the incident we use the following expression:

$$f_3 = \frac{\sum_{c=1}^{nATC} \begin{cases} 1, & \text{if } S_{ic} = S_{i(c-1)} \\ 0, & \text{if } S_{ic} \neq S_{i(c-1)} \end{cases}}{nATC}, \tag{9}$$

   where S_{ic} is the contents of the slot in the shift i and the incident slot c, whereas $S_{i(c-1)}$ is the content of the slot in the shift i and the slot before the incident slot.

4. Maintain solution-like structure: The aim is to derive a solution as compact as possible. For this, we use the following expression, where each is time slot is compared against the one on his right and under it:

$$v = \sum_{k=1}^{nATC-1} \sum_{j=1}^{nSlot-1} \begin{cases} 0.2, & \text{if } S_{kj} = S_{(k+1)j} \\ 1, & \text{if } S_{kj} = S_{k(j+1)}. \end{cases} \tag{10}$$

In Equation (10) is checked for each S_{kj}, if the slot on its right ($S_{k(j+1)}$) has the same content and if so, add 1, also checks if the slot below ($S_{(k+1)j}$) has the same content and if so adds to 0.2. If both are different, no value will be added. To get the theoretical maximum value we use the Equation (11).

$$v_{max} = (nSlot - 1) \times (nATC - 1) \times 2. \tag{11}$$

This is a theoretical maximum value without regard to constraints. Once we have this value we can normalize the target with the Equation (12).

$$f_4 = 1 - \frac{v_{max} - v}{v_{max}}. \tag{12}$$

Once all four objectives are calculated, a weighting is made between all objectives to obtain the target function:

$$f_t = \mu_1 f_1 + \mu_2 f_2 + \mu_3 f_3 + \mu_4 f_4. \tag{13}$$

CRIDA experts ranked the above objectives by importance, and the weights were derived using a rank-order centroid (ROC) method [33].

The following sections describe the adaptation of SA and VNS to our ATC work shift scheduling problem and the corresponding parameter tuning are described.

3.1. Simulated Annealing Adaptation

Simulated annealing (SA) [34,35] is a trajectory-based metaheuristic which is named for and inspired by annealing in metallurgy. SA is one of the oldest metaheuristics and has been adapted to solve many combinatorial optimization problems. Over the years, many authors have proposed both general and problem-specific improvements and variants of SA [36]. Different variants of scheduling problems have been tackled using SA [37], such as the job-shop scheduling problem [38–41], university course timetabling problems [42], or sports scheduling problems [43].

SA pseudocode for a minimization optimization problem is shown in Algorithm 1. The basic idea of SA is as follows. An initial feasible solution, x_0, is randomly generated. Then, in each iteration i, a new solution (y) is randomly generated from the neighborhood, $N(x_i)$, of the solution considered in that iteration, x_i. If the new solution is better than the current one, then the algorithm moves to that solution. Otherwise, there is some probability of it moving to a worse solution. The acceptance of worse solutions makes for a broader search for the optimal solution and avoids trapping in local optima in early iterations.

The search is initially very diversified, since practically all moves are allowed. As the temperature drops, the probability of accepting a worse moves decreases, and only better moves will be accepted when it is zero. This makes SA work like hill climbing.

Some elements in the above algorithm require clarification. Numerous studies propose techniques to calculate initial temperature (t_0) that allow an acceptance rate of about 95% of solutions. However, in this case we prefer to reduce the acceptance rate to 60% due to the nature of the problem to avoid an unnecessary scan that entails a high computation time. This allows us to perform a slower temperature drop without wasting exploration time in unpromising search spaces.

Algorithm 1 Basic SA.

Do $x^* = x_0$, $f^* = f(x_0)$, $i = 0$. Select the initial temperature t_0 (t_i temperature in step i)
repeat
  Randomly generate $y_i \in N(x_i)$
  if $(f(y_i) - f(x_i)) \leq 0$ **then**
    $x_{i+1} = y_i$
    if $(f(x^*) > f(y_i)$ **then**
      $x^* = y_i, f^* = f(y_i)$
    end if
  else
    $p \sim U(0,1)$
    if $(p \leq e^{-(f(y_i)-f(x_i))/t_i})$ **then**
      $x_{i+1} = y_i$
    else
      $x_{i+1} = x_i$
    end if
  end if
  Update temperature, $i = i + 1$
until stopping criterion

Regarding the temperature update, after running different tests looking for the best cooling function, we found that, in this case, the classic geometric function is the most promising function:

$$t_i = \alpha t_{i-1}. \tag{14}$$

The number of iterations in which the temperature is kept constant influences the evolution of the temperature. We use the cutoff method, which is to incorporate a counter of the number of improving iterations and, when it exceeds a threshold (1000 iterations), the temperature may drop before the established.

Besides, choosing solutions from a neighborhood ($y_i \in N(x_i)$) is one of the most important points for the proper operation of SA, and, for this reason, different neighborhood definitions have been tested. Several movements have been proposed for this problem and, after a comparison process, the following movement was chosen.

Starting with the initial solution, the solution is divided into different sets of slots using a grid. This grid sets the cuts at different points where there are changes from work to rest or from one sector to another. Figure 4 shows an example of how a grid would be created from an initial solution.

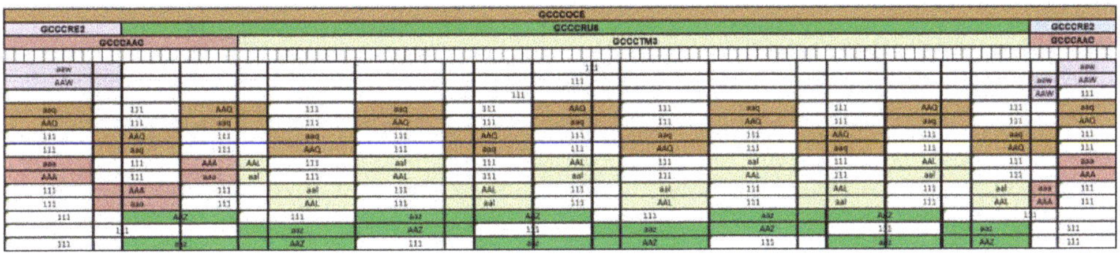

Figure 4. Example of a grid set.

The grid sets the start and end points of the different sets of slots that allow permutations along the iterations of the algorithm, and the movement is defined as the exchange of the set of slots between two distinct rows. We have an example of movement in Figure 5.

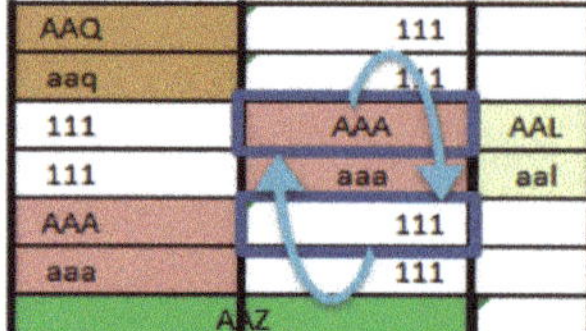

Figure 5. Two solutions of the same neighborhood.

Throughout the execution, in order to promote exploitation, the grid is modified reducing the size of the spaces to a minimum number of slots. This initially achieves greater exploration and increased exploitation as iterations progress.

The acceptance function is a method that allows us to move towards worse solutions than the current one with some probability. There are several functions to determine the likelihood of acceptance p of these solutions. Equation (15) shows the chosen function.

$$p \leq e^{-(f(y_i)-f(x_i))/t_i}, \quad (15)$$

where y_i is the a random solution in the neighborhood of actual solution x_i and t_i is the temperature in step i.

Finally, the stopping condition sets when the algorithm should stop. We use a solution improvement-related condition: the algorithm is stopped when the best solution found is not improved within a certain number of iterations. This threshold has been set at 50,000 iterations.

3.2. Variable Neighborhood Search Adaptation

Variable neighborhood search (VNS) is based on the idea of successively exploring a set of neighborhoods to solve optimization problems [44]. VNS explores neighborhoods either at random or systematically in search of local optima. Local searches of different neighborhoods should generate different local optima, and the global optimum will be a local optimum for one such neighborhood.

Algorithm 2 illustrates the basic version of VNS. There are many other variants of VNS in the literature, including variable neighborhood descent, reduced variable neighborhood search and variable neighborhood decomposition search. They differ according to:

- Neighborhood order: forward VNS starts with $k = 1$ and increases k by one if no better solutions are found; otherwise it sets $k \leftarrow 1$; backward VNS starts with $k = k_{max}$ and decreases k by one if no better solutions are found, and extended VNS uses parameters k_{min} and k_{step}, sets $k \leftarrow k_{min}$ and increases k by k_{step} if no better solution is found.
- Solution acceptance: skewed VNS accepts worse solutions if $f(x'') - \alpha d(x, x'') < f(x)$, where $d(x, x'')$ is the distance between candidate solutions.

Algorithm 2 Basic VNS.

Require: N_k: set of neighborhood structures, $k = 1, ..., k_{max}$
Ensure: Best solution found.

$k = 1$
Generate an initial solution x
repeat
  Randomly generate $x' \in N_k(x)$
  x''= Local-search(x',N_k)
  if $(f(x'') < f(x))$ **then**
    $x = x''$
    $k = 1$
  else
    $k = k + 1$
  end if
until stopping criterion

We used variable neighborhood descent (VND) [45]. This deterministic algorithm selects an initial solution and iterates through all the selected neighborhoods in the specified order, running a search process to find a local optimum in each iteration. If this local optimum is found to be better than the incumbent best solution, it becomes the new initial solution for a local search. If not, the algorithm explores the next specified neighborhood.

It includes four types of neighborhoods, see Figure 6.

- Neighborhood 1: Time slot exchange between to ATCs subject two the following constraints:
  1. ATCs remain in the same sector and position (planner or executive) for approximately 45 min.
  2. Optimally, ATCs should work for 90 min between breaks.
  3. ATCs should spend from 40% to 60% of working time in executive positions.
- Neighborhood 2: Time slot exchange between two ATCs subject to the following constraints
  1. ATCs remain in the same sector and position (planner or executive) for approximately 45 min.
  2. Optimally, ATCs should work for 90 min between breaks.
  3. ATCs should spend from 40 to 60% of working time in executive positions.
  4. Exchanged time slots must be work periods.
- Neighborhood 3: Time slot exchange between two ATCs subject to the following constraints:
  1. ATCs remain in the same sector and position (planner or executive) for approximately 45 min.
  2. Optimally, ATCs should work for 90 min between breaks.
  3. ATCs should spend from 40 to 60% of working time in executive positions.
  4. One ATC was working in the same sector and position immediately before or after the exchange time slot (no work period extension is required).
- Neighborhood 4: Time slot exchange between two ATCs subject to the following constraints:
  1. ATCs remain in the same sector and position (planner or executive) for approximately 45 min.
  2. Optimally, ATCs should work for 90 min between breaks.
  3. ATCs should spend from 40 to 60% of working time in executive positions.
  4. Exchanged time slots must be work periods.
  5. One ATC was working in the same sector and position immediately before or after the exchange time slot (no work period extension is required).

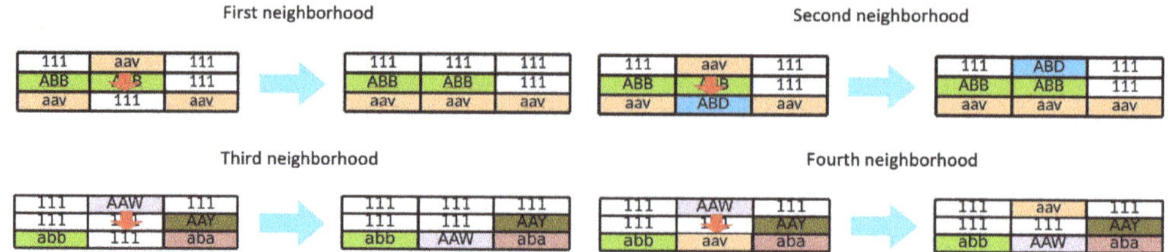

Figure 6. Neighborhood definitions.

4. An Illustrative Example

We now illustrate the operation of the problem-solving methodology explain above using a real example of application from the Barcelona control center originated during an afternoon shift, referred to henceforth as Instance 1.

Note that a control center may be responsible for managing one or more cores. Each core should be solved separately, unless there are sectors belonging to more than one core. In this case, ATCs should be assigned to the respective cores. The Barcelona control center manages two cores, the *western route* and the *eastern route* cores, and the sectors under consideration in this instance belong to only one of these cores. Thus, each core can be solved separately.

The shift lasts from 15:00 to 21:35, the sectorization established is shown in Figure 7, a 04A configuration (sectors adc, aee, acy, adu) for the *western route* core and a 05A configuration (sectors adq, adl, adp, adr, aea) for the *eastern route* core.

Figure 8 shows the solution (ATCs shifts) covering the sectorization established before the incident. The number of available ATCs is 25, 11 (C1–C11) belonging to the eastern route core and 14 (C12–C25) to the western route core.

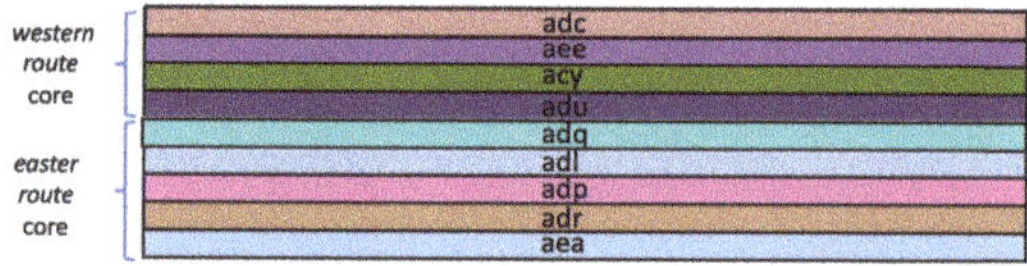

Figure 7. Sectorization of Barcelona control center (Instance 1).

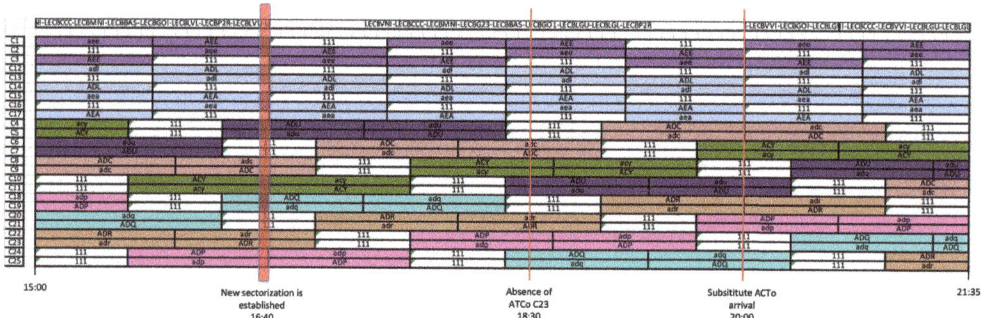

Figure 8. Initial solution before the incident of Barcelona control center (Instance 1).

The incident in question is as follows. An ATC has to leave his/her workplace, leaving a vacant slot at 18:30, that is, three and a half hours after starting the shift, see Figure 8. The manager reports the incident and notifies a substitute ATC who should provide support to his/her colleague. However, the substitute cannot get up until one and half hours later (20:00). In addition, one of the open sectors is closed as a result of an expected decline in air traffic. In response to the incident, a new sectorization is established at 16:40, as shown in Figure 9.

In the new sectorization, there are six open sectors in the Barcelona western route core: a 04A configuration (sectors adc, aee, acy, adu) from 15:00 to 20:00, and a 03A configuration (sectors adc, adt, aef) from 20:00 to 21:35; and 10 sectors open in the Barcelona eastern route core: a 05A configuration (sectors adl, adp, adr, adq, aea) from 15:00 to 16:40, a 05C configuration (sectors aea, adi, adn, adh, adm) from 16:40 to 20:00, a 04A configuration (sectors adl, aea, adn, adm) from 20:00 to 20:40, and a 03A configuration (sectors adn, adm, aeb) from 20:40 to 21:35 (see Figure 9).

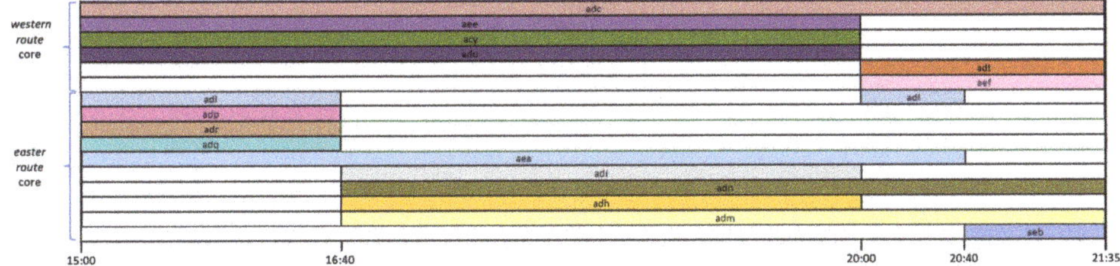

Figure 9. Sectorization after the incident of Barcelona control center (Instance 1).

The first phase of the problem-solving methodology modifies the current ATC working time (shown in Figure 8) to output a new schedule including all the modifications required to deal with the unforeseen event.

The solution derived from the first phase is shown in Figure 10. Even though it includes four more than the available ATCs (denoted as C0, see the first four lines in Figure 10) and is not feasible, this solution accounts for all the necessary changes to meet the new working hour requirements and will be the starting point of the second phase. The solution shows that ATC C23 (ATC absent from his/her position at 18:30) and C26 (substitute ATC getting up at 20:00) are not working the shift, neither will be able to assume workload from 16:30 to 20:00. ATCs not working the shift are represented by the string "000".

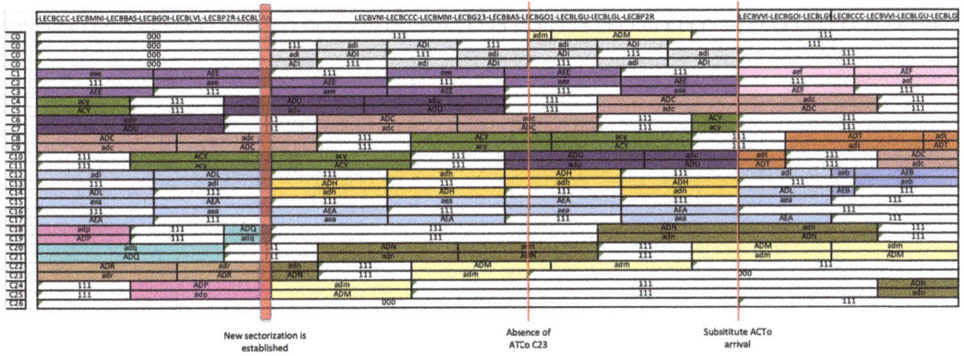

Figure 10. Solution derived from Phase 1 for Instance 1.

In the second phase, SA and VNS are used to derived a final solution.

In SA, the parameters are set as follows. Initial temperature is $t_0 = 0.035$, which stands for a low acceptance rate. This is because the time available to solve the problem is limited, and it is inconvenient to break the built-in structures of the solution. The proposed cooling function uses a cooling ratio of $\alpha = 0.95$, and the temperature is reduced every $L = 3000$ iterations. Besides, the cutoff method lowers the temperature without having to wait for the specific number of iterations if 1500 iterations consecutively improve the current solution.

Finally, SA and VNS both use the following stopping criteria: both stop the execution if there is no improvement of 0.02% in the target value corresponding to the best solution found for 50,000 iterations.

At the end of the second phase, we have the best solution so far. Figures 11 and 12 show the solution reached by SA and VNS, respectively, where the four artificial ATCs (C0) are no longer present as the other ATCs have taken on their workload. In addition, Table 1 shows the values for the different objective functions and the respective computation times.

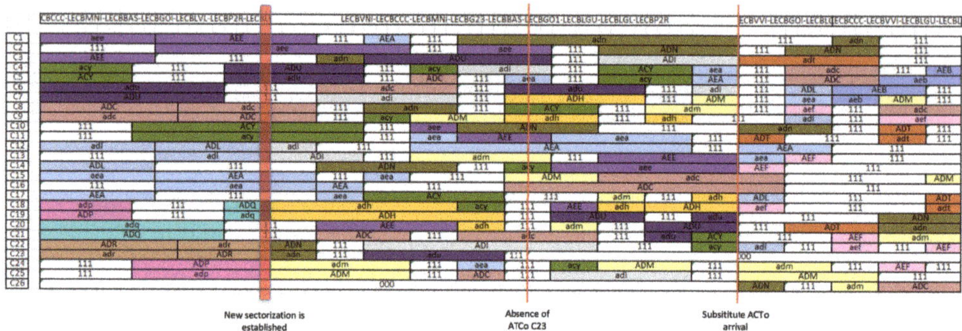

Figure 11. Optimal solution derived from Phase 2 using simulated annealing (SA).

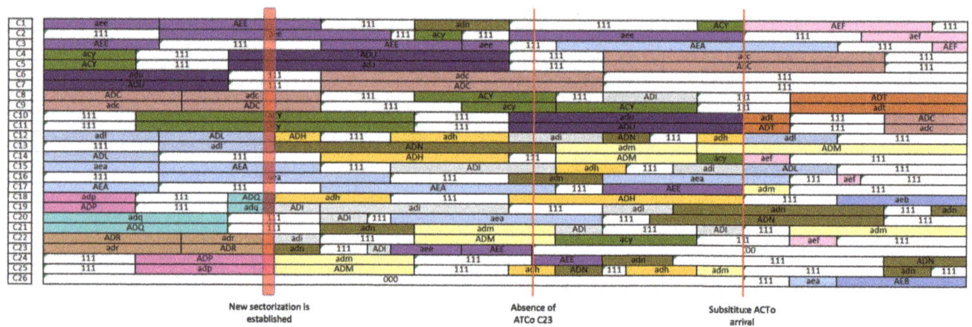

Figure 12. Optimal solution derived from Phase 2 using variable neighborhood search (VNS).

Table 1. Results for Instance 1 using SA and VNS.

	Metaheuristic	f	f_1	f_2 (Violated Constraints)	f_3 (Changes)	f_4	Computation Time (min.)
Instance 1	SA	0.9316	1.0000	1.0000 (0)	0.6923 (8)	0.6293	10.45
	VNS	0.9037	1.0000	0.9219 (26)	0.6923 (8)	0.5165	1.02

Note that a value of $f_1 = 1$ means that the solution uses no more than the number of available ATCs, whereas $f_2 = 1$ means that all the ATC working conditions are met, and, consequently, the solution is feasible. Thus, the solution reached by SA is feasible, whereas the solution derived by VNS uses the available number of ATCs but violates 26 constraints.

The proportion of ATCs who do not change their position during the incident slot is the same in both solutions (0.6923). Eighteen out of the 26 ATCs are kept in the same position just after the sectorization change (incident slot). The SA solution outperforms the solution derived by VNS with regard to f_4, that is, the SA solution is more compact, which makes it easier to understand for the NMOC (see Figures 11 and 12).

Finally, it takes considerable less time for VNS to reach the solution (1.02 min) than SA (10.45 min). We can conclude that SA outperforms VNS with regard to the solution quality (f_2 and f_4), but VNS reaches the final solution faster. However, the SA computation time also satisfies CRIDA experts.

Table 2 shows information about the constraint violation distribution using VNS. Priority constraints are highlighted in bold. Four out the fourteen constraint types are not met, and two priority constraints (LC5 and LC12) are not met four and eight times, respectively.

Table 2. Constraint violation distribution for Instance 1 using VNS.

	LC1	LC2	LC3	LC4	LC5	LC6	LC7	LC8	LC9	LC10	LC11	LC12	LC13	LC14	Total
Instance 1	0	0	0	0	4	0	0	4	0	0	10	8	0	0	26

5. Numerical Analysis

Simulated annealing (SA) and variable neighborhood search (VNS) performance has been tested on a set of representative instances provided by CRIDA experts (see Table 3).

Table 3. Instance description.

	Control Center	Cores	Shift	Available ATCs	Incident
Instance 1	Barcelona	2	Afternoon (15:00–21:35)	25	Closure of sectors in both cores + Absent ATC + substitute ATC
Instance 2	Barcelona	2	Morning (7:30–15:00)	24	Additional sector opened at 10:30
Instance 3	Madrid	1	Afternoon (15:30–22:30)	19	Different sectors are closed throughout the shift
Instance 4	Madrid	1	Afternoon (15:30–22:30)	19	Different sectors are closed throughout the shift + absent ATC + substitute ATC
Instance 5	Madrid	1	Afternoon (15:30–22:30)	19	Different sectors are opened and closed throughout the shift
Instance 6	Gran Canaria	1	Morning (7:00–15:00)	22	Absent ATC at 10:10
Instance 7	Gran Canaria	1	Morning (7:00–15:00)	22	Absent ATC + substitute ATC

Instance 1 was used to illustrate the proposed methodology in Section 3. Instance 2 also refers to the Barcelona control center but is centered on a morning shift with a different number of available ATCs and a different incident. Instances 3, 4 and 5 focus on the Madrid control center on an afternoon shift with the same number of available ATCs but different incidents. Finally, instances 6 and 7 address the Gran Canaria control center during a morning shift with 25 available ATCs and one absent ATC absence, in the first case, plus a substitute ATC in the second instance.

In the following sections, we describe Instances 2–9 in detail, show the solution before the incident, plus the solutions derived from Phases 1 and 2 and compare the solutions reached using SA and VNS.

5.1. Instance 2

Instance 2 is a real case of a morning shift turn at Barcelona control center from 7:30 to 15:00. Two cores are managed simultaneously, as in Instance 1. The number of available ATCs is 24, seven belonging to the eastern route core and 17 to the western route core.

In the new sectorization, the western route core does not undergo any change with three open sectors throughout the shift. However, the eastern route core moves from five to six open sectors (sector adl is closed and sectors adg and adk are opened) at 10:30, which remains so for four and a half hours until the end of the shift.

Figure 13 shows the solution before the incident, the solution derived in Phase 1 and the solution derived in Phase 2 using SA for Instance 2.

The first phase modifies the initial working time of ATCs to output a new schedule incorporating all modifications to deal with the unforeseen event. Three more ATCs than are available (denoted as

C0, see the first three lines in Figure 13) are used to cover the new open sector adk. Therefore, the solution is not feasible.

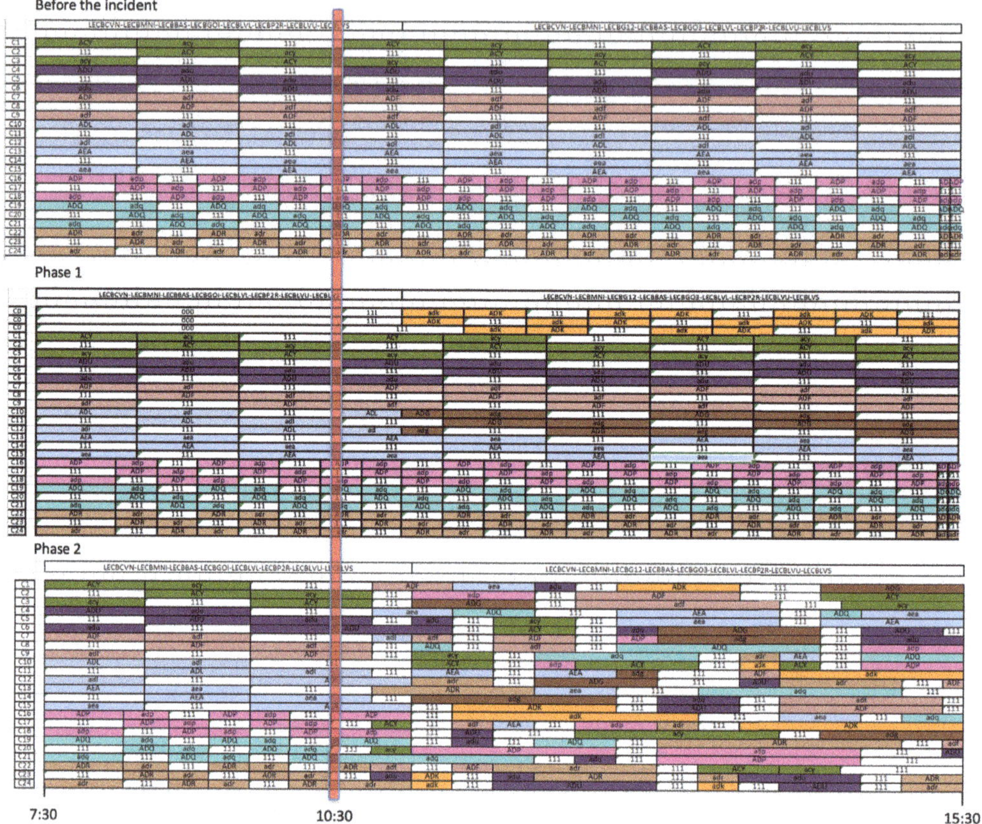

Figure 13. Solution before the incident, solution derived in Phase 1 and solution for Instance 2.

Table 4 shows the values of the different objective functions for the SA and VNS solutions in the seven instances under consideration and their computation times.

As mentioned for Instance 1, a value of $f_1 = 1$ means that the solution uses no more than the available number of ATCs, whereas $f_2 = 1$ means that all the ATC working conditions are met, and, consequently, the solution is feasible. Thus, the solutions reached by SA and VNS in Instance 2 use the number of available ATCs but violate six and 49 constraints, respectively. All ATCs remain in the same position during the incident slot in both solutions ($f_2 = 1$), and the SA solution outperforms the solution derived using VNS with regard to f_4, that is, the SA solution is more compact. This makes it easier to understand for the NMOC.

Although Instance 2 looks simple, the incident is very complex to manage. The solution before the incident is composed of 24 ATCs covering 9 (3 $\times$ (3 $\times$ 8)) sectors, and the percentage break time for ATCs is 25%, which exactly matches a labor constraint. Thus, ATCs do not have even one extra minute of break time. Besides, it accounts for two cores simultaneously, and the different ATCs can only be assigned to the sectors belonging to the core in which they are working. This further restricts the problem resulting in very few feasible solutions, as the working margin is very small. Consequently, it should be no surprise if a feasible solution is not reached.

Finally, the time it takes VNS to reach the solution (1.83 min) is considerably less than for SA (13.32 min). The latter is considered very satisfactory by CRIDA experts.

Table 4. Results for the instances under consideration using SA and VNS.

		f	f_1	f_2 (Violated Constraints)	f_3 (Changes)	f_4	Computation Time
Instance 1	SA	0.9316	1.0000	1.0000 (0)	0.6923 (8)	0.6293	10.45
	VNS	0.9037	1.0000	0.9219 (26)	0.6923 (8)	0.5165	1.02
Instance 2	SA	0.9752	1.0000	0.9918 (6)	1.0000 (0)	0.6238	13.32
	VNS	0.9191	1.0000	0.8188 (49)	1.0000 (0)	0.4670	1.83
Instance 3	SA	0.9115	1.0000	0.9945 (2)	0.4736 (10)	0.8664	6.35
	VNS	0.9018	1.0000	0.9945 (2)	0.4736 (10)	0.7045	0.13
Instance 4	SA	0.9439	1.0000	1.0000 (0)	0.7894 (4)	0.5660	6.21
	VNS	0.9503	1.0000	0.9680 (4)	0.8947 (3)	0.5660	0.68
Instance 5	SA	0.9280	1.0000	0.9379 (6)	0.7894 (4)	0.6062	11.18
	VNS	0.9105	1.0000	0.8805 (9)	0.7894 (4)	0.5737	0.47
Instance 6	SA	0.9545	1.0000	0.9660 (8)	0.9047 (2)	0.6223	11.52
	VNS	0.9128	1.0000	0.8295 (18)	0.9047 (2)	0.5426	0.53
Instance 7	SA	0.9593	1.0000	1.0000 (0)	0.9047 (2)	0.5398	7.68
	VNS	0.9136	1.0000	0.8240 (18)	0.9047 (2)	0.5691	0.63

Table 5 shows information about the constraint violation distribution using both SA and VNS in the instances under consideration. Priority constraints are highlighted in bold.

In Instance 2, as mentioned above, the SA and VNS solutions failed to met six and 49 labour constraints, respectively. The six constraints are not met by the SA solution are priority constraints (LC9 three times and LC12 three times). Twenty-five out of the 46 constraints not met by the VNS solution are priority constraints (LC3 and LC5 five times each, LC9 once and LC12 14 times). SA clearly outperforms VNS with respect to this objective.

Table 5. Constraint violation distribution.

		LC1	LC2	LC3	LC4	LC5	LC6	LC7	LC8	LC9	LC10	LC11	LC12	LC13	LC14	Total
Instan. 1	SA	0	0	0	0	0	0	0	0	0	0	0	0	0	0	0
	VNS	0	0	0	0	4	0	0	4	0	0	10	8	0	0	26
Instan. 2	SA	0	0	0	0	0	0	0	0	3	0	0	3	0	0	6
	VNS	0	0	5	0	5	0	0	4	1	3	17	14	0	0	49
Instan. 3	SA	0	0	0	0	0	0	0	0	0	0	2	0	0	0	2
	VNS	0	0	0	0	0	0	0	0	0	0	2	0	0	0	2
Instan. 4	SA	0	0	0	0	0	0	0	0	0	0	0	0	0	0	0
	VNS	0	0	1	0	1	0	0	1	0	1	0	0	0	0	4
Instan. 5	SA	0	0	0	0	4	0	0	0	0	2	0	0	0	0	6
	VNS	0	0	1	0	7	0	0	0	0	0	1	0	0	0	9
Instan. 6	SA	0	0	1	0	1	0	0	0	0	4	0	2	0	0	8
	VNS	0	0	5	0	8	0	0	0	0	0	0	5	0	0	18
Instan. 7	SA	0	0	0	0	0	0	0	0	0	0	0	0	0	0	0
	VNS	0	0	6	0	8	0	0	1	0	1	1	1	0	0	18
Total	SA	0	0	1	0	5	0	0	0	3	6	2	5	0	0	22
	VNS	0	0	32	0	48	0	0	18	2	5	41	42	1	0	189

5.2. Instance 3

Instance 3 refers to a real case of an afternoon shift at Madrid control center from 15:30 to 22:30. Only one core is managed, and the number of available ATCs is 19.

The incident involves the closure of several sectors, moving from a sectorization with seven open sectors throughout the shift to seven sectors open until 20:40. At this time, two sectors are closed and the other five sector are kept open until 21:00, when another sector is closed. Then, the four sector are kept open until 21:40, when another sector is closed until the end of the shift (three open sectors).

Figure 14 shows the solution before the incident, the solution derived in Phase 1, and the solution derived in Phase 2 using SA on Instance 3. Artificial ATCs (C0) do not have to be used in Phase 1.

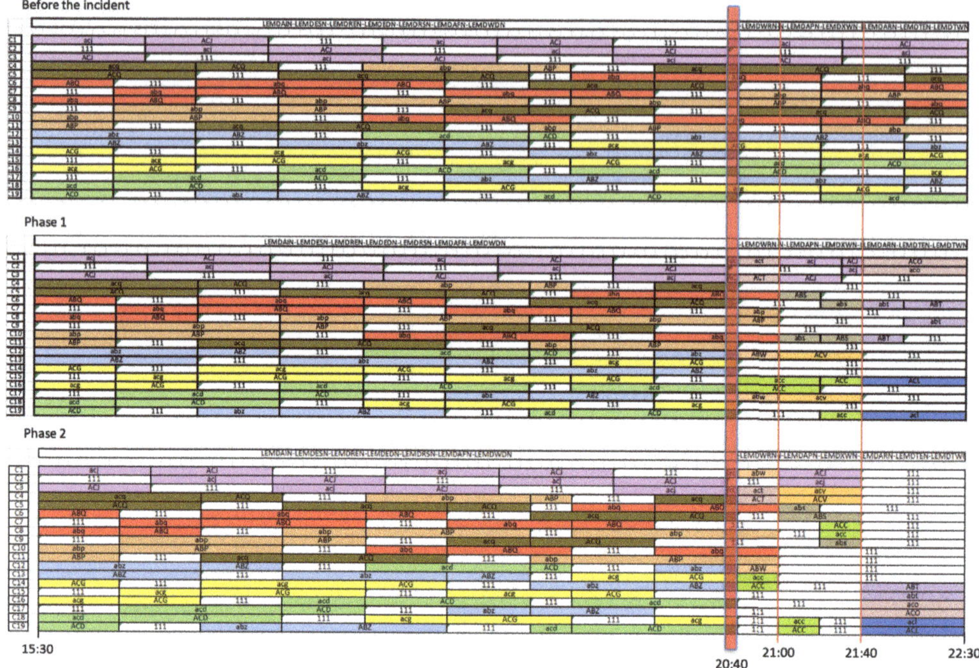

Figure 14. Solution before the incident, solution derived in Phase 1 and solution for Instance 3.

Looking at Table 4, we find that the solutions reached by SA and VNS for Instance 3 both use the available number of ATCs, but violate two labor constraints, corresponding to LC11 (twice). LC11 is not a priority constraint (Table 5).

Note that a feasible solution can never be reached for Instance 3 due to the position of the incident slot. The minimum consecutive working time of ATCs working in the sector that is closed as a consequence of the incident is irreparably violated, and the problem-solving method cannot modify any slot in the solution before the incident slot. The proportion of ATCs who remain in the same position during the incident slot is the same in both solutions (0.4736). Nine out of the 19 ATCs remain in the same position just after the sectorization change (incident slot).

The SA solution outperforms the solution derived by VNS with regard to f_4, that is, the SA solution is more compact, which makes it easier to understand for the NMOC. Finally, the time it takes for VNS to reach the solution (0.13 min) is slightly lower than for SA (6.35 min)—The latter is considered very satisfactory by CRIDA experts.

The solutions derived by SA and VNS for Instance 3 are equal with regard to objectives f_1, f_2 and f_3, where the SA solution is better in terms of compactability, and VNS has a lower computation time (SA computation time also satisfies the CRIDA experts).

5.3. Instance 4

Instance 4 refers to a real case of an afternoon shift at Madrid control center from 15:30 to 22:30. Only one core is managed, and the number of available ATCs is 19.

The incident involves the closure of several sectors, like Instance 3. It passes from a sectorization with seven open sectors throughout the shift to seven sectors open until 20:40. At this time, two sectors are closed, and five sectors are kept open until 21:00, when another sector is closed. Then, the four sectors are kept open until 21:40, when another sector is closed until the end of the shift (three open sectors).

Additionally, an ATC (C8) has to leave his/her position, leaving a vacant slot at 17:00, that is, two and a half hours after starting the shift. The manager reports the incident and notifies a substitute ATC who should provide support to his/her colleague, but the substitute cannot get up until three hours and 20 min later (20:20), Figure 15.

Figure 15 shows the solution before the incident, the solution derived in Phase 1, and the solution derived in Phase 2 using SA on Instance 4.

The first phase modifies the initial working time of ATCs to output a new schedule incorporating all modifications to deal with the unforeseen event. One more than the available ATCs (denoted as C0, see the first line in Figure 15) is used, and it is, therefore, a feasible solution.

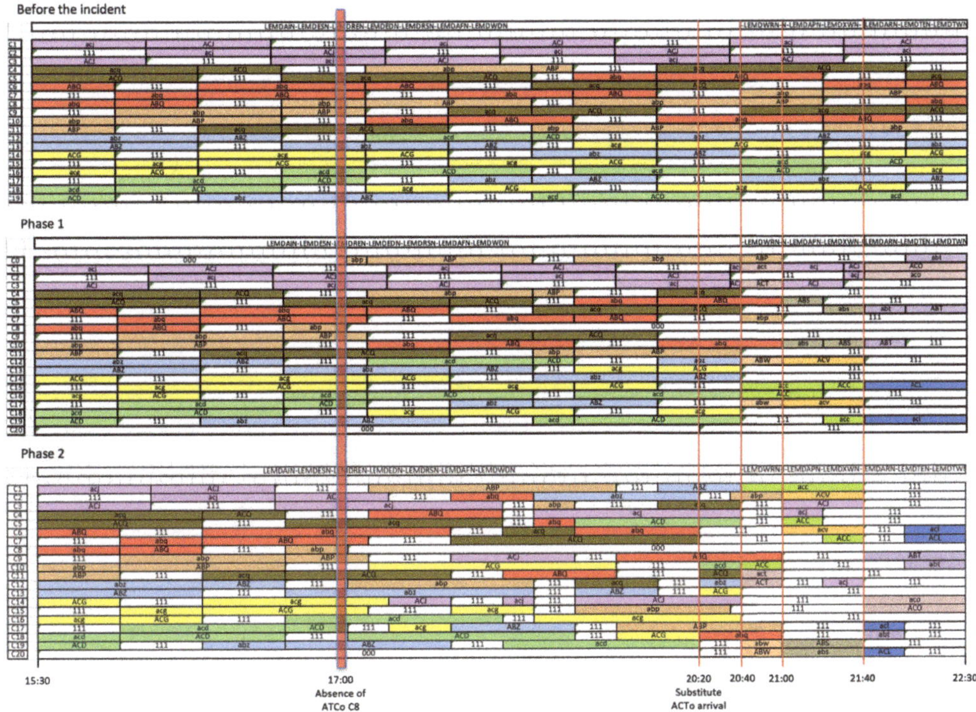

Figure 15. Solution before the incident, solution derived in Phase 1 and solution for Instance 4.

Looking at Table 4, we find that the solution reached by the SA is feasible, whereas the solution derived by the VNS uses the number of available ATCs but does not met four constraints, violating LC3, LC5, LC8 and LC11 once each, where the first two are priority constraints (Table 5). The proportion of ATCs who remain in the same position during the incident slot is higher in the VNS solution, where 16 out of the 19 ATCs remain in the same position just after the sectorization change (incident slot). There is an additional change in the SA solution.

The SA and VNS solutions are equal in terms of compactability, f_4. However, VNS is again faster at reaching the solution (0.6 vs. 6.21 min). Given that the SA computation time also satisfies the CRIDA experts, the SA solution involves no more than an additional control center change (in the incident slot) but is a feasible solution (the VNS solution violates four of the labor constraints), the SA solution was selected for Instance 4 by CRIDA experts.

5.4. Instance 5

Instance 5 refers to a real case from the afternoon shift at Madrid control center from 15:30 to 22:30. Only one core is managed, and the number of available ATCs is 19.

The incident in question involves the closure and opening of several sectors. It passes from a sectorization with seven open sectors throughout the shift to seven open sectors until 17:40. At this time, an additional sector is opened, and the eight sectors are kept open until 20:40, when a sector is closed. Then, the seven sectors are kept open until 21:00, when the sectorization changes and only four open sectors are established. A sector is closed 20 min later (21:20), and the three sectors are kept open until the end of the shift.

Figure 16 shows the solution before the incident, the solution derived in Phase 1, and the solution derived in Phase 2 using SA on Instance 5.

In the first phase, three more ATCs than available (denoted as C0, see the first three lines in Figure 16) are incorporated to manage the newly opened sector, abv.

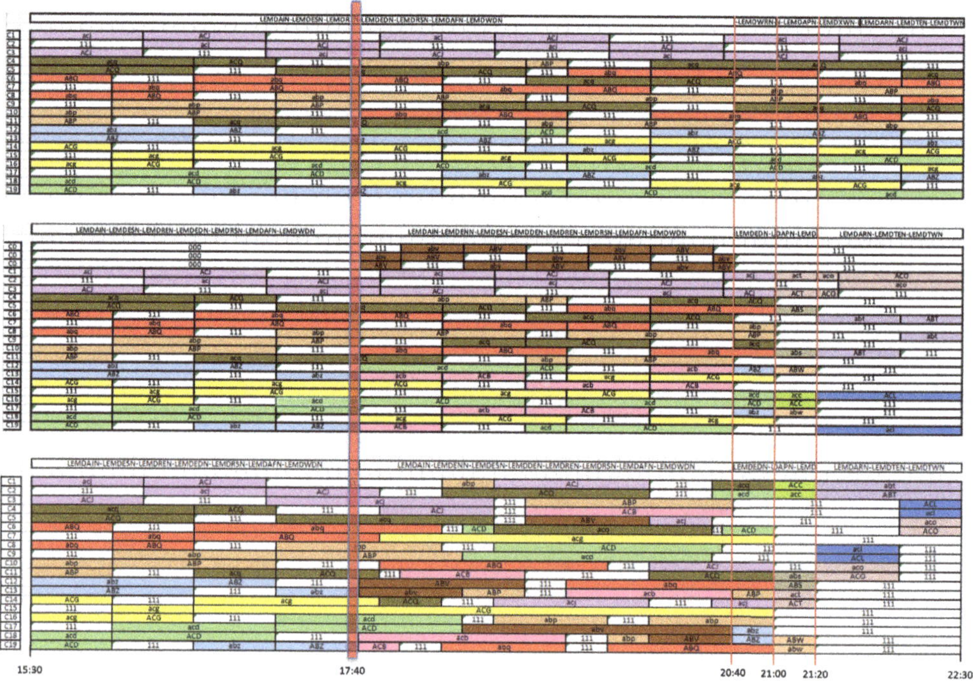

Figure 16. Solution before the incident, solution derived in Phase 1 and solution for Instance 5.

Looking at Table 4, we find the solutions reached by SA and VNS for Instance 2 use the available number of ATCs but violate six and nine constraints, respectively. Four out the six labor constraints not met by the SA solution are priority constraints (LC5 four times). C12 is also violated twice (Table 5). Eight out the nine labor constraints not met by the VNS solution are priority constraints (LC3 once and LC5 seven times).

The incident in Instance 5 involves a high workload. Therefore, it is impossible to find a feasible solution because the difference between the workload (8 sectors $\times$ 2 positions $\times$ 36 slots (3 h) = 576) and the ATC-assumed work ((19 ATCs $\times$ 36 slots $-$ 6 slots (mandatory timeout)) = 570) is 576 $-$ 570 $=$ 6 slots, without counting the ATC labor constraints. The proportion of ATCs who do not change their position during the incident slot is 0.7894 in both solutions. Fifteen out of the 19 ATCs remain in the same position just after the sectorization change (incident slot).

The SA solution outperforms the solution derived by VNS with regard to f_4, that is, the SA solution is more compact, which makes it easier to understand for the NMOC. However, VNS is again faster at reaching the solution (0.47 vs. 11.18 min). SA outperforms VNS with regard to objectives f_2 and f_4, but VNS has a lower computation time. As SA computing time also satisfies the CRIDA experts, the SA solution is chosen for Instance 5.

5.5. Instance 6

Instance 6 is a real case from the morning shift at Gran Canaria control center. An 8A sectorization is used throughout the shift, from 7:00 to 15:00. There are 22 available ATCs to cover this sectorization, but ATC C16 is relieved at 10:10, and no substitute is available. Therefore, the remainder of the shift is covered by 21 ATCs. Figure 17 shows the solution before the incident, the solution derived in Phase 1, and the solution derived in Phase 2 using SA.

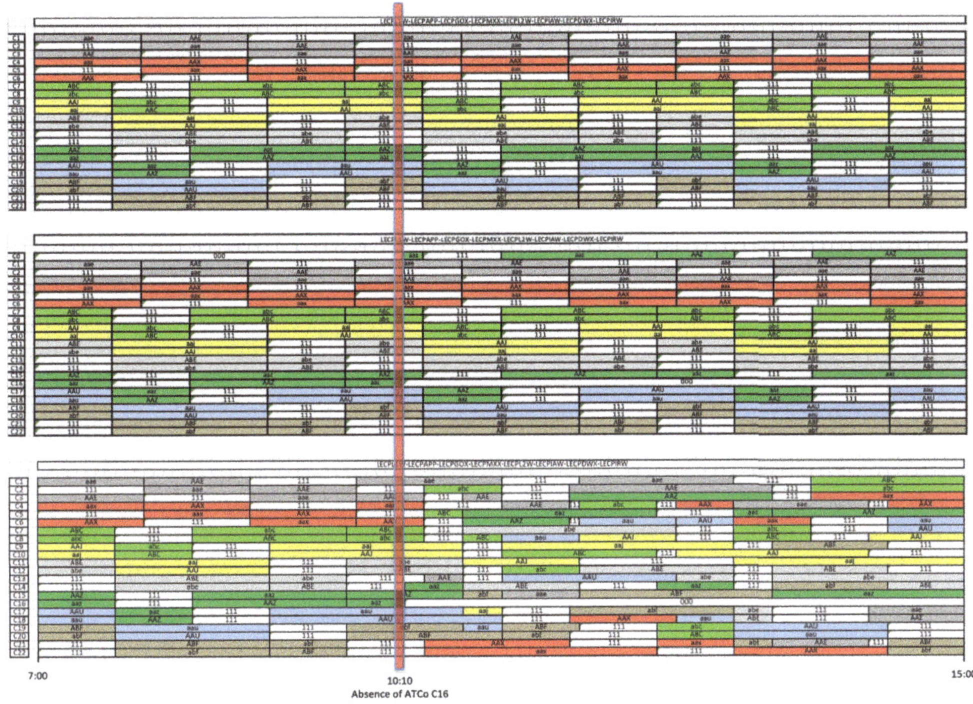

Figure 17. Solution before the incident, solution derived in Phase 1 and solution for Instance 6.

In the Phase 1 solution, the workload of ATC 16 is assigned as of 10:10 to ATC C0 (artificial ATC). This workload is distributed among the other ATCs in Phase 2, and the artificial ATC is no longer necessary.

In Instance 6, the solutions reached by SA and VNS for Instance 2 use the number of available ATCs but violate eight and 18 constraints, respectively, see Table 4. Four out of the eight labor constraints not met by the SA solution are priority constraints (LC3 once, LC5 once and LC10 four times). LC10 is also violated four times (Table 5). The 18 labor constraints not met by the VNS solution are priority constraints (LC3 five times, LC5 five times and LC12 five times).

The solution for Instance 6 involves a high workload. Therefore, it is impossible to find a feasible solution because the difference between the workload (8 sectors × 2 positions × 96 slots (8 h) = 1536) and the ATC-assumed work ((22 ATCs × 38 slots + 21 ATCs × 58 slots) × 3/4 (25% mandatory timeout)

= 1540) is 1540 − 1536 = 4 slots. The proportion of ATCs who do not change their position during the incident slot is 0.9047 in both solutions. Nineteen out of the 21 ATCs remain in the same position when ATC 16 is relieved.

The SA solution outperforms the solution derived from VNS with regard to f_4, that is, the SA solution is more compact, which makes it easier to understand for the NMOC. However, VNS is again faster at reaching the solution (0.53 vs. 11.52 min). Given that the SA computation time also satisfies the CRIDA experts, and the SA solution violates fewer labor constraints and is more compact, this solution was selected for Instance 6 by CRIDA experts.

5.6. Instance 7

Instance 7 is very similar to Instance 6. It is a real case of a morning shift at Gran Canaria control center. An 8A configuration is used throughout the shift, from 7:00 to 15:00. There are 22 available ATCs to cover this sectorization, but ATC 16 is relieved at 10:10, and the substitute (ATC 23) is not available until 12:40. Figure 18 shows the solution before the incident, the solution derived in Phase 1, and the solution derived in Phase 2 using SA for Instance 7.

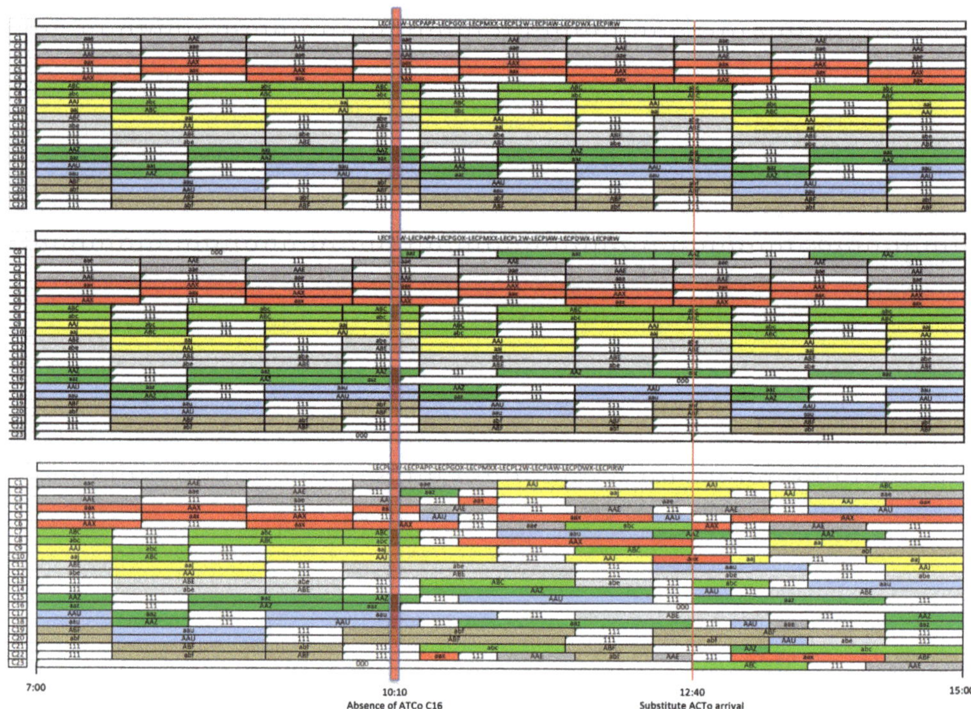

Figure 18. Solution before the incident, initial solution, and solution for Instance 7.

In the Phase 1 solution, the workload of ATC 16 is assigned as of 10:10 to ATC C0 (artificial ATC). In addition, ATC 23 is entered at 12:40 as resting. This workload is distributed among the other ATCs in Phase 2, the artificial ATC is no longer necessary, and ATC 23 has been assigned the work of other ATCs as of 12:40.

In Instance 7, the solution reached by SA is feasible, whereas the solution derived by VNS uses the available number of ATCs but violates 18 priority constraints (Table 4), corresponding to LC3 (six times), LC5 (eight times) and LC12 (once), and LC8, LC10 and LC11 once, see Table 5. The proportion of ATCs who do not change their position during the incident slot is 0.9047 in both solutions. Nineteen out of the 21 ATCs remain in the same position when the ATC 16 is relieved.

The VNS solution outperforms the solution derived by SA with regard to f_4, that is, the VNS solution is more compact, which makes it easier to understand for the NMOC. Moreover, VNS is again faster at reaching the solution (0.63 vs. 7.68 min). Although the VNS solution is faster and slightly more compact than the SA solution, the CRIDA experts prefer the SA solution since it is feasible and the computation time is satisfactory.

6. Conclusions

This paper proposes a methodology for reassigning available ATCs to air sectors after the occurrence of an incident, such as one or more ATCs having to be relieved and/or a change in sectorization due to an increase in air traffic.

The methodology has been applied to a set of seven representative instances corresponding to three different control centers in Spain, provided by CRIDA experts. Although solutions including the available number of ATCs are achieved for all the instances using the SA and VNS metaheuristics, the solutions derived by SA are feasible in three out of the seven instances, whereas no feasible solutions are reached by VNS in any of the seven instances under consideration.

The SA-derived solutions outperform VNS solutions in terms of both the number of violated constraints in all seven instances and solution compactability in six out the seven instances, and both are very similar with regard to the number of control center changes at the time of the incident. VNS computation times are clearly better than for SA. However, CRIDA experts were also satisfied with SA computation times. Thus, SA solutions are preferred.

The proposed methodology constitutes an improvement on the current approach adopted by the network managers, who in an attempt to find a solution as quickly as possible, forsake optimization, often leading to systematic violation of ATC labor conditions. Using the proposed methodology, an optimization process outputs a solution, aimed at meeting as many labor conditions as possible and taking into account the relative importance of the violation of each constraint, in a relatively short time (and the stopping criterion could be modified to output the best solution found in a specified time).

The next step before the methodology can be applied is its integration into the information systems at the network manager operations centers to analyze other types of incidents and compare, in batch mode, the quality of the real solutions applied by the network managers against solutions provided by the proposed methodology.

Besides, network managers will be asked to evaluate the solutions derived by the proposed methodology, and their opinions will be useful for tuning some parameters of the methodology like the relative importance of objectives or how they are measured.

Author Contributions: Conceptualization, A.J.-M., F.T., and A.M.; methodology, F.T.; software, F.T.; validation, F.T., A.J.-M. and A.M.; formal analysis, A.J., F.T. and A.M.; investigation, A.J.-M., F.T. and A.M.; writing—original draft preparation, A.J.-M. and A.M.; writing—review and editing, A.J.-M. and A.M.; visualization, F.T., A.J.-M. and A.M.; supervision, A.J.-M. and A.M.; project administration, A.J.-M. and A.M.; funding acquisition, A.J.-M. and A.M. All authors have read and agreed to the published version of the manuscript.

Funding: This research was funded by Spanish Ministry of Economy and Competitiveness project grant number MTM2017-86875-C3-3-R. The authors are also grateful to Sergio Flórez, who implemented the VNS algorithm and applied simulation techniques for his M.S. final project, and José Miguel de Pablo, Andrés Perillo and Álvaro Rodríguez, from the Reference Center for Research, Development and Innovation in ATM, whose expertise enhanced the reported research.

Conflicts of Interest: The authors declare no conflict of interest.

References

1. IATA Forecasts Passenger Demand to Double Over 20 Years. Available online: http://www.iata.org/pressroom/pr/Pages/2016-10-18-02.aspx (accessed on 17 July 2019).

2. Niarchakou, S. *NMOC ATFCM Operations Manual: Network Operations Handbook*, 21th ed.; EUROCONTROL: Brussels, Belgium, 2017.
3. Ivanov, N.; Jovanovic, R.; Fichert, F., Strauss, A.; Starita, S.; Babic, O.; Pavlovic, G. Coordinated capacity and demand management in a redesigned Air Traffic Management value-chain. *J. Air Trans. Manag.* **2019**, *75*, 139–152. [CrossRef]
4. European Commission. *Accelerating the Implementation of the Single European Sky, Communication from the Commission to the European parlament, the Council, the European Economic and Social Committee and the Committee of the Regions*; European Commission: Strasbourg, France, 2013.
5. Eurocontrol. 2017 overview & 2018 outlook airspace user perspective. In Proceedings of the Network Manager User Forum (NM UF 2018), Brussels, Belgium, 24th January 2018.
6. Eurocontrol. *PRC Performance Review Report: an Assessment of Air Traffic Management in Europe during the Calendar Year 2016*; EUROCONTROL: Brussels, Belgium, 2017.
7. Malakis, S.; Psaros, P.; Kontogiannis, T.; Malaki, C. Classification of air traffic control scenarios using decision trees: insights from a field study in terminal approach radar environment. *Cogn. Technol. Work* **2020**, *22*, 159–179. [CrossRef]
8. Eurocontrol. *Strategic, Pre-Tactical, Tactical and Post-Ops Air Traffic Flow and Capacity Management*; Technical Report; EUROCONTROL: Brussels, Belgium, 2017.
9. Di Gaspero, L.; Gartner, J.; Kortsarz, G.; Musliu, N.; Schaerf, A.; Slany, W. The minimum shift design problem. *Ann. Oper. Res.* **2007**, *155*, 79–105. [CrossRef]
10. Musliu, N.; Schaerf, A.; Slany, W. Local search for shift design, *Eur. J. Oper. Res.* **2004**, *153*, 51–64. [CrossRef]
11. Aykin, T. A comparative evaluation of modeling approaches to the labor shift scheduling problem, *Eur. J. Oper. Res.* **2000**, *25*, 381–397. [CrossRef]
12. Bechtold, S.E.; Jacobs, L.W. Implicit modeling of flexible break assignments in optimal shift scheduling. *Man. Sci.* **1990**, *36*, 1339–1351. [CrossRef]
13. Rekik, M.; Cordeau, J-F.; Soumis, F. Implicit shift scheduling with multiple breaks and work stretch duration restrictions. *J. Schedul.* **2010**, *13*, 49–75. [CrossRef]
14. Tellier, P.; White, G. Generating personnel schedules in an industrial setting using a Tabu Search algorithm. In *Proceedings of the 6th international Conference on Practice and Theory of Automated Timetabling*; Springer-Verlag: Berlin Heidelberg, Germany, 2006, 293–302.
15. Dechter, R.; Meiri, I.; Pearl, J. Temporal constraint networks. *Artif. Intell.* **1991**, *49*, 61–95. [CrossRef]
16. Musliu, N.; Schafhauser, W.; Widl, M. A memetic algorithm for a break scheduling problem. In *Hybrid Metaheuristics (Lecture Notes in Computer Science)*; Springer: Berlin/Heidelberg, Germany, 2010; pp. 133–147.
17. Widl, M.; Musliu, N. An improved memetic algorithm for break scheduling. In *Hybrid Metaheuristics (Lecture Notes in Computer Science)*; Springer: Berlin/Heidelberg, Germany, 2010; p. 6373.
18. Fonseca, G.H.; Santos, H.G.; Carrano, E.G. Integrating matheuristics and metaheuristics for timetabling. *Comput. Oper. Res.* **2016**, *74*, 108–117. [CrossRef]
19. Telhada, J. Alternative MIP formulations for an integrated shift scheduling and task assignment problem. *Discret. Appl. Math.* **2014**, *164*, 328–343. [CrossRef]
20. Brucker, P.; Qu, R.; Burke, E. Personnel scheduling: models and complexity. *Eur. J. Oper. Res.* **2011**, *210*, 467–473. [CrossRef]
21. Kletzander, L.; Musliu, N. Solving the general employee scheduling problem, *Comput. Oper. Res.* **2020**, *113*, 460–473.
22. Arnving, M.; Beermann, B.; Koper, B.; Maziul, M.; Mellett, U.; Niesing, C.; Vogt, J. *Managing shiftwork in European ATM, Literature Review*; European Organization for the Safety of Air Navigation, EATM Infocentre: Brussels, Belgium, 2006.
23. Eurocontrol. *Shiftwork Practices Study ATM and Related Industries*; EUROCONTROL: Brussels, Belgium, 2006.
24. Eurocontrol. *EATCHIP Human Resources Team. ATM Manpower Planning in Practice: Introduction to a Qualitative and Quantitative Staffing Methodology*; EUROCONTROL: Brussels, Belgium, 1998.
25. Stojadinovic, M. Air traffic controller shift scheduling by reduction to CSP, SAT and SAT-related Problems. In *Principles and Practice of Constraint Programming 20th Int. Conf. (CP 2014)*; Springer: Berlin/Heidelberg, Germany 2015; pp. 886–902.
26. Biere, A.; Heule, M.; van Maaren, H. *Handbook of Satisfiability; Frontiers in Artificial Intelligence and Applications*; IOS Press: Amsterdam, The Netherlands, 2009; p. 185.

27. Stojadinovic, M. Hybrid of hill climbing and SAT solving for air traffic controller shift scheduling. *J. Inform. Techn. Appl.* **2015**, *5*, 81–87. [CrossRef]
28. Conniss, R.; Curtis, T.; Petrovic, S. Scheduling air traffic controllers. In Proceedings of the 10th Int. Conf of the Practice and Theory of Automated Timetabling, York, UK, 26–29 August 2014; pp. 465–468.
29. Tello, F.; Mateos, A.; Jiménez-Martín, A. ATC work shift scheduling using multistart simulated annealing and regular expressions. In *Proc. Int. Conf. Decis. Support Syst. Tech*; IEEE Xplore: Thessaloniki, Greece, 2017; pp. 169–175.
30. Friedl, J.E.F. *Mastering Regular Expressions*; O'Reilly and Associates: Sebastopol, CA, USA, 1997.
31. Tello, F.; Mateos, A.; Jiménez-Martín, A. The air traffic controller work-shift scheduling problem in Spain from a multiobjective perspective: A metaheuristic and regular expression-based approach. *Math. Probl. Eng.* **2018**, *2018*, 15. [CrossRef]
32. Tello, F.; Jiménez-Martín, A.; Mateos, A.; Lozano, P. A Comparative Analysis of Simulated Annealing and Variable Search in the ATC Work-Shift Scheduling Problem. *Mathematics* **2018**, *7*, 636. [CrossRef]
33. Butler, J.; Olson, D.L. Comparison of centroid and simulation approaches for selection sensitivity analysis. *J. Multi-Criteria Decis. Mak.* **1999**, *8*, 146–161. [CrossRef]
34. Kirkpatrick, S.; Gelatt., C.D.; Vecchi, M.P. Optimization by simulated annealing. *Science* **1983**, *220*, 671–680. [CrossRef] [PubMed]
35. Cerny, V. Thermodynamical approach to the traveling salesman problem: An efficient simulation algorithm. *J. Optim. Theory Appl.* **1985**, *45*, 41–51. [CrossRef]
36. Franzin, A.; Stutzle, T., Revisiting simulated annealing: A component-based analysis. *Comput. Oper. Res.* **2019**, *104*, 191–206. [CrossRef]
37. Gallo, C.; Capozzi, V. A Simulated annealing algorithm for scheduling problems. *J. Appl. Math. Phys.* **2019**, *7*, 2579–2594. [CrossRef]
38. Chakraborty, S.; Bhowmik, S. An efficient approach to job shop scheduling problem using Simulated Annealing. *Int. J. Hybrid Inf. Technol.* **2015**, *8*, 273–284. [CrossRef]
39. Cruz-Chávez, M.A.; Martínez-Rangel, M.G.; Cruz-Rosales, M.H. Accelerated Simulated Annealing algorithm applied to the flexible job shop scheduling problem. *Int. Trans. Oper. Res.* **2017**, *24*, 1119–1137. [CrossRef]
40. Krishnaraj, J.; Pugazhendhi, S.; Rajendran, C.; Thiagarajan, S. Simulated Annealing Algorithms to minimise the completion time variance of jobs in permutation flowshops. *Int. J. Ind. Syst. Eng.* **2019**, *31*, 425–451. [CrossRef]
41. Sel, C.; Hamzadayi, A. A Simulated Annealing approach based simulation-optimization to the dynamic job-shop scheduling problem. *J. Eng. Sci.* **2015**, *24*, 665–674.
42. Babaei, H.; Karimpour, J.; Hadidi, A. A survey of approaches for university course timetabling problem. *Comput. Ind. Eng.* **2015**, *86*, 43–59. [CrossRef]
43. Ribeiro, C.S. Sports scheduling: Problems and applications. *Int. Trans. Oper. Res.* **2012**, *19*, 201–226. [CrossRef]
44. Mladenovic, N.; Hansen, P. Variable neighborhood search. *Comput. Oper. Res.* **1997**. *24*, 1097–1100. [CrossRef]
45. Hertz, A.; Mittaz, M. A variable neighborhood descent algorithm for the undirected capacitated arc routing problem. *J. Algorithms* **2001**, *35*, 425–434. [CrossRef]

Article

A Spectral Conjugate Gradient Method with Descent Property

Jinbao Jian [1], Lin Yang [1], Xianzhen Jiang [1,*], Pengjie Liu [2] and Meixing Liu [3]

1 College of Science, Guangxi University for Nationalities, Nanning 530006, Guangxi, China; jianjb@gxu.edu.cn (J.J.); yanglinlin1125@163.com (L.Y.)
2 College of Mathematics and Information Science, Guangxi University, Nanning 530004, Guangxi, China; liupengjie2019@163.com
3 Guangxi Colleges and Universities Key Laboratory of Complex System Optimization and Big Data Processing, Yulin Normal University, Yulin 537000, Guangxi, China; liumeixing2011@126.com
* Correspondence: yl2811280@163.com

Received: 6 January 2020; Accepted: 13 February 2020; Published: 19 February 2020

Abstract: Spectral conjugate gradient method (SCGM) is an important generalization of the conjugate gradient method (CGM), and it is also one of the effective numerical methods for large-scale unconstrained optimization. The designing for the spectral parameter and the conjugate parameter in SCGM is a core work. And the aim of this paper is to propose a new and effective alternative method for these two parameters. First, motivated by the strong Wolfe line search requirement, we design a new spectral parameter. Second, we propose a hybrid conjugate parameter. Such a way for yielding the two parameters can ensure that the search directions always possess descent property without depending on any line search rule. As a result, a new SCGM with the standard Wolfe line search is proposed. Under usual assumptions, the global convergence of the proposed SCGM is proved. Finally, by testing 108 test instances from 2 to 1,000,000 dimensions in the CUTE library and other classic test collections, a large number of numerical experiments, comparing with both SCGMs and CGMs, for the presented SCGM are executed. The detail results and their corresponding performance profiles are reported, which show that the proposed SCGM is effective and promising.

Keywords: unconstrained optimization; spectral conjugate gradient method; Wolfe line search; global convergence

1. Introduction

Conjugate gradient method (CGM) is one class of the prevailing methods commonly used for solving large-scale optimization problems, since it possesses simple iterations, fast convergence properties and low memory requirements. In this paper, we consider the following unconstrained optimization problem:

$$\min\{f(x)|\ x \in R^n\}, \tag{1}$$

where $f : R^n \to R$ is a continuously differentiable function and its gradient is denoted by $g(x) = \nabla f(x)$. The iterates of the classical CGM can be formulated as

$$x_{k+1} = x_k + \alpha_k d_k, \tag{2}$$

and

$$d_k = \begin{cases} -g_k, & \text{if } k = 1, \\ -g_k + \beta_k d_{k-1}, & \text{if } k > 1, \end{cases} \tag{3}$$

where $g_k = g(x_k)$, d_k is the search direction, β_k is the so-called conjugate parameter, and α_k is the steplength which obtained by a suitable exact or inexact line search. However, as the high cost of the exact line search, α_k is usually generated by an inexact line search, such as the Wolfe line search

$$\begin{cases} f(x_k + \alpha_k d_k) \leq f(x_k) + \delta \alpha_k g_k^T d_k, \\ g(x_k + \alpha_k d_k)^T d_k \geq \sigma g_k^T d_k, \end{cases} \tag{4}$$

or the strong Wolfe line search

$$\begin{cases} f(x_k + \alpha_k d_k) \leq f(x_k) + \delta \alpha_k g_k^T d_k, \\ |g(x_k + \alpha_k d_k)^T d_k| \leq \sigma |g_k^T d_k|. \end{cases} \tag{5}$$

The parameters δ and σ above generally are required to satisfy $0 < \delta < \sigma < 1$. As we know, different choice for β_k would generate different CGM. The most well-known CGMs are the Hestenes-Stiefel (HS, 1952) method [1], Fletcher-Reeves (FR, 1964) method [2], Polak-Ribière-Polyak (PRP, 1969) method [3,4] and the Dai-Yuan (DY, 1999) method [5], where the corresponding formulas for β_k are

$$\beta_k^{\mathrm{HS}} = \frac{g_k^T (g_k - g_{k-1})}{d_{k-1}^T (g_k - g_{k-1})}, \quad \beta_k^{\mathrm{FR}} = \frac{\|g_k\|^2}{\|g_{k-1}\|^2}, \quad \beta_k^{\mathrm{PRP}} = \frac{g_k^T (g_k - g_{k-1})}{\|g_{k-1}\|^2}, \quad \beta_k^{\mathrm{DY}} = \frac{\|g_k\|^2}{d_{k-1}^T (g_k - g_{k-1})},$$

respectively, where $\|\cdot\|$ is the standard Euclidean norm. Generally, these four methods are often referred to the classical CGMs. Under the corresponding assumptions, the authors analysed the convergence properties and tested the numerical performance of the four CGMs in References [2–6], respectively.

It is well known that the FR CGM and DY CGM possess nice convergence properties. However, their numerical performances are not so good. Inversely, the PRP and HS methods have excellent performance in practical computation. But their convergence properties are hardly to be obtained. Thus, to overcome these shortcomings of the classical CGMs, many researchers pay great attention to improve the CGMs. As a result, many improvements with excellent theoretical properties and numerical performance of the CGMs results were proposed, for example, References [7–20]. Where, the spectral conjugate gradient method (SCGM) proposed by Birgin and Martinez [12] can be seen as an important development of CGM. The main difference between SCGM and CGM lies in the computation of the search direction. The search direction of SCGM is usually yielded as follows:

$$d_k = \begin{cases} -g_k, & \text{if } k = 1, \\ -\theta_k g_k + \beta_k d_{k-1}, & \text{if } k > 1, \end{cases} \tag{6}$$

where θ_k is a spectral parameter. Obviously, for a SCGM, the selection techniques for the spectral parameter θ_k and the conjugate parameter β_k are core work and very important.

In Reference [12], after giving the concrete conjugate parameter β_k, Birgin and Martinez required the spectral search direction yielded by (6) satisfying $g_k^T d_k = -\|g_k\|^2$ (a special sufficient descent property), and then obtained the corresponding spectral parameter:

$$\theta_k = \frac{s_k^T s_k}{s_k^T y_k}, \text{ where } s_k = x_k - x_{k-1},\ y_k = g_k - g_{k-1}. \tag{7}$$

Under suitable line search, the SCGM yielded by (2) and (6) as well as (7) performs superiorly to the PRP CGM, the FR CGM and the Perry method [21].

Yielding the conjugate parameter β_k by a modified DY formula, based on the Newton's direction and the quasi-Newton equation as well as the conjugate conditions, respectively, Andrei [13] considered two approaches to generate the spectral parameter θ_k, namely,

$$\theta_k^{\mathrm{N}} = \frac{1}{y_k^T g_k}(\|g_k\|^2 - \frac{\|g_k\|^2 s_k^T g_k}{y_k^T s_k} + s_k^T g_k),$$

and

$$\theta_k^{\mathrm{C}} = \frac{1}{y_k^T g_k}(\|g_k\|^2 - \frac{\|g_k\|^2 s_k^T g_k}{y_k^T s_k}).$$

The two SCGMs associated with θ_k^{N} and θ_k^{C} are all sufficient descent without depending on any line search, and are global convergent with the Wolfe line search (4). Also, the numerical results show that the SCGM associated with θ_k^{N} is more encouraging.

Recently, by requiring the spectral direction d_k defined by (6) satisfying the special sufficient descent condition $g_k^T d_k = -\|g_k\|^2$ for general β_k, Liu et al. [15] proposed a class of choice for θ_k as follows:

$$\theta_k^{\mathrm{LFZ}} = -\frac{g_{k-1}^T d_{k-1}}{\|g_{k-1}\|} + \beta_k \frac{g_k^T d_{k-1}}{\|g_k\|^2}.$$

Under the conventional assumptions and request $|\beta_k| \leq |\beta_k^{\mathrm{FR}}|$ as well as the Wolfe line search (4), the SCGM developed by θ_k^{LFZ} is global convergent, and implemented with good computation performance.

On the other hand, Jiang et al. [19] considered to improve both the FR method and the DY method by utilizing the strong Wolfe line search (5), and achieved their good numerical effect. As a result, two schemes for the conjugate parameter are proposed, namely,

$$\beta_k^{\mathrm{IFR}} = \frac{|g_k^T d_{k-1}|}{-g_{k-1}^T d_{k-1}} \cdot \frac{\|g_k\|^2}{\|g_{k-1}\|^2}, \quad \beta_k^{\mathrm{IDY}} = \frac{|g_k^T d_{k-1}|}{-g_{k-1}^T d_{k-1}} \cdot \frac{\|g_k\|^2}{d_{k-1}^T (g_k - g_{k-1})}.$$

Interestingly, it is found, from formulas (3) and (6), that the SCGM can lead to more decrease than the classical CGM for a same β_k and any $\theta_k > 1$. Therefore, in this work, motivated by the ideas of the modified FR method and DY method [19], and making full use of the second condition of the strong Wolfe line search (5), we first introduce a new approach for yielding the spectral parameter as follows:

$$\theta_k^{\mathrm{JYJLL}} = 1 + \frac{|g_k^T d_{k-1}|}{-g_{k-1}^T d_{k-1}}. \tag{8}$$

Obviously, $\theta_k^{\mathrm{JYJLL}} \geq 1$ if the previous d_{k-1} is a descent direction.

Secondly, based on the scheme of the conjugate parameter in Reference [20] with the form

$$\beta_k^{\mathrm{JYJ}} = \frac{\|g_k\|^2 - \frac{(g_k^T d_{k-1})^2}{\|d_{k-1}\|^2}}{\|g_{k-1}\|^2 + u|d_{k-1}^T g_k|},$$

and fully absorbing the hybrid idea of Reference [11], we propose a new conjugate parameter in the following manner

$$\beta_k^{\mathrm{JYJLL}} = \frac{\|g_k\|^2 - \frac{(g_k^T d_{k-1})^2}{\|d_{k-1}\|^2}}{\max\{\|g_{k-1}\|^2, d_{k-1}^T (g_k - g_{k-1})\}}. \tag{9}$$

So far, a basic conception of our SCGM has been formed. As a result, a new SCGM is proposed, and the theoretical features and numerical performance is analysed and reported.

2. Algorithm and the Descent Property

Based on the formulas (2) and (6) as well as (9), we establish the new SCGM as follows (Algorithm 1).

Algorithm 1: JYJLL-SCGM

Step 0. Given any initial point $x_1 \in R^n$, parameters δ and σ satisfying $0 < \delta < \sigma < 1$, and an accuracy tolerance $\epsilon > 0$. Let $d_1 = -g_1$, set $k := 1$.
Step 1. If $\|g_k\| \leq \epsilon$, terminate. Otherwise, go to Step 2.
Step 2. Determine a steplength α_k by an inexact line search.
Step 3. Generate the next iterate by $x_{k+1} = x_k + \alpha_k d_k$, compute gradient $g_{k+1} := g(x_{k+1})$ and the spectral parameter $\theta_{k+1} := \theta_{k+1}^{\text{JYJLL}}$ by (8), as well as the conjugate parameter $\beta_{k+1} := \beta_{k+1}^{\text{JYJLL}}$ by (9).
Step 4. Let $d_{k+1} = -\theta_{k+1}g_{k+1} + \beta_{k+1}d_k$. Set $k := k+1$, repeat Step 1.

The following lemma indicates that the JYJLL-SCGM always satisfies the descent condition without depending on any line search, and the conjugate parameter β_k^{JYJLL} has the similar properties as the DY formula.

Lemma 1. *Suppose that the search direction d_k is generated by JYJLL-SCGM. Then, we have $g_k^T d_k < 0$ for $k \geq 1$, that is, the search direction satisfies the descent condition. Furthermore, we obtain $0 \leq \beta_k^{\text{JYJLL}} \leq \frac{g_k^T d_k}{g_{k-1}^T d_{k-1}}$.*

Proof. We first prove the former claim by induction. For $k = 1$, it is easy to see that $g_1^T d_1 = -\|g_1\|^2 < 0$. Assume that $g_{k-1}^T d_{k-1} < 0$ holds for $k-1$ $(k \geq 2)$. Now, we prove that $g_k^T d_k < 0$ holds for k. Let $\hat{\theta}_k$ be the angle between g_k and d_{k-1}. The proof is divided into two cases as follows.

(a) If $d_{k-1}^T(g_k - g_{k-1}) > \|g_{k-1}\|^2$, then $d_{k-1}^T(g_k - g_{k-1}) > 0$. It follows from (8) and (9) that

$$\begin{aligned} g_k^T d_k =& g_k^T(-\theta_k^{\text{JYJLL}} g_k + \beta_k^{\text{JYJLL}} d_{k-1}) \\ =& -(1 + \frac{|g_k^T d_{k-1}|}{-g_{k-1}^T d_{k-1}})\|g_k\|^2 + \frac{\|g_k\|^2(1-\cos^2\hat{\theta}_k)}{d_{k-1}^T(g_k - g_{k-1})} g_k^T d_{k-1} \\ =& -\frac{|g_k^T d_{k-1}|}{-g_{k-1}^T d_{k-1}}\|g_k\|^2 - \|g_k\|^2\cos^2\hat{\theta}_k + \frac{\|g_k\|^2(1-\cos^2\hat{\theta}_k)}{d_{k-1}^T(g_k - g_{k-1})} g_{k-1}^T d_{k-1} \\ =& -(\cos^2\hat{\theta}_k + \frac{|g_k^T d_{k-1}|}{-g_{k-1}^T d_{k-1}})\|g_k\|^2 + \beta_k^{\text{JYJLL}} g_{k-1}^T d_{k-1} < 0. \end{aligned} \tag{10}$$

(b) If $d_{k-1}^T(g_k - g_{k-1}) \leq \|g_{k-1}\|^2$, then $g_k^T d_{k-1} \leq \|g_{k-1}\|^2 + g_{k-1}^T d_{k-1}$, and hence by (8) and (9), we have

$$\begin{aligned} g_k^T d_k =& g_k^T(-\theta_k^{\text{JYJLL}} g_k + \beta_k^{\text{JYJLL}} d_{k-1}) \\ =& -\theta_k^{\text{JYJLL}}\|g_k\|^2 + \frac{\|g_k\|^2(1-\cos^2\hat{\theta}_k)}{\|g_{k-1}\|^2} g_k^T d_{k-1} \\ \leq& -(1 + \frac{|g_k^T d_{k-1}|}{-g_{k-1}^T d_{k-1}})\|g_k\|^2 + \frac{\|g_k\|^2(1-\cos^2\hat{\theta}_k)}{\|g_{k-1}\|^2}(\|g_{k-1}\|^2 + g_{k-1}^T d_{k-1}) \\ =& -(\cos^2\hat{\theta}_k + \frac{|g_k^T d_{k-1}|}{-g_{k-1}^T d_{k-1}})\|g_k\|^2 + \beta_k^{\text{JYJLL}} g_{k-1}^T d_{k-1} < 0. \end{aligned} \tag{11}$$

Thus, $g_k^T d_k < 0$ holds for $k \geq 1$.

Now we prove the second assertion. From (10) and (11), it follows that $g_k^T d_k \leq \beta_k^{\text{JYJLL}} g_{k-1}^T d_{k-1}$. This together with $g_k^T d_k < 0$ implies that $\beta_k^{\text{JYJLL}} \leq \frac{g_k^T d_k}{g_{k-1}^T d_{k-1}}$, furthermore, we deduce that $0 \leq \beta_k^{\text{JYJLL}}$ from (9), and the proof is complete. □

3. Convergence Analysis

To analyze and ensure the global convergence of the JYJLL-SCGM, we choose the Wolfe line search (4) to yield the steplength α_k. Further, a basic assumption about the objective function as follows is needed.

Assumption 1. (H1) *For any initial point $x_1 \in R^n$, the level set $\Lambda = \{x \in R^n \mid f(x) \leq f(x_1)\}$ is bounded;* (H2) *$f(x)$ is continuously differentiable in a neighborhood U of Λ, and its gradient $g(x)$ is Lipschitz continuous, namely, there exists a constant $L > 0$ such that $\|g(x) - g(y)\| \leq L\|x - y\|, \forall\, x, y \in U$.*

In the following lemma, we review the well-known Zoutendijk condition [6], which plays an important role in the convergence analysis of CGMs. Also, the Zoutendijk condition is suitable for the convergence analysis of the JYJLL-SCGM.

Lemma 2. *Suppose that Assumption 1 holds. Consider a general iterative method $x_{k+1} = x_k + \alpha_k d_k$, where d_k is a descent direction such that $g_k^T d_k < 0$, and the steplength α_k satisfies the Wolfe line search condition (4). Then $\sum_{k=1}^{\infty} \frac{(g_k^T d_k)^2}{\|d_k\|^2} < \infty$.*

Based on Lemmas 1 and 2, we can establish the global convergence of the JYJLL-SCGM.

Theorem 1. *Suppose that Assumption 1 holds, and let the sequence $\{x_k\}$ be generated by the JYJLL-SCGM with the wolfe line search (4). Then $\liminf_{k\to\infty} \|g_k\| = 0$.*

Proof. By contradiction, suppose that the conclusion is not true. Then there exists a positive constant $\gamma > 0$ such that $\|g_k\|^2 \geq \gamma, \forall\, k$. Again, from (8), we obtain $d_k + \theta_k^{\text{JYJLL}} g_k = \beta_k^{\text{JYJLL}} d_{k-1}$. Combining this equation and Lemma 1, we have

$$\begin{aligned} \|d_k\|^2 &= (\beta_k^{\text{JYJLL}})^2 \|d_{k-1}\|^2 - 2\theta_k^{\text{JYJLL}} g_k^T d_k - (\theta_k^{\text{JYJLL}})^2 \|g_k\|^2 \\ &\leq \left(\frac{g_k^T d_k}{g_{k-1}^T d_{k-1}}\right)^2 \|d_{k-1}\|^2 - 2\theta_k^{\text{JYJLL}} g_k^T d_k - (\theta_k^{\text{JYJLL}})^2 \|g_k\|^2. \end{aligned}$$

Next, dividing both sides of the above inequality by $(g_k^T d_k)^2$, we obtain

$$\begin{aligned} \frac{\|d_k\|^2}{(g_k^T d_k)^2} &\leq \frac{\|d_{k-1}\|^2}{(g_{k-1}^T d_{k-1})^2} - \frac{2\theta_k^{\text{JYJLL}}}{g_k^T d_k} - \frac{(\theta_k^{\text{JYJLL}})^2 \|g_k\|^2}{(g_k^T d_k)^2} \\ &= \frac{\|d_{k-1}\|^2}{(g_{k-1}^T d_{k-1})^2} + \frac{1}{\|g_k\|^2} - \left(\frac{1}{\|g_k\|} + \frac{\theta_k^{\text{JYJLL}} \|g_k\|}{g_k^T d_k}\right)^2 \\ &\leq \frac{\|d_{k-1}\|^2}{(g_{k-1}^T d_{k-1})^2} + \frac{1}{\|g_k\|^2}. \end{aligned}$$

In terms of $\frac{\|d_1\|^2}{(g_1^T d_1)^2} = \frac{1}{\|g_1\|^2}$, together with the above relations and $\|g_k\|^2 \geq \gamma$, we have

$$\begin{aligned} \frac{\|d_k\|^2}{(g_k^T d_k)^2} &\leq \frac{\|d_{k-1}\|^2}{(g_{k-1}^T d_{k-1})^2} + \frac{1}{\|g_k\|^2} \\ &\leq \frac{\|d_{k-2}\|^2}{(g_{k-2}^T d_{k-2})^2} + \frac{1}{\|g_k\|^2} + \frac{1}{\|g_{k-1}\|^2} \\ &\leq \cdots \leq \sum_{i=1}^{k} \frac{1}{\|g_i\|^2} \leq \frac{k}{\gamma}, \end{aligned}$$

that is, $\frac{(g_k^T d_k)^2}{\|d_k\|^2} \geq \frac{\gamma}{k}$. Hence $\sum_{k=1}^{\infty} \frac{(g_k^T d_k)^2}{\|d_k\|^2} = \infty$, which contradicts Lemma 2. Therefore, the proof is complete. □

4. Numerical Results

In this section, we test the numerical performance of our method (denoted by JYJLL for short) via 108 test problems, and compare it with the four methods HZ [7], KD [8], AN1 [13] and LFZ [15]. The HZ and KD methods belong to the CGMs with excellent effect, and the AN1 and LFZ methods are the SCGMs with more efficient performance. The first 53 (from bard to woods) test problems are taken from the CUTE library in N. I. M. Gould et al. [22], and the last 55 are from References [23,24], their dimensions ranging from 2 to 1,000,000. All codes were written in Matlab 2016a and run on a DELL PC with 4GB of memory and windows 10 operating system. All the steplength α_k is generated by the Wolfe line search with $\sigma = 0.1$ and $\delta = 0.01$.

In the experiments, notations Itr, NF, NG, and Tcpu and $\|g_*\|$ denote the number of iteration, function evaluation, gradient evaluation, computing time of CPU and gradient values, respectively. We stop the iteration, if one of the following two cases is satisfied: (i) $\|g_k\| \leq 10^{-6}$; (ii) Itr > 2000. When case (ii) appears, the method is deemed to be invalid and is denoted by "F".

To show the numerical performance of the tested methods, we observed and reported the values of Itr, NF, NG, Tcpu and $\|g_*\|$ generated by the five tested methods for each test instance, see Tables 1 and 2 below. On the other hand, to visually characterize and compare the numerical results in Tables 1 and 2, we use the performance profiles introduced by Dolan and Morè [25] to describe the performance of the five tested methods according to Itr, NF, NG and Tcpu, respectively, see Figures 1–4 below.

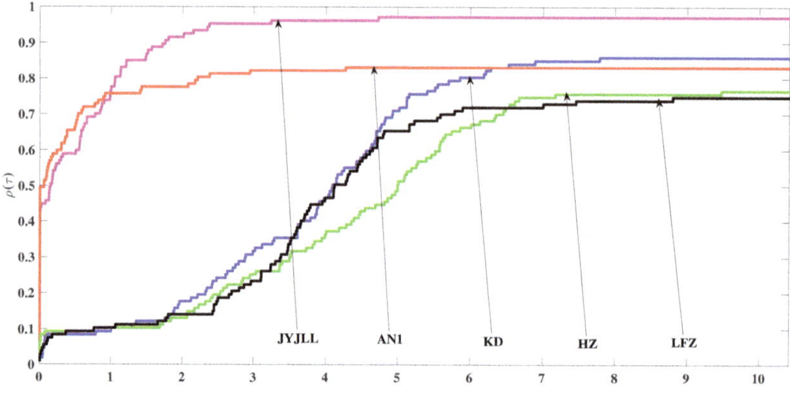

Figure 1. Performance profiles on Tcpu.

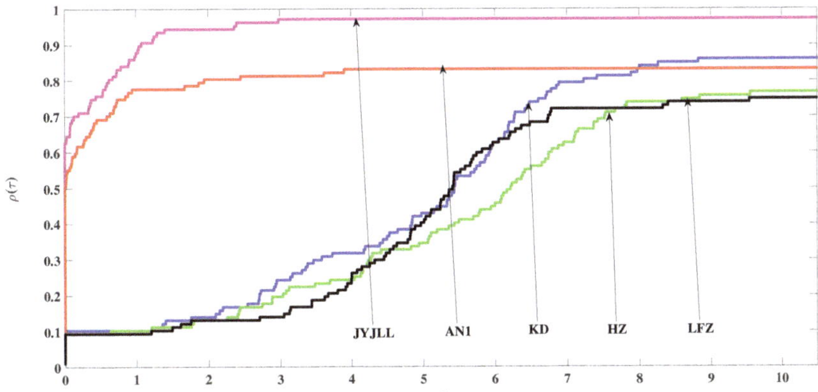

Figure 2. Performance profiles on NF .

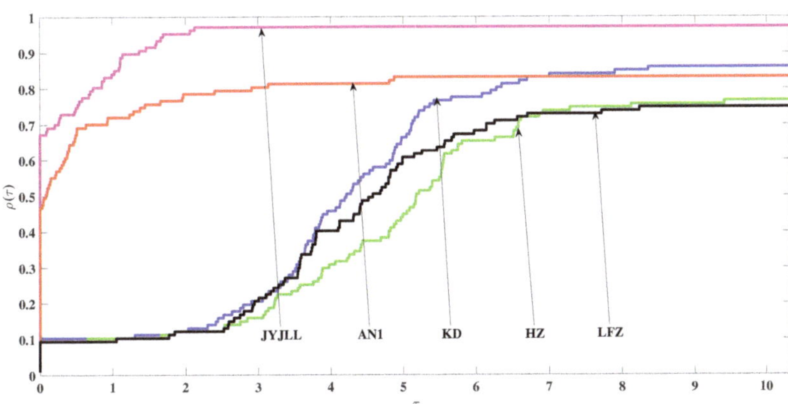

Figure 3. Performance profiles on NG

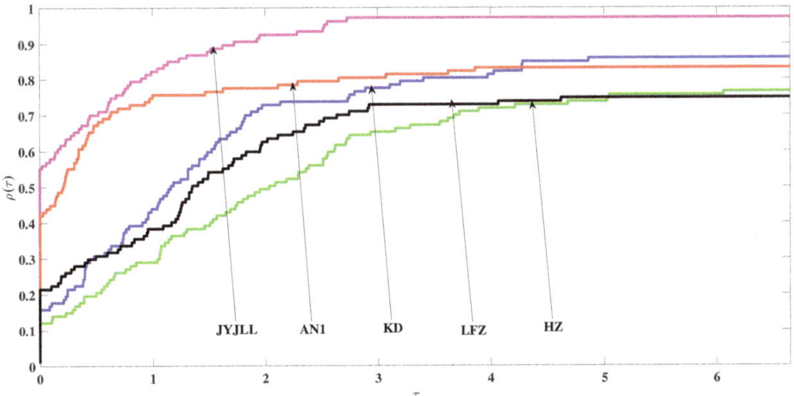

Figure 4. Performance profiles on Itr.

Table 1. Numerical test reports for the five tested methods.

Problems	JYJLL	KD	AN1	HZ	LFZ
Name/n	Itr/ NF/ NG/ Tcpu/ $\|g_*\|$	Itr/ NF/ NG/ Tcpu/ $\|g_*\|$	Itr/ NF/ NG/ Tcpu/ $\|g_*\|$	Itr/ NF/ NG/ Tcpu/ $\|g_*\|$	Itr/ NF/ NG/ Tcpu/ $\|g_*\|$
bard 3	178/218/259/0.171/8.74 × 10^{-7}	1565/48,617/19,915/4.657/8.01 × 10^{-7}	81/206/156/0.051/8.35 × 10^{-7}	620/18,622/7190/1.732/3.95 × 10^{-7}	265/6647/3284/0.691/1.72 × 10^{-7}
beale 2	67/93/87/0.015/9.28 × 10^{-7}	326/9668/4385/0.429/7.71 × 10^{-7}	44/150/92/0.012/9.17 × 10^{-7}	251/7325/3130/0.318/6.17 × 10^{-7}	171/4757/2347/0.214/4.31 × 10^{-8}
box 3	139/193/207/0.026/6.64 × 10^{-7}	330/10,108/4330/0.504/5.98 × 10^{-7}	36/104/64/0.009/7.52 × 10^{-7}	475/14,246/5816/0.702/8.94 × 10^{-7}	68/1569/764/0.083/3.17 × 10^{-7}
cosine 2000	17/55/19/0.104/2.48 × 10^{-7}	493/13,783/6221/3.593/8.15 × 10^{-7}	247/673/556/0.438/9.68 × 10^{-7}	435/11,148/5281/3.126/4.62 × 10^{-7}	F/F/F/F/1.88 × 10^{3}
cosine 4000	20/61/24/0.050/9.50 × 10^{-8}	F/F/F/F/6.04 × 10^{-3}	246/908/671/0.964/5.96 × 10^{-7}	178/4537/2189/3.495/8.83 × 10^{-7}	F/F/F/F/7.43 × 10^{-4}
cosine 20,000	26/65/32/0.228/6.76 × 10^{-7}	F/F/F/F/3.13 × 10^{-3}	163/207/240/1.029/7.51 × 10^{-7}	1730/48,495/21,578/161.976/6.59 × 10^{-7}	F/F/F/F/1.85 × 10^{4}
dixmaana 3000	21/63/22/0.318/8.90 × 10^{-7}	17/266/117/0.683/3.09 × 10^{-7}	20/60/22/0.178/9.35 × 10^{-7}	26/520/206/1.282/1.33 × 10^{-7}	24/393/169/1.012/1.85 × 10^{-7}
dixmaanb 3000	12/59/12/0.159/3.75 × 10^{-7}	12/150/49/0.370/3.26 × 10^{-7}	14/59/14/0.145/8.48 × 10^{-7}	49/1246/565/3.231/9.61 × 10^{-8}	14/180/72/0.458/1.48 × 10^{-7}
dixmaanc 3000	15/64/15/0.197/8.85 × 10^{-7}	25/458/217/1.207/5.54 × 10^{-7}	19/61/19/0.174/9.87 × 10^{-7}	33/678/323/1.829/1.92 × 10^{-7}	35/659/315/1.718/8.03 × 10^{-7}
dixmaand 3000	19/68/19/0.194/1.05 × 10^{-8}	25/389/167/0.977/3.58 × 10^{-7}	21/67/21/0.191/3.53 × 10^{-7}	54/1304/627/3.484/2.30 × 10^{-7}	45/822/409/2.179/6.68 × 10^{-7}
dixmaane 3000	564/525/797/3.316/8.67 × 10^{-7}	787/22,182/9876/60.145/9.62 × 10^{-7}	406/249/500/2.203/9.08 × 10^{-7}	1239/35,277/14,918/96.673/7.67 × 10^{-7}	369/9463/4622/26.732/7.85 × 10^{-7}
dixmaanf 3000	486/470/691/3.060/5.26 × 10^{-7}	720/20,789/8946/56.355/7.80 × 10^{-7}	395/238/483/2.094/7.12 × 10^{-7}	1387/40,775/17,299/111.617/9.92 × 10^{-7}	283/7443/3631/21.630/8.83 × 10^{-7}
dixmaang 3000	291/286/403/1.872/8.98 × 10^{-7}	557/15,559/6999/43.071/8.50 × 10^{-7}	430/266/531/2.370/6.61 × 10^{-7}	1492/43,815/18,382/120.388/6.52 × 10^{-7}	441/11,502/5595/32.348/6.66 × 10^{-7}
dixmaanh 3000	501/499/719/3.300/8.91 × 10^{-7}	933/26,886/11971/73.607/9.35 × 10^{-7}	428/245/518/2.273/9.72 × 10^{-7}	F/F/F/F/2.23 × 10^{-5}	478/12,359/6015/35.150/9.39 × 10^{-7}
dixmaanj 3000	874/829/1258/5.304/7.87 × 10^{-7}	1201/34,580/15,144/93.734/9.58 × 10^{-7}	1038/575/1295/5.617/7.20 × 10^{-7}	F/F/F/F/3.16 × 10^{-6}	642/16,203/7988/45.658/9.24 × 10^{-7}
dixmaank 3000	891/876/1297/5.757/9.74 × 10^{-7}	1158/33,535/14,587/89.969/7.82 × 10^{-7}	978/551/1222/5.291/6.98 × 10^{-7}	F/F/F/F/2.66 × 10^{-5}	1899/49,488/24,316/139.479/8.81 × 10^{-7}
dixon3dq 5	51/54/54/0.011/5.58 × 10^{-7}	136/3776/1687/0.156/4.24 × 10^{-7}	52/58/58/0.007/5.73 × 10^{-7}	77/1803/786/0.072/7.91 × 10^{-7}	168/4217/2066/0.165/4.14 × 10^{-7}
dixon3dq 40	631/600/908/0.076/9.98 × 10^{-7}	1164/34,979/14,945/1.418/5.17 × 10^{-7}	394/114/427/0.041/7.07 × 10^{-7}	1789/52,134/21,597/1.999/7.37 × 10^{-7}	477/12,218/5965/0.492/3.93 × 10^{-7}
dqdrtic 10,000	105/159/144/0.174/8.38 × 10^{-7}	275/7800/3415/3.965/2.96 × 10^{-7}	78/124/99/0.159/7.25 × 10^{-7}	911/28,054/11,907/15.545/7.25 × 10^{-7}	191/5281/2547/3.215/3.10 × 10^{-7}
dqdrtic 100,000	118/176/164/1.740/9.70 × 10^{-7}	455/13,662/6077/53.011/3.67 × 10^{-7}	104/232/178/2.033/8.42 × 10^{-7}	685/20,674/8804/82.583/5.80 × 10^{-7}	293/7575/3761/30.040/3.73 × 10^{-7}
dqdrtic 1,000,000	67/143/96/13.649/2.81 × 10^{-7}	256/7211/3161/354.291/7.07 × 10^{-7}	84/185/132/19.420/9.45 × 10^{-7}	854/26,185/11,120/1264.582/5.93 × 10^{-7}	214/5856/2872/297.431/3.91 × 10^{-7}
dqrtic 480	33/89/33/0.043/6.19 × 10^{-7}	F/F/F/F/4.77 × 10^{9}	43/100/51/0.037/2.83 × 10^{-7}	F/F/F/F/4.77 × 10^{9}	F/F/F/F/4.77 × 10^{9}
dqrtic 510	28/89/28/0.028/2.08 × 10^{-7}	F/F/F/F/5.90 × 10^{9}	F/F/F/F/1.16 × 10^{9}	F/F/F/F/5.90 × 10^{9}	F/F/F/F/5.90 × 10^{9}
edensch 4000	39/326/155/0.683/9.42 × 10^{-7}	45/855/380/1.666/6.67 × 10^{-7}	F/F/F/F/7.59 × 10^{-6}	51/1123/490/2.195/8.67 × 10^{-7}	F/F/F/F/1.22 × 10^{-5}
edensch 5000	34/196/83/0.573/6.12 × 10^{-7}	66/1491/715/3.760/6.48 × 10^{-7}	F/F/F/F/1.99 × 10^{-6}	48/1035/475/2.590/6.56 × 10^{-7}	F/F/F/F/2.25 × 10^{-5}
edensch 7500	35/259/96/0.956/2.94 × 10^{-7}	68/1673/797/4.458/3.76 × 10^{-7}	F/F/F/F/2.64 × 10^{-6}	F/F/F/F/2.13 × 10^{-6}	F/F/F/F/9.86 × 10^{-6}
eg2 30	66/84/79/0.036/6.82 × 10^{-7}	F/F/F/F/3.05 × 10^{-1}	62/138/102/0.016/7.12 × 10^{-7}	F/F/F/F/3.20 × 10^{-3}	F/F/F/F/2.61 × 10^{-5}
eg2 100	111/156/157/0.025/7.73 × 10^{-7}	156/4392/1951/0.320/5.40 × 10^{-7}	45/289/160/0.031/5.13 × 10^{-7}	F/F/F/F/7.42 × 10^{-6}	F/F/F/F/2.74 × 10^{-5}
fletchcr 50	51/107/81/0.028/4.96 × 10^{-7}	88/2309/1016/0.239/4.70 × 10^{-7}	63/108/89/0.025/7.14 × 10^{-7}	140/3679/1711/0.383/1.92 × 10^{-7}	181/4599/2284/0.539/3.65 × 10^{-7}
fletchcr 100	59/75/69/0.016/5.93 × 10^{-7}	157/4341/1989/0.306/4.38 × 10^{-7}	F/F/F/F/4.08 × 10^{-6}	86/2125/978/0.127/5.68 × 10^{-7}	67/1635/810/0.093/9.77 × 10^{-7}
fletchcr 200	59/75/72/0.012/7.79 × 10^{-7}	117/3079/1414/0.205/7.30 × 10^{-7}	75/288/195/0.032/4.41 × 10^{-7}	F/F/F/F/8.57 × 10^{-6}	F/F/F/F/1.00 × 10^{-5}

Table 1. *Cont.*

Problems	JYJLL	KD	AN1	HZ	LFZ
freuroth 5	269/390/421/0.061/9.04×10^{-7}	941/28,582/12,159/1.563/3.39×10^{-8}	F/F/F/F/1.56×10^{-5}	1580/48,118/19,484/2.562/6.79×10^{-7}	F/F/F/F/5.26×10^{-6}
genrose 5000	283/292/397/0.537/7.01×10^{-7}	371/10,726/47,41/6.456/3.94×10^{-7}	219/194/283/0.441/7.91×10^{-7}	1070/31,988/13,201/16.842/4.83×10^{-7}	616/16,065/7855/5.567/8.05×10^{-7}
genrose 100,000	142/170/193/2.068/9.01×10^{-7}	416/12,351/5418/62.603/9.70×10^{-7}	166/203/233/2.965/9.99×10^{-7}	494/14,196/6016/71.947/9.73×10^{-7}	361/9600/4616/54.420/9.84×10^{-7}
genrose 1000	233/266/335/0.077/8.18×10^{-7}	295/8492/3729/0.942/8.13×10^{-7}	224/197/291/0.087/7.28×10^{-7}	701/20,368/8502/2.625/4.17×10^{-7}	393/10,221/5053/1.526/9.81×10^{-7}
gulf 3000	2/1/2/0.008/0.00×10^{0}	2/1/2/0.001/0.00×10^{0}	2/1/2/0.000/0.00×10^{0}	2/1/2/0.000/0.00×10^{0}	2/1/2/0.000/0.00×10^{0}
helix 20,000	133/203/199/0.064/7.70×10^{-7}	282/8168/3481/0.733/4.20×10^{-7}	119/258/212/0.044/8.96×10^{-7}	731/21,643/8941/2.214/6.74×10^{-7}	F/F/F/F/4.06×10^{1}
himmelbg 30,000	3/6/7/0.051/8.71×10^{-28}	3/6/7/0.043/7.27×10^{-28}	3/6/7/0.043/8.73×10^{-28}	3/6/7/0.042/6.57×10^{-28}	3/6/7/0.043/8.75×10^{-28}
himmelbg 50,000	3/6/7/0.075/1.12×10^{-27}	3/6/7/0.067/9.38×10^{-28}	3/6/7/0.081/1.13×10^{-27}	3/6/7/0.066/8.48×10^{-28}	3/6/7/0.072/1.13×10^{-27}
himmelbg 10	3/6/7/0.001/1.59×10^{-29}	3/6/7/0.001/1.33×10^{-29}	3/6/7/0.001/1.59×10^{-29}	3/6/7/0.001/1.20×10^{-29}	3/6/7/0.001/1.60×10^{-29}
kowosb 10	449/470/668/0.097/9.29×10^{-7}	817/24,298/10,391/1.768/4.66×10^{-7}	118/285/243/0.042/6.03×10^{-7}	F/F/F/F/5.22×10^{-4}	312/8413/4176/0.666/7.72×10^{-7}
liarwhd 15	59/95/73/0.018/9.69×10^{-7}	72/1725/753/0.127/4.53×10^{-7}	35/129/68/0.013/7.67×10^{-7}	229/6565/2858/0.417/5.51×10^{-7}	47/1028/474/0.072/6.32×10^{-7}
liarwhd 1000	258/421/426/0.094/6.91×10^{-7}	F/F/F/F/1.33×10^{-4}	39/211/102/0.027/7.23×10^{-7}	1267/38,967/15,799/3.867/9.84×10^{-7}	224/5956/2888/0.612/5.39×10^{-7}
nondquar 10	385/461/586/0.085/9.98×10^{-7}	F/F/F/F/2.49×10^{-4}	F/F/F/F/2.76×10^{-3}	F/F/F/F/1.28×10^{-3}	F/F/F/F/1.99×10^{1}
penalty1 1000	14/75/14/0.279/6.45×10^{-8}	15/253/94/1.084/1.21×10^{-7}	14/75/14/0.279/7.70×10^{-8}	15/253/94/0.996/1.21×10^{-7}	15/253/94/0.954/1.21×10^{-7}
penalty1 8000	17/79/17/13.954/3.44×10^{-7}	114/3407/1633/545.410/1.79×10^{-7}	17/79/17/11.996/3.00×10^{-7}	114/3407/1624/568.693/1.79×10^{-7}	114/3407/1607/573.841/1.77×10^{-7}
quartc 50	23/67/23/0.010/5.57×10^{-7}	28/516/226/0.034/1.42×10^{-8}	21/68/23/0.006/1.03×10^{-7}	26/449/192/0.028/8.61×10^{-7}	85/2319/1093/0.157/6.36×10^{-7}
quartc 300	25/86/26/0.019/4.33×10^{-7}	34/562/249/0.100/3.74×10^{-7}	29/87/32/0.020/2.30×10^{-7}	37/633/284/0.116/8.92×10^{-8}	108/2821/1345/0.494/4.26×10^{-7}
tridia 400	1196/1131/1723/0.225/8.86×10^{-7}	F/F/F/F/1.07×10^{-3}	865/482/1067/0.152/6.41×10^{-7}	F/F/F/F/3.76×10^{-5}	832/22,141/10,798/1.437/5.76×10^{-7}
tridia 500	1163/1073/1661/0.230/9.65×10^{-7}	1655/47,555/20,693/3.330/4.97×10^{-7}	F/F/F/F/1.52×10^{3}	F/F/F/F/3.97×10^{-3}	991/25,621/12,586/1.797/5.03×10^{-7}
sinquad 3	202/284/320/0.032/6.77×10^{-8}	272/7998/3343/0.334/8.95×10^{-7}	86/220/173/0.018/6.56×10^{-8}	501/14,834/5870/0.639/2.96×10^{-7}	166/4182/2052/0.201/9.27×10^{-7}
vardim 2	10/54/10/0.014/5.74×10^{-7}	10/124/36/0.005/9.89×10^{-8}	10/54/10/0.003/5.74×10^{-7}	10/124/36/0.009/9.89×10^{-8}	10/124/36/0.006/9.89×10^{-8}
woods 10,000	662/760/998/1.323/9.70×10^{-7}	1851/54,675/25,702/47.380/9.00×10^{-7}	215/321/331/0.640/5.16×10^{-7}	1122/32,157/13,925/29.130/8.10×10^{-7}	515/13,196/6550/12.415/1.79×10^{-7}

Table 2. Numerical test reports for the five tested methods (Continue).

Problems	JYJLL	KD	AN1	HZ	LFZ
Name/n	Itr/ NF/ NG/ Tcpu/ $\|g_*\|$	Itr/ NF/ NG/ Tcpu/ $\|g_*\|$	Itr/ NF/ NG/ Tcpu/ $\|g_*\|$	Itr/ NF/ NG/ Tcpu/ $\|g_*\|$	Itr/ NF/ NG/ Tcpu/ $\|g_*\|$
bdexp 50,000	3/2/3/0.079/1.33 × 10^{-109}	3/2/3/0.075/1.28 × 10^{-109}	3/2/3/0.072/1.32 × 10^{-109}	3/2/3/0.073/1.25 × 10^{-109}	3/2/3/0.074/1.33 × 10^{-109}
bdexp 100,000	3/2/3/0.134/1.73 × 10^{-109}	3/2/3/0.129/1.70 × 10^{-109}	3/2/3/0.132/1.73 × 10^{-109}	3/2/3/0.123/1.68 × 10^{-109}	3/2/3/0.123/1.73 × 10^{-109}
bdexp 1,000,000	3/2/3/1.245/5.09 × 10^{-109}	3/2/3/1.276/5.08 × 10^{-109}	3/2/3/1.341/5.09 × 10^{-109}	3/2/3/1.239/5.08 × 10^{-109}	3/2/3/1.259/5.09 × 10^{-109}
exdenschnf 100,000	24/84/25/0.751/8.51 × 10^{-7}	69/1886/922/16.019/6.00 × 10^{-7}	74/130/97/2.004/8.07 × 10^{-7}	59/1490/692/12.027/2.21 × 10^{-7}	123/3284/1568/26.773/8.35 × 10^{-7}
exdenschnf 1,000,000	29/86/30/8.772/3.60 × 10^{-7}	72/1974/940/155.062/9.90 × 10^{-8}	57/114/73/16.513/3.88 × 10^{-7}	123/3517/1677/281.678/5.75 × 10^{-8}	220/6391/3128/520.133/6.92 × 10^{-7}
mccormak 2	23/58/30/0.024/7.24 × 10^{-7}	19/364/163/0.021/8.11 × 10^{-7}	33/55/39/0.005/7.39 × 10^{-7}	39/900/404/0.039/1.21 × 10^{-7}	F/F/F/F/2.89 × 10^{4}
exdenschnb 15,000	24/62/25/0.067/3.34 × 10^{-7}	28/600/263/0.496/8.71 × 10^{-7}	32/61/33/0.101/7.18 × 10^{-7}	38/787/367/0.764/9.24 × 10^{-8}	134/3726/1766/3.688/8.57 × 10^{-7}
exdenschnb 120,000	23/62/24/0.497/7.31 × 10^{-7}	30/632/282/3.333/2.86 × 10^{-7}	33/63/34/0.638/8.21 × 10^{-7}	28/523/223/2.688/2.98 × 10^{-7}	89/2344/1134/12.149/4.31 × 10^{-7}
genquartic 120,000	35/78/41/0.793/9.37 × 10^{-7}	46/929/437/6.281/1.34 × 10^{-7}	47/83/57/1.178/5.73 × 10^{-7}	59/1355/643/9.081/9.61 × 10^{-7}	166/4182/2080/28.307/9.70 × 10^{-7}
genquartic 100,000	28/67/31/0.556/5.69 × 10^{-7}	37/736/306/4.120/5.17 × 10^{-7}	39/74/44/0.786/6.59 × 10^{-7}	81/2048/984/11.667/8.80 × 10^{-7}	174/4464/2191/25.852/9.57 × 10^{-7}
biggsb1 110	790/738/1136/0.118/7.29 × 10^{-7}	F/F/F/F/2.37 × 10^{-4}	522/143/571/0.065/9.63 × 10^{-7}	1868/55,264/23,430/2.526/9.82 × 10^{-7}	594/15,762/7618/0.728/4.96 × 10^{-7}
biggsb1 200	1717/1618/2504/0.255/8.47 × 10^{-7}	1809/52,361/22,385/2.674/9.04 × 10^{-7}	1148/205/1228/0.163/9.09 × 10^{-7}	F/F/F/F/2.38 × 10^{-3}	866/22,409/11,027/1.192/7.31 × 10^{-7}
sine 750,000	111/147/156/31.089/8.79 × 10^{-7}	141/3661/1775/379.752/3.73 × 10^{-11}	F/F/F/F/1.65 × 10^{-3}	101/2714/1330/318.993/5.72 × 10^{-7}	F/F/F/F/5.77 × 10^{-3}
sine 1,000,000	238/407/412/110.321/2.67 × 10^{-7}	72/2052/994/297.446/2.56 × 10^{-7}	89/198/160/47.856/2.98 × 10^{-7}	150/3798/1858/538.100/9.23 × 10^{-8}	F/F/F/F/4.35 × 10^{1}
fletcbv3 55	667/488/895/0.408/2.07 × 10^{-7}	1381/26,076/12,787/1.507/8.07 × 10^{-8}	F/F/F/F/2.03 × 10^{-4}	1388/26,927/13,220/1.712/8.48 × 10^{-7}	F/F/F/F/1.59 × 10^{-4}
fletcbv3 85	1654/911/2080/0.348/5.69 × 10^{-7}	F/F/F/F/7.89 × 10^{-4}	F/F/F/F/7.43 × 10^{-4}	F/F/F/F/8.15 × 10^{-4}	772/13,968/7106/1.214/2.55 × 10^{-7}
nonscomp 30,000	F/F/F/F/2.14 × 10^{-4}	147/3574/1604/6.135/8.57 × 10^{-7}	95/124/119/0.409/4.82 × 10^{-7}	927/27600/11616/41.504/5.89 × 10^{-7}	F/F/F/F/4.08 × 10^{-3}
nonscomp 25000	60/78/62/0.217/6.47 × 10^{-7}	1164/35,210/14,738/47.678/6.09 × 10^{-7}	93/125/118/0.381/9.82 × 10^{-7}	199/4988/2234/6.920/4.66 × 10^{-7}	1006/26,503/13,011/38.180/4.78 × 10^{-7}
power1 100	1588/1500/2295/0.277/7.70 × 10^{-7}	F/F/F/F/8.27 × 10^{-5}	1496/908/1907/0.181/9.52 × 10^{-7}	F/F/F/F/2.49 × 10^{-1}	1178/31,067/15,281/1.268/7.06 × 10^{-7}
power1 90	1614/1548/2346/0.201/9.42 × 10^{-7}	F/F/F/F/5.18 × 10^{-5}	F/F/F/F/1.03 × 10^{-2}	F/F/F/F/5.28 × 10^{-2}	1043/27,237/13,510/1.104/8.22 × 10^{-7}
raydan1 1000	320/362/470/0.141/6.35 × 10^{-7}	527/14,585/6962/1.225/8.97 × 10^{-7}	F/F/F/F/1.91 × 10^{-5}	1029/29,815/13,278/2.239/7.75 × 10^{-7}	F/F/F/F/4.19 × 10^{-5}
raydan1 1200	404/391/568/0.101/2.46 × 10^{-7}	601/17,005/7837/1.484/7.66 × 10^{-7}	1971/2141/3002/0.527/8.77 × 10^{-7}	850/24,008/10,762/1.970/9.80 × 10^{-7}	F/F/F/F/1.31 × 10^{-4}
raydan2 5000	17/52/17/0.067/9.00 × 10^{-7}	21/456/230/0.265/7.22 × 10^{-8}	17/52/17/0.033/9.35 × 10^{-7}	38/972/472/0.461/2.67 × 10^{-7}	21/458/205/0.207/8.34 × 10^{-7}
raydan2 10,000	17/56/20/0.069/1.63 × 10^{-7}	18/364/149/0.374/1.55 × 10^{-7}	23/60/28/0.089/6.20 × 10^{-7}	17/303/143/0.326/3.56 × 10^{-7}	33/780/348/0.819/8.82 × 10^{-7}
raydan2 20,000	13/50/14/0.100/7.46 × 10^{-7}	29/662/305/1.224/3.55 × 10^{-7}	20/83/44/0.185/6.68 × 10^{-7}	14/238/81/0.602/8.97 × 10^{-7}	21/413/163/0.867/5.83 × 10^{-7}
diagonal1 50	69/67/72/0.047/6.89 × 10^{-7}	152/4438/2120/0.247/8.73 × 10^{-7}	79/244/169/0.017/7.22 × 10^{-7}	204/5701/2669/0.239/9.74 × 10^{-7}	188/4914/2287/0.218/5.44 × 10^{-7}
diagonal1 80	100/321/228/0.063/8.82 × 10^{-7}	227/6698/3133/0.355/5.49 × 10^{-7}	F/F/F/F/1.81 × 10^{-5}	215/5707/2704/0.267/8.70 × 10^{-7}	F/F/F/F/1.11 × 10^{-5}
diagonal2 5000	821/775/1183/1.415/7.04 × 10^{-7}	1614/47,941/22,089/25.459/9.28 × 10^{-7}	470/415/652/0.717/7.50 × 10^{-7}	F/F/F/F/1.33 × 10^{-1}	619/15,686/7753/8.985/9.65 × 10^{-7}
diagonal2 10,000	1043/974/1505/3.503/5.89 × 10^{-7}	F/F/F/F/1.93 × 10^{-4}	873/775/1235/5.941/9.42 × 10^{-7}	F/F/F/F/2.20 × 10^{-1}	819/21,318/10,576/30.265/7.97 × 10^{-7}
diagonal3 150	164/575/422/0.058/7.53 × 10^{-7}	194/5161/2419/0.294/7.94 × 10^{-7}	F/F/F/F/2.95 × 10^{-5}	F/F/F/F/5.99 × 10^{-6}	F/F/F/F/2.13 × 10^{-5}
diagonal3 200	144/655/421/0.056/9.54 × 10^{-7}	F/F/F/F/3.24 × 10^{-6}	F/F/F/F/2.96 × 10^{-5}	F/F/F/F/2.85 × 10^{-5}	F/F/F/F/6.48 × 10^{-5}
bv 1000	19/13/23/0.152/8.69 × 10^{-7}	29/805/356/3.956/6.69 × 10^{-7}	82/25/90/0.705/7.51 × 10^{-7}	33/931/383/4.416/9.42 × 10^{-7}	143/4195/1754/19.534/8.70 × 10^{-7}
bv 2000	5/5/5/0.125/5.84 × 10^{-7}	14/385/172/5.830/3.11 × 10^{-7}	42/9/44/0.964/8.45 × 10^{-7}	9/225/93/3.519/4.59 × 10^{-7}	123/3714/1509/55.763/5.77 × 10^{-7}

Table 2. *Cont.*

Problems	JYJLL	KD	AN1	HZ	LFZ
ie 215	13/40/13/0.977/8.36 × 10^{-7}	13/183/64/3.470/7.39 × 10^{-7}	15/42/15/1.083/9.11 × 10^{-7}	17/294/120/6.959/9.65 × 10^{-8}	40/889/394/21.664/3.31 × 10^{-7}
singx 900	F/F/F/F/1.46 × 10^{-5}	1479/44741/16,996/180.509/2.36 × 10^{-7}	223/580/477/3.497/8.63 × 10^{-7}	F/F/F/F/4.08 × 10^{-5}	366/9279/4616/38.993/4.12 × 10^{-7}
band 3	20/64/22/0.024/1.48 × 10^{-7}	67/1838/860/0.089/6.29 × 10^{-7}	40/96/57/0.007/4.87 × 10^{-7}	77/2156/1006/0.097/6.95 × 10^{-7}	46/1025/467/0.048/8.94 × 10^{-8}
gauss 3	23/38/28/0.029/4.72 × 10^{-7}	47/1263/533/0.085/6.39 × 10^{-7}	12/30/13/0.003/6.97 × 10^{-7}	14/245/114/0.018/8.13 × 10^{-7}	30/743/361/0.052/7.35 × 10^{-7}
jensam 2	69/113/98/0.024/8.86 × 10^{-7}	119/3545/1697/0.155/6.69 × 10^{-7}	37/139/82/0.009/2.89 × 10^{-7}	211/6152/2894/0.271/6.99 × 10^{-7}	188/5415/2603/0.242/3.45 × 10^{-8}
lin 200	2/2/2/0.012/1.71 × 10^{-13}	2/2/2/0.006/1.71 × 10^{-13}	2/2/2/0.006/1.71 × 10^{-13}	2/2/2/0.006/1.71 × 10^{-13}	2/2/2/0.007/1.71 × 10^{-13}
lin 500	2/2/2/0.021/9.93 × 10^{-14}	2/2/2/0.021/9.93 × 10^{-14}	2/2/2/0.020/9.93 × 10^{-14}	2/2/2/0.021/9.93 × 10^{-14}	2/2/2/0.020/9.93 × 10^{-14}
lin 500	2/2/2/0.023/9.93 × 10^{-14}	2/2/2/0.021/9.93 × 10^{-14}	2/2/2/0.022/9.93 × 10^{-14}	2/2/2/0.020/9.93 × 10^{-14}	2/2/2/0.020/9.93 × 10^{-14}
osb2 11	1266/1180/1833/0.415/9.27 × 10^{-7}	F/F/F/F/2.54 × 10^{-3}	615/483/831/0.200/5.98 × 10^{-7}	F/F/F/F/6.43 × 10^{-4}	F/F/F/F/5.58 × 10^{-2}
pen1 50	121/233/192/0.035/8.32 × 10^{-7}	222/6619/2668/0.414/7.03 × 10^{-7}	58/152/89/0.016/2.20 × 10^{-7}	691/21,027/8471/1.298/7.18 × 10^{-7}	21/427/184/0.027/7.53 × 10^{-7}
pen1 90	260/411/417/0.069/9.04 × 10^{-7}	340/10,061/4115/0.750/8.46 × 10^{-7}	104/284/194/0.035/2.08 × 10^{-7}	348/10,351/4128/0.768/4.68 × 10^{-7}	93/2475/1171/0.190/5.09 × 10^{-9}
pen2 105	135/360/265/0.116/5.37 × 10^{-7}	224/6471/3041/1.603/7.59 × 10^{-7}	F/F/F/F/4.85 × 10^{-5}	F/F/F/F/7.28 × 10^{-5}	F/F/F/F/2.24 × 10^{-4}
pen2 100	F/F/F/F/1.09 × 10^{-5}	274/7792/3688/1.809/6.18 × 10^{-7}	F/F/F/F/8.89 × 10^{-6}	430/12,042/5778/3.055/8.95 × 10^{-7}	F/F/F/F/7.94 × 10^{-6}
rose 2	166/261/263/0.063/4.08 × 10^{-7}	911/28,228/11,815/1.219/8.56 × 10^{-7}	58/239/145/0.016/4.35 × 10^{-7}	1075/32,570/13,545/1.405/9.49 × 10^{-7}	148/3833/1917/0.174/8.42 × 10^{-8}
rosex 100	377/471/577/0.089/1.30 × 10^{-7}	1077/33,420/13,972/2.948/8.43 × 10^{-7}	66/317/190/0.041/8.67 × 10^{-7}	811/23,841/10,191/2.349/2.55 × 10^{-7}	148/4021/1954/0.423/8.91 × 10^{-8}
rosex 500	292/415/463/2.101/3.95 × 10^{-7}	1160/36,016/15,090/88.764/7.03 × 10^{-7}	60/241/146/0.751/6.09 × 10^{-7}	891/26,552/11,144/65.695/7.84 × 10^{-7}	150/4066/1960/10.359/4.00 × 10^{-7}
sing 4	896/1018/1373/0.142/7.31 × 10^{-7}	380/11052/4591/0.502/6.68 × 10^{-7}	153/388/314/0.032/5.24 × 10^{-7}	F/F/F/F/5.50 × 10^{-6}	360/8971/4386/0.427/9.01 × 10^{-7}
trid 50	82/79/93/0.029/9.78 × 10^{-7}	201/5551/2510/0.475/5.26 × 10^{-7}	81/89/96/0.019/9.39 × 10^{-7}	196/5383/2389/0.451/4.84 × 10^{-7}	317/8411/4103/0.723/9.60 × 10^{-7}
trid 90	87/87/101/0.026/3.19 × 10^{-7}	240/6330/2957/0.737/9.36 × 10^{-7}	103/105/126/0.033/8.56 × 10^{-7}	182/4501/2012/0.531/4.11 × 10^{-7}	207/5447/2632/0.654/8.16 × 10^{-7}
wood 4	351/391/509/0.074/8.86 × 10^{-7}	631/17,914/8171/0.865/7.97 × 10^{-7}	210/342/341/0.037/7.14 × 10^{-7}	1284/36,519/15,935/1.768/7.91 × 10^{-7}	375/9721/4715/0.479/4.73 × 10^{-7}

5. Discussion of Results

First, it is known, from the characters of the performance profiles, that the higher the curve in the figures, the better the associated method. Second, by summarizing the convergence analysis and numerical reports in Tables 1 and 2 and Figures 1–4, the proposed JYJLL-SCGM shows the following three advantages.

(i) It has good global convergence under mild assumptions.

(ii) It is practically effective, at least for the 108 tested instances.

(iii) It is the most effective in the five tested methods. In addition, the numerical performance of AN1 [13] method and the JYJLL-SCGM is relatively stable.

Of cause, the advantages above are attributed to the choice techniques (8) and (9) for the spectral parameter and the conjugate parameter.

6. Conclusions

The contributions of this work are two aspects. The one is to design a new computing schemes for the spectral parameter which ensures the $\theta_k > 1$. The other is to propose a new computing method for the conjugate parameter. These two techniques such that the search directions always possess descent property independent of the line search technique. As a result, the presented JYJLL-SCGM possesses global convergence if using the Wolfe line search to yield the steplength. A lot of numerical experiments in comparison with relative methods show that our SCGM is promising.

As further works, we think the following two problems are interesting and worth studying. The one is to design new approaches for the spectral parameter to guarantee $\theta_k > 1$, such as, combining the Newton direction and some new quasi-Newton equations, or the new conjugate conditions. The other is to find new computation techniques for the conjugate parameter with the help of the existing approaches, for example, the hybrid parameter, the three–term conjugate parameter et al.

Author Contributions: Conceptualization, J.J. and X.J.; methodology, J.J. and X.J.; formal analysis, L.Y. and P.L.; numerical experiments, L.Y. and P.L.; writing—original draft preparation, M.L. All authors have read and agreed to the published version of the manuscript.

Funding: This research was supported by Natural Science Foundation of China under Grant No. 11771383, Natural Science Foundation of Guangxi Province under Grant No. 2016GXNSFAA380028, Research Foundation of Guangxi University for Nationalities under Grant No. 2018KJQD02, and Middle-aged and Young Teachers' Basic Ability Promotion Project of Guangxi Province under Grant No. 2017KY0537.

Conflicts of Interest: The authors declare no conflict of interest.

References

1. Hestenes, M.R.; Stiefel, E. Method of conjugate gradient for solving linear equations. *J. Res. Nati. Bur. Stand.* **1952**, *49*, 409–436.
2. Fletcher, R.; Reeves, C. Function minimization by conjugate gradients. *Comput. J.* **1964**, *7*, 149–154.
3. Polak, E.; Ribière, G. Note surla convergence de directions conjugèes. *Rev. Fr. Inform. Rech. Oper. 3e. Ann.* **1969**, *16*, 35–43.
4. Polyak, B.T. The conjugate gradient method in extreme problems. *USSR Comput. Math. Math. Phys.* **1969**, *9*, 94–112.
5. Dai, Y.H.; Yuan Y.X. A nonlinear conjugate gradient method with a strong global convergence property. *SIAM J. Optim.* **1999**, *10*, 177–182.
6. Zoutendijk, G. Nonlinear programming computational methods. In *Integer and Nonlinear Programming*; Abadie, J., Ed.; North-Holland: Amsterdam, The Netherlands, 1970; pp. 37–86.
7. Hager, W.W.; Zhang, H. A new conjugate gradient method with guaranteed descent and an efficient line search. *SIAM J. Optim.***2005**, *16*, 170–192.
8. Kou, C.X.; Dai, Y.H. A modified self-scaling memoryless Broyden-Fletcher-Goldfarb-Shanno method for unconstrained optimization. *J. Optimiz. Theory. App.* **2015**, *165*, 209–224.

9. Dai, Y.H.; Yuan, Y.X. An efficient hybrid conjugate gradient method for unconstrained optimization. *Ann. Oper. Res.* **2001**, *103*, 33–47.
10. Andrei, N. Hybrid conjugate gradient algorithm for unconstrained optimization. *J. Optimiz. Theory. Appl.* **2009**, *141*, 249–264.
11. Jian, J.B.; Han, L.; Jiang, X.Z. A hybrid conjugate gradient method with descent property for unconstrained optimization. *J. Comput. Appl. Math.* **2015**, *39*, 1281–1290.
12. Birgin, E.G.; Martine, J.M. A spectral conjugate gradient method for unconstrained optimization. *Appl. Math. Optim.* **2001**, *43*, 117–128.
13. Andrei, N. New acceleration conjugate gradient algorithms as a modification of Dai-Yuan's computational scheme for unconstrained optimization. *J. Comput. Appl. Math.* **2010**, *234*, 3397–3410.
14. Lin, S.H.; Huang, H. A new spectral congjugate gradient method. *Chin. J. Eng. Math. (Chin. Ser.)* **2014**, *31*, 837–846.
15. Liu, J.K.; Feng, Y.M.; Zou, L.M. A spectral conjugate gradient method for solving large-scale unconstrained optimization. *Comput. Math. Appl.* **2019**, *77*, 731–739
16. Zhang, L.; Zhou, W.; Li, D.H. A descent modified Polak-Ribiere-Polyak conjugate gradient method and its global convergence. *IMA J. Numer. Anal.* **2006**, *26*, 629–640.
17. Sun, M.; Liu, J. Three modified Polak-Ribiere-Polyak conjugate gradient methods with sufficient descent property. *IMA J. Numer. Anal.* **2015**, *2015*, 125–129.
18. Li, X.L.; Shi, J.J.; Dong, X.L.; Yu, J.L. A new conjugate gradient method based on Quasi-Newton equation for unconstrained optimization. *J. Comput. Appl. Math.* **2019**, *350*, 372–379.
19. Jiang, X.Z.; Jian, J.B. Improved Fletcher-Reeves and Dai-Yuan conjugate gradient methods with the strong Wolfe line search. *J. Comput. Appl. Math.* **2019**, *348*, 525–534.
20. Jian, J.B.; Yin, J.H.; Jiang, X.Z. An efficient conjugate gradient method with sufficient descent property. *Math. Num. Sin.(Chin. Ser.)* **2015**, *11*, 415–424.
21. Perry, A. A modified conjugate gradient algorithm. *Oper. Res.* **1978**, *26*, 1073–1078.
22. Gould, N.I.M.; Orban, D.; Toint, P.L. CUTEr and SifDec: A constrained and unconstrained testing environment, revisited. *ACM Trans. Math. Softw.* **2003**, *29*, 373–394.
23. Morè, J.; Garbow, B.S.; Hillstrome, K.E. Testing unconstrained optimization software. *ACM Trans. Math. Softw.* **1981**, *7*, 17–41
24. Andrei, N. An unconstrained optimization test functions collection. *Adv. Model. Optim.* **2008**, *10*, 147–161.
25. Dolan, E.D.; Morè, J. Benchmarking optimization software with performance profiles. *Math. Program.* **2002**, *91*, 201–213.

Article

Relevant Aspects for an EF3-Evaluation of E-Cognocracy

José María Moreno-Jiménez [1,*], Cristina Pérez-Espés [2] and Pilar Rivera-Torres [3]

1 Grupo Decisión Multicriterio Zaragoza (GDMZ), Facultad de Economía y Empresa, Universidad de Zaragoza, 50005 Zaragoza, Spain
2 HARMONIA-UAM Research Group, Facultad de Ciencias Económicas y Empresariales, Universidad Autónoma de Madrid, 28049 Madrid, Spain; cristina.perez@uam.es
3 CREVALOR Research Group, Facultad de Economía y Empresa, Universidad de Zaragoza, 50005 Zaragoza, Spain; privera@unizar.es
* Correspondence: moreno@unizar.es

Received: 2 January 2020; Accepted: 15 February 2020; Published: 19 February 2020

Abstract: The search for an appropriate response to the new challenges and needs posed by the Knowledge Society in the area of public decisions has led to the development of a number of participation models whose value must be assessed and analysed in an integral manner. Using a theoretical model based on structural equations, the present work identifies the relevant factors for an EF3-approach to the democracy model named e-Cognocracy: it comprises a conjoint evaluation of its effectiveness (doing what is right), efficacy (achieving goals) and efficiency (doing things correctly). The model was applied to a real-life e-Cognocracy experience undertaken in the municipality of Cadrete, Zaragoza. The evaluation resulted in the extraction and identification of a series of relationships that allow the advancement of an EF3-participation acceptance model, in line with the TAM model of Davis and the work of Delone and MacLean, which can be used for the integral evaluation of any e-participation model.

Keywords: e-Cognocracy; citizen participation; integral evaluation; effectiveness; structural equations

1. Introduction

As exemplified by the social protest movements of 2010 (e.g., the revolution in Arabic countries) and 2011 (the Spanish '15M Real Democracy, Now!'), there is growing pressure on governments to increase civic participation in the management of society. The traditional democratic model encounters significant difficulties when it is expected to react effectively within complex, uncertain and dynamic environments. The democratic legitimacy of public institutions is being questioned by a better educated, more reflexive and more critical citizenry.

There is a need for new models of participation that can make use of the potential of the Knowledge Society (KS) and respond to the challenges (transparency, participation, control, etc.) that it generates. The determination of a model of participation that is most appropriate for a given epoch is by no means an original topic of debate and discussion; as Ibn Jaldun concluded more than six centuries ago [1], this problem may only be resolved if the dynamic of the system is understood and the system is self-organised and adaptive [2].

Participation must have a practical outcome; citizens must be able to see its impact and results. Joint and in-house evaluation procedures should be developed in order to measure the scope and impact of e-Participation experiences. Evaluation allows the visualisation of the results of an initiative and the degree of achievement in terms of the proposed objectives; it is also an expression of rigour, transparency, analysis and continuous improvement which can reinforce the consistency and credibility of participatory experiences.

Evaluation should aim to undertake the most rigorous analysis possible of the different stages and the results achieved by the e-Participation experiences. It should provide an assessment of the scientific rigour of the methodology: the effectiveness, efficacy and efficiency (EF3-approach), as well as the economic, social and environmental impacts of the actions.

Macintosh and White [3] put forward three arguments for a rigorous evaluation framework: (i) the increasing amount of information available on the internet requires new knowledge and information management systems; (ii) the range of stakeholders demands personalised communication integrated with the delivery of relevant information; and (iii) information systems design must move towards more collaborative working environments to support a partnership of government and civil society.

Aichholzer and Westholm [4] argue that the evaluation of e-participation is indispensable if knowledge of greater precision and objectivity is wanted about the effectiveness, the value, the success of an e-participation project, initiative or programme.

In the last 15 years, a number of e-participation experiences have taken place and many of them have made extensive use of information and communication technologies (ICTs). Some examples and descriptions of e-participation experiences can be found in [5–9]. Many of these projects have made a contribution to the revitalisation of democracy by increasing transparency in governance and creating new political spaces for communication and participation.

Since the emergence of e-Cognocracy as a new model of citizen participation [10–12] in 2003, it has been widely studied from a variety of perspectives and viewpoints: political, sociological, scientific, technological and economic [13–17]. However, an integral evaluation of the impact of the application of e-Cognocracy in real-life situations has yet to be undertaken.

This paper presents a theoretical framework for selecting the most appropriate participation model, based on structural equations, which identifies the relevant factors for an integral evaluation of e-Cognocracy; the evaluation simultaneously considers (EF3-approach) its effectiveness (doing what is right), efficacy (achieving goals) and efficiency (doing things correctly). The framework was applied to the evaluation of a real-life practical initiative on the design of cultural and sporting policies for the Municipal Council of Cadrete, Zaragoza. The implementation resulted in the extraction and identification of a series of relationships that allow the advancement of an EF3-participation acceptance model (EF3-PAM), in line with the Technology Acceptance Model (TAM) [18] and the work of Delone and MacLean [19].

The structure of the article is as follows: after this brief introduction, Section 2 deals with the evaluation of e-Participation; Section 3 offers an evaluation of E-Cognocracy and presents the case study; and Section 4 details the main conclusions that can be drawn from the work.

2. The Evaluation of e-Participation

2.1. e-Participation

In the context of governance, participation is a basic democratic right. Sartori [20] describes it as "personally taking part in something; self-activated, willed taking part, not caused by others or mobilised from above". In general, the term participation includes a range of concepts, from formal participation (exercising the right to vote and the reception of administrative actions) to active participation or political participation in the resolution of conflicts and making public decisions (governance).

The electronic government (e-Government) of society can be defined as the utilisation of ICTs in Public Administration with the aim of improving public services and democratic processes, thereby strengthening support of public policies [21]; e-Government can be applied to the two broad areas of Public Administration activity: (i) the provision of services (e-Administration) and (ii) political participation in democratic processes (e-Governance). Nevertheless, the term e-Participation has, traditionally, been exclusively associated with political participation or e-Governance in its widest sense (participation in public decision making relative to the governance of society).

Given this interpretation of e-Participation (or e-Governance), e-Cognocracy can be considered as a new model of participation which, in the context of e-Governance, provides cognitive collaborative public decision making [22] and addresses [16] the demand of Michels and De Graaf [9] for e-participation models: " ... to think about smart ways of combining participatory processes and formal decision-making, for example, by splitting the process into different phases and coupling participatory and deliberative phases to the phases of decision-making."

2.2. e-Cognocracy

E-Cognocracy [10–12,23–25] is a new system of democratic representation that combines liberal (representative) democracy and direct (participative) democracy to cognitive ends. It seeks the creation and social diffusion of knowledge and the construction of a more open, transparent, cultured, educated and freer society; a society that is more cohesive and connected, more participative, egalitarian and cooperative.

The new system uses multicriteria decisions as its methodological support, the internet as its communication support and the democratic system as a catalyst for learning. As explained in [16,26], the e-Cognocracy methodology consists of 16 stages grouped in four blocks (Figure 1): (1) Problem Formulation, Stages 1 and 2; (2) Problem Resolution, Stages 3 to 10; (3) Knowledge Extraction and Democratisation, Stages 11 to 14; and (4) Evaluation and Documentation, Stages 15 and 16.

Block 1, Problem Formulation, refers to the initial problem proposed by the representatives or the citizens, its presentation to the actors involved in its resolution and its final setting.

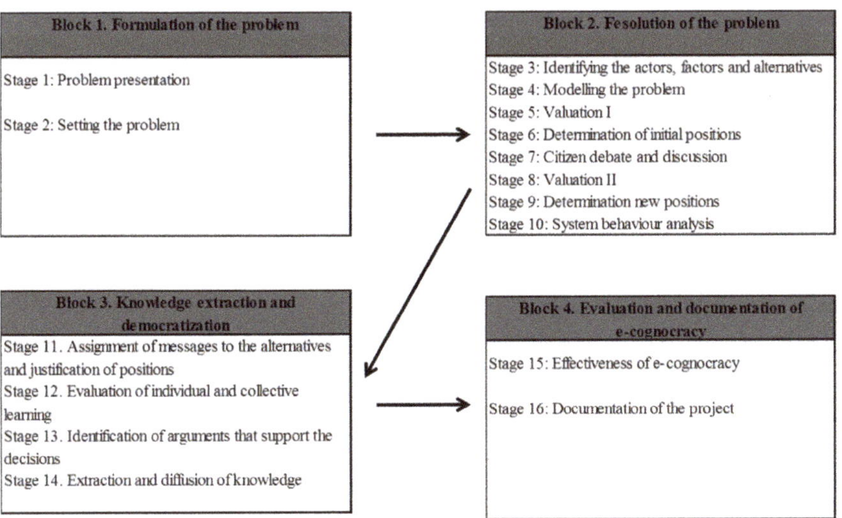

Figure 1. Blocks and stages of the e-Cognocracy methodology [16].

Block 2, Problem Resolution, includes the two voting rounds, in which the scientific resolution of the problem is obtained, using the Internet as communication support and the analytic hierarchy process (AHP) as the multicriteria decision making technique. The rounds are interspersed with an e-discussion process, in which the actors justify their preferences. The online discussion is the step prior to the extraction of knowledge that takes place in third block. The discussion allows the creative capacity of all individuals interested in the resolution of the problem to be incorporated into the decision making process and this accords with Dahl's proposals [27,28] on the improvement of democracy based on the use of ICTs. The e-discussion allows the active citizens, a *minipopulus*, to complement the institutions in order to reduce the gap between the representatives/politicians, and the represented/citizens. As argued by Habermas [29], the e-discussion fosters the co-creation of a more cohesive, fair, educated and effective society.

Block 3, Knowledge Democratisation, provides the arguments that support the different positions [14,15], identifies the social leaders (the citizens whose arguments are followed by the majority of the citizens) and shares this knowledge in order to generate individual and social learning.

Finally, and as recommended for any procedure that makes use of public funds, Block 4 analyses the effectiveness (doing what is right), the efficacy (achieving goals) and the efficiency (doing things correctly) of the e-Cognocracy public policy making process.

The methodology for the conjoint design (politicians and citizens) of local public policies comprises the following phases or steps [11,23]: Step1: Project presentation; Step 2: Problem presentation; Step 3: Identification of the actors, factors and alternatives; Step 4: Modelling the problem; Step 5: Evaluation; Step 6: Identification of the initial positions; Step 7: Citizen debate and discussion. Step 8: Evaluation II; Step 9: Identification of new positions; Step 10: System behaviour; Step 11: Assignation of messages to alternatives and justification of positions; Step 12: Evaluation of individual and collective learning; Step 13: Identification of arguments that support the decisions; Step 14: Extraction and diffusion of knowledge; Step 15: Effectiveness of e-Cognocracy; and Step 16: Project documentation (final report).

The new methodology for the evaluation of the behaviour of models of citizen participation in the taking of public decisions uses an EF3 framework that allows the simultaneous evaluation of the model's efficiency, efficacy and effectiveness.

2.3. The Evaluation of e-Participation

E-participation is still evolving and there have only been a few proposals on evaluation. Bagozzi and Warshaw [30] advanced a secure and stable technology model for predicting users' acceptance of a range of new technologies that has been widely employed and studied in the last decades. Fred Davis [18] suggested a technology acceptance model (TAM) that explains the process of acceptation of information technology at an individual level. The Davis model is probably the most recognised and utilised in the scientific literature, it is commonly referenced and has been the inspiration behind a number of other, similar, models.

Delone and MacLean [19] designed the 'Information System Success' model as a conceptual framework for measuring the complex dependent variable in research on information technology systems. Ten years later, the same authors updated the original model based on changes in the management of information systems [31].

Rowe and Frewer [32] put forward a framework to evaluate participation in general (it was not specific to e-participation); they defined a number of theoretical criteria which are essential for effective public participation and divided them into two types: acceptance criteria and process criteria. Acceptance criteria refer to representativeness, independence, early involvement, influence and transparency that offer a measure of acceptability to the wider public; process criteria refer to resource accessibility, task definition, structured decision making and cost-effectiveness that offer a measure of effectiveness.

Henderson and Henderson [33] constructed a model for evaluating on-line consultations, e-petitions and internet live broadcasting of parliamentary initiatives. There are seven evaluation dimensions and indicators: Effectiveness; Equity; Quality; Efficiency; Appropriateness; Sustainability; and Process.

Macintosh and Whyte [3] developed an evaluation methodology with criteria that cover three perspectives of an e-participation experience: (i) democratic—the overarching democratic criteria addressed by the experience; (ii) project—the identification of the aims and objectives; and (iii) socio-technical—the extent to which the ICTs directly affect the outcomes. Each evaluation perspective is linked to a number of criteria:

- Democratic criteria: representation, engagement, transparency, conflict and consensus, political equality, community control;

- Project criteria: engaging with a wider audience, obtaining better informed opinions, enabling more in-depth consultation, cost-effective analysis of contributions, providing feedback to citizens;
- Socio-technical criteria: social acceptability, usefulness, usability.

In 2009, [4] published an e-Participation model as part of the DEMO-net project, undertaken in cooperation with other European researchers [34]. They reviewed and analysed applied methods for the evaluation of e-participation and gave core criteria and indicators relevant to e-participation activities such as consultation and deliberation.

Mamaqui and Moreno-Jiménez [35] formulated a methodology based on the utilisation of Structural Equation Models (SEM) for the evaluation of the effectiveness of e-Cognocracy.

Luna-Reyes, Gil-García and Romero [36] devised a multidimensional system for measuring and evaluating electronic government. The model seeks to incorporate the approaches currently used to measure electronic government and ideas from published literature related on the issue.

Finally, Wimmer and Bicking [37] gave us the impact evaluation framework, based on evaluation methods of empirical research thereby reflecting the programmatic contexts of the projects. Evaluation is based on the interaction of the elements of a holistic e-participation solution: the participation process; the topics to be discussed; the policy; and the technology and tools employed.

This current paper proposes a theoretical framework based on the three dimensions for evaluating e-participation processes: effectiveness, efficacy and efficiency. The framework is designed for the evaluation of citizen participation experiences and projects, particularly those that involve e-Cognocracy.

3. The Evaluation of e-Cognocracy

3.1. Background

Most researchers, administrators and users of electronic public services agree that the adoption and use of the services depends, to a great extent, on the benefits that are perceived by the citizens. To foment the use of electronic public services by citizens, public administrations need to know the key aspects that encourage their adoption. Given that the online delivery of public services is based on the use of technology, the variables and models of technological utilisation have acquired a fundamental role beyond the factors traditionally employed in the previous literature.

Al-Adawi, Yousafzai and Pallister [38] produced a model that follows the TAM and explains intention to use e-Government websites by postulating four direct determinants: perceived usefulness, perceived ease-of-use, trust, and perceived risk. The TAM offers a promising theoretical foundation for examining the factors contributing to the acceptance of new technologies; it has been successfully applied in customer behaviour, technology take-up, system use and in a variety of studies human behaviour.

As previously mentioned, at the end of the 1980s, Bagozzi and Warshaw [30] created a secure and stable technology model for predicting user acceptance of a range of new technologies. In 1986, in his doctoral thesis, Fred Davis [18,39] defined a technology acceptance model (TAM) that explains the process of acceptance of information technology at an individual level. It is in line with the tradition of previous research into information systems [40–42] that aimed to identify the attributes that lead to the success of information systems in business, taking user satisfaction as its measure.

The TAM is based on the Theory of Reasoned Action (TRA) [43] and its methodology of expected values. The authors of the model had already used the TRA in their research. Bagozzi [44] used it in a work on blood donation, Warshaw and Davis [45] in a number of studies and Warshaw [46] in research on brand selection. The TAM permits evaluation measurements of the quality of information technology systems and their adjustment to the requirements of the tasks that are to be executed; it is used to predict the level of acceptance and use of new technologies. The model (Figure 2) assumes that attitudes toward the use of an information system are based on two antecedent variables: (i) Perceived usefulness (PU) and (ii) perceived ease-of-use (PEOU).

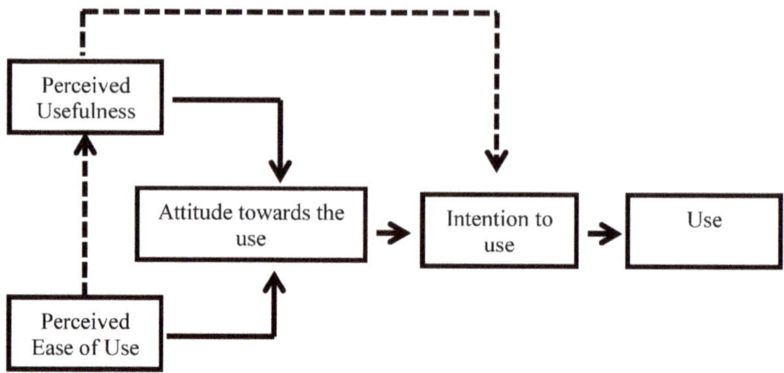

Figure 2. Basic TAM model [18].

This is similar to Bandura's [47] concept of self-efficacy. Perceived usefulness is defined as "the degree to which a person believes that using a particular system would enhance his or her job performance" [18]. The items employed by Davis [18] as indicators of perceived usefulness are directed at determining people's evaluations of the consequences that the use of an information system may have on their productivity at work.

Another of the fundamental constructs of the TAM is the perceived ease-of-use of a technology, which is based on the self-efficacy of Bandura [47], defined by Davis [18] as "the degree to which a person believes that using a particular system would be free from effort". The items that measure this concept are: flexibility; ease-of-use; control; and the simplicity of becoming an expert in its use.

In the TAM, a direct link is made between beliefs (perceived usefulness) and intentions. This is a significant difference with respect to the Theory of Reasoned Action [43], where beliefs only have an impact on attitudes.

In 1992, DeLone and McLean [19] established an Information System Success Model, based on research on communications carried out by Shannon and Weaver [48]. They specified two quality dimensions: System quality, that measures technological success, and, Information quality, that measures semantic success (the content). The model was founded on an analysis of the literature from 1980 to 1990, consisting of around 100 publications. The original model was applied and its results published in more than 300 papers, guaranteeing the applicability of the model.

In 2003, they updated this model to include three dimensions of quality that affect use and user satisfaction: Information quality, system quality and service quality that measure the quality of service delivery. In the updated model, Intention to use refers to attitude; use refers to behaviour (Figure 3).

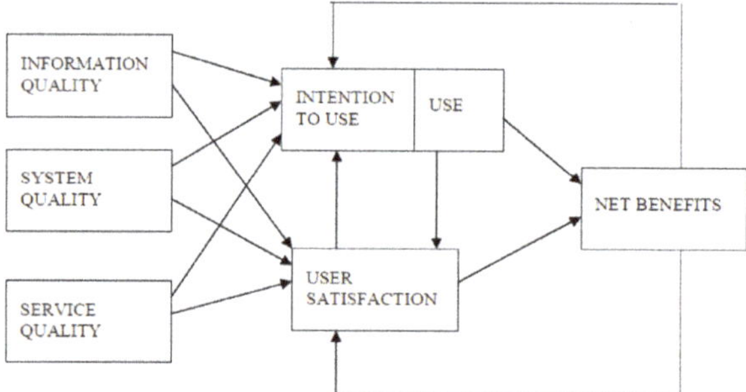

Figure 3. The DeLone and McLean model [31].

The two models were utilised for the development of a new approach for the evaluation of e-Cognocracy: the Theoretical EF3-framework.

The Theoretical EF3-Framework

Considering the evaluation of enterprise behaviour [11], the EF3-approach contemplates the following ideas:

(a) Effectiveness: a political criterion associated with strategic planning or long-term behaviour, related to aspects relevant to the resolution process (doing what is right);
(b) Efficacy: an administrative criterion concerning tactical planning or medium-term behaviour, related to measuring how well the goals that are set are achieved (achieving goals); and
(c) Efficiency: an economic criterion linked with operational planning or short-term behaviour that measures the best possible allocation of public resources (doing things correctly).

The theoretical EF3-framework for the integration of effectiveness, efficacy and efficiency can be seen as an extension of the TAM and DeLone and McLean approaches: The perceptions and behaviour of citizens are used to evaluate the processes of citizen participation and the adoption of technology.

The EF3-framework (Figure 4) was designed for the identification of the relevant aspects required for evaluating e-Cognocracy, but it can also be employed in the evaluation of any e-participation model.

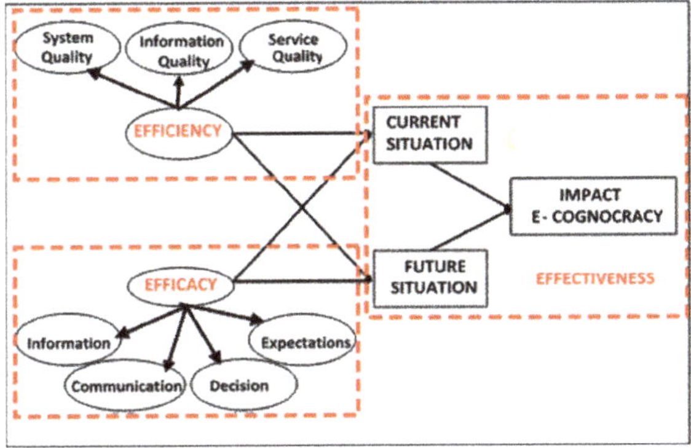

Figure 4. Theoretical EF3-framework for the Evaluation of e-Cognocracy (source: the authors).

In the theoretical EF3-framework: *efficiency* is "the operational improvement of the current democratic system"; *efficacy* is the capacity of the current democratic system to "defend the interests of the citizens through their representatives"; and, *effectiveness* is "the conjoint creation of a better society".

The relevant aspects determining efficiency are based on the three constructs contemplated by the DeLone and McLean model [31]: the IT application (system quality), the information that is obtained (information quality) and the human resources support (service quality).

Four constructs are considered for the evaluation of efficacy: information, communication, decision and expectations. Information is a unidirectional flow of interaction (usually from the administration to the citizens); communication is understood as two-way interaction: debate and discussion. In addition to the bi-directional flow of information; decision includes the production of a co-decision between the administration and citizens, and expectations refer to the identification of the characteristics that participation experiences should have in the future.

Effectiveness is studied through the analysis of two scenarios as latent intermediate variables: the current situation and the future situation (ideal), and an endogenous variable that captures the impact of e-Cognocracy (the creation of a better society).

The following section of this work presents the application of the theoretical EF3-framework through a survey implemented in the real-life experience of Cadrete (e-Cognocracy) using SEM or the covariance structure analysis approach [49–52].

3.2. Relevant Aspects: The Estimated Structural Model

The identification of relevant aspects for the EF3 evaluation of e-Cognocracy, using structural equation models, was by means of a survey undertaken after the implementation of the project in Cadrete.

3.2.1. Case Study

In April 2010, the Cadrete Municipal Council, in collaboration with the University of Zaragoza, undertook a citizen participation project aimed at giving the residents of the municipality a voice in public decision making. The project was financed by the Government of Aragon and organised by the Zaragoza Multicriteria Decision Making Group. The issue in question was the design of cultural and sporting policies. There were two main objectives: (i) That decisions on the budget assigned to the aforementioned policies would be conjointly made by the politicians and the citizenry; and (ii) that citizens would be encouraged to involve themselves in the debate and take part in the decision making process and the arguments that supported the decisions would be publicly disseminated.

Participation was encouraged by the incorporation of a new group of actors: the neighbourhood associations. There were therefore three groups of actors that were given different weightings: (i) The politicians, with a weighting of 40%; (ii) the citizens, of 44%; and (iii) the local associations, of 16%.

The participants were local residents (on the electoral register) of over 18 years of age (politicians, citizens and representatives of the local associations). There were two voting options: with National Identity Card or with username and password. As was the case with Mamaqui and Moreno-Jiménez [35], the analytic hierarchy process [53,54] was used as a methodological support and the Internet as the communication support.

The census of actors that fulfilled the requirements necessary to participate and the weights awarded to each group are shown in Table 1. The voters could determine what proportion of economic and financial resources should be allocated to each of the four segments of the population considered: children (0–14 years old), the young (15–29 years old), adults (30–64 years old) and the elderly (over 65 years old).

Table 1. Voters and Weightings.

Participants	Census	Weight
Associations	15	16
Citizens	1949 *	44
Politicians	11	40
Total	1975	100

* Over 18s with the right to vote in 2008 (data from the Aragon Statistics Institute) (source: the authors).

A hierarchy with two criteria (Cultural and Sports) and six sub-criteria was constructed (Figure 5). Within the criterion associated with cultural aspects, three sub-criteria could be selected: Education, Leisure and Identity. Within the sports criterion, the selection sub-criteria were: Entertainment, Physical Development and Social Relations.

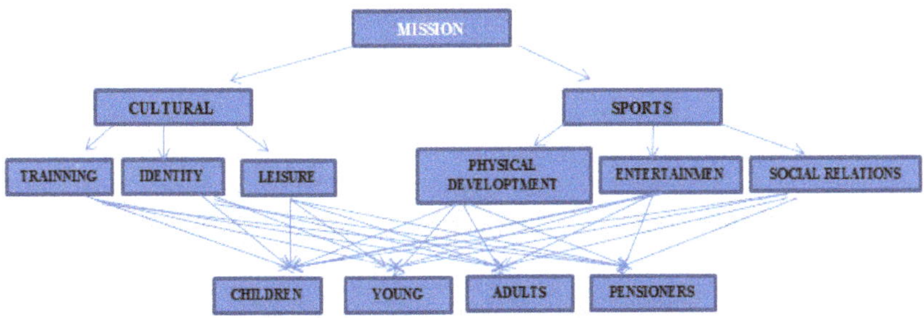

Figure 5. Hierarchy of the Cadrete experience (source: the authors)

3.2.2. Phases of the Process

Following the structure of the e-Cognocracy democracy model [11,12,23–26], the process of participation was:

1. Problem formulation.
2. Information and training.
3. Modelling the problem: following the methodology of the analytic hierarchy process [53], one of the most widespread multicriteria techniques.
4. First round of voting: Those inscribed in the citizen participation census could vote on their (cultural or sports) preferences between 1:00 p.m. and 10:00 p.m., on the 8 April 2010.
5. Discussion: From the 8th to the 16th of April, a forum was opened on the Internet. Any citizen, even if they had not logged in for the voting process, could freely express their opinions, proposals, suggestions and ideas, and respond to the comments of the other citizens.
6. Second round of voting: As in the first vote, those inscribed in the citizen participation census could vote electronically on their preferences (cultural or sports) from 12:00 midday to 7:00 p.m. on the 16 April 2010.
7. Presentation of the results and closing ceremony: This took place on the 23 April 2010. The results were announced and a prize draw was held.

In order to evaluate the experience, at the end of the elicitation process (on the same day and at the same location), the participants were asked to complete a questionnaire (see Appendix A) comprising 51 questions grouped into seven sections: (i) The System of Citizen Participation; (ii) The Creation of a Better Society; (iii) Motivation; (iv) Evaluation of the Technology Support and Applications; (v) Evaluation of the Information; (vi) Evaluation of the Support Personnel; and (vii) Overall Evaluation. The information from the questionnaire was used to analyse the effectiveness of e-Cognocracy.

3.2.3. Methodology

The methodological support used in the voting experiment was the analytic hierarchy process (AHP), a multicriteria technique proposed by T.L. Saaty [53]. First, a hierarchy (Figure 5) that contains the relevant aspects of the problem was constructed by the research team responsible for the Cadrete project. Before eliciting the citizens' preferences, a brief presentation of the AHP methodology was made to the citizens by the research team. Then, following a top-down procedure, the individuals incorporate their preferences by means of the pairwise comparisons of the elements considered, following Saaty's fundamental scale (Table 2). Finally, the methodology aggregates the values throughout the hierarchy to obtain the total priority of each alternative with respect to the objective of the problem [54].

Table 2. Fundamental scale [53].

Intensity	Definition	Explanation
1	Equal importance	Two activities contribute equally to objective
3	Moderate importance	Experience and trial slightly favour one activity over the other
5	Strong importance	Experience and trial strongly favour one activity over the other
7	Very strong importance	An activity is much more favoured than the other and its dominance is demonstrated in practice
9	Extreme importance	Evidence in favour of an alternative over another has the highest possible order for his claim

3.2.4. Results

In the first round of voting (Table 3), 43 participants voted electronically; 37 were citizens, three were politicians and there were three from associations. This represented 2.17% of the total census and 14.96% of the weighted participation. In this round, the participants opted for the cultural criteria (52.99%) over sports (47.01%) (Table 4). The priorities of both rounds can be seen in [16].

Table 3. The two rounds of voting.

Participants	Census	Electronic Voting in the 1st Voting	Percentage	Electronic Voting in the 2nd Voting	Percentage
Associations	15	3	20%	2	13.3%
Citizen	1949 *	37	1.9%	35	1.8%
Politicians	11	3	27.3%	4	36.7%
Total weighted	1975	43	2.17% (14.96%)	41	2.08% (17.60%)

* Citizens over 18 in the 2008 Census (data from the Aragon Statistics Institute) (source: the authors)

Table 4. Priorities of the criteria of each group of actors.

Criteria		Associations	Citizens	Politicians	Total Vote
1st vote	Cultural	39.33%	57.64%	53.35%	52.99%
	Sports	60.67%	42.36%	46.65%	47.01%
2nd vote	Cultural	54.88%	62.56%	50.47%	56.58%
	Sports	45.12%	37.44%	49.53%	43.42%

(Source: the authors).

Between the first round and the second round, a forum of debate was opened on the Internet, participants expressed their concerns and preferences (Table 5). A total of 61 messages were posted, 37 belonged to the cultural criteria and 24 to sports. There were 195 comments about these messages, 114 belonged to the cultural criteria and 81 to sports. In the second round of voting (Table 3), there were 41 participants, 35 were citizens, four were politicians and two were from associations, a weighted participation of 17.60% (2.08% of the total census).

Table 5. Messages and comments on the Cadrete forum.

	Total Messages	Total Comments	Total
Sport	24	81	105
Culture	37	114	151
Total	61	195	156

(Source: the authors).

In the two rounds, the variations in absolute terms were minimal, although, due to the small number of voters, the relative variation, at least with respect to politicians and associations, shows significant modifications. The percentage of politicians voting increased (from the first to the second round) by 33.3% and that of associations fell by the same figure (33.3%). The voters again opted for cultural criteria (56.58%) over sports (43.42%), by an increased margin (Table 4).

The low citizen participation in the electronic experience contrasts with the municipal (2007) and the general (2004 and 2008) elections whose levels of participation were 69.9% and 76.6%, respectively (Table 6).

Table 6. Participation in electoral processes in Cadrete.

Elections	Participation Rate
General Elections 2004	80.2%
Municipal Elections 2007	69.9%
General Elections 2008	76.6%
Electronic consultation 2010	2.08% (17.60% weighted)

(Source: the authors).

3.3. Estimated Structural Model

After the e-participation experience in Cadrete, participants were asked to complete an online questionnaire; 24 residents responded though only 20 replies were valid. The questionnaire was considered as invalid if less than 80% of the questions were answered and/or there was zero variability with regards to the total number of questions.

The measurement scale was from 0 to 10 (0 = total disagreement, 10 = total agreement). A total of 51 questions were grouped into seven sections: (i) The System of Citizen Participation; (ii) The Creation of a Better Society; (iii) Motivation; (iv) Evaluation of the Technological Support and Applications; (v) Evaluation of the Information; (vi) Evaluation of the Support Personnel; and, (vii) Overall Evaluation.

The theoretical EF3-framework was first evaluated through a survey in Cadrete using the SEM or the Covariance Structure Analysis approach [49–52]. The first analysis was a conjoint descriptive analysis of the variables, using the measurements of their position, dispersion and correlation. These measurements led to the identification of groups of interrelated indicators that could be expected to define the utilised constructs; the analysis was complemented by an examination of the main components. The study was completed with structural equation models with latent variables, or covariance structure analysis [51].

This methodological approach was chosen as it allows the researcher to formulate and evaluate the existence of latent variables from the reflected indicators [50], that is to say, variables that are not susceptible to direct observation. The software used was EQS 6.1 [52]. Table 7 shows the constructs and indicators used to describe the effectiveness, efficacy and efficiency in the estimation of the structural model for an EF3-evaluation of e-Cognocracy.

The indicators concerning *information*, *communication* and *decision*, revealed evaluations that were just above or just below an acceptable level of satisfaction (5.0, 4.7, 5.5, 5.0, 5.2 and 5.1, respectively); *expectations* reached a high level (7.5 and 7.2), which suggests that citizens are not convinced about actively participating in the system. Therefore, efficacy is defined by the indicators *information*, *communication*, *decision* and *expectations*.

Table 7. Constructs and Indicators.

EFFICACY			*
INFORMATION	X1	The Administration informs society about the existing mechanisms of citizen participation	5.00
	X2	The Administration informs society about the decision taken	4.70
COMMUNICATION	X3	Political powers take citizens' opinions into account for the design public policies	5.00
	X4	Political powers take associations' opinions into account for the design public policies	4.95
DECISION	X5	The citizens have weight in political decision making	5.15
	X6	Associations have weight in political decisions making	5.05
EXPECTATIONS	X7	Citizens should participate in the design of public policies	7.50
	X8	Citizens should decide the design of public policies in conjunction with the elected representatives	7.15
EFFICIENCY			
INFORMATION QUALITY	X9	In general, I am satisfied with the information that I have received	6.90
SYSTEM QUALITY	X10	In general, I liked the design of the software application	5.95
EFFECTIVENESS			
CURRENT SYSTEM	Y1	With the current system of citizen participation, the representatives defend my interests	5.50
FUTURE SYSTEM	Y2	E-Cognocracy improves the current democratic system	7.90
E-COGNOCRACY	Y3	E-Cognocracy contributes to create a better society	7.70

* Average indicator evaluation. (Source: the authors).

In general, the average valuations of the indicators of efficiency were relatively high. Participants gave a more positive valuation to *service quality* (8.50) than to *information* and *system quality* (7.0 and 6.0, respectively). The correlation was over 0.7, which indicates that the structure should be maintained. Therefore, efficiency is defined by the participants' satisfaction with the *quality of information and of the system.* In our empirical study, the third construct (personal support) considered by DeLone and McLean was centred on informing about the AHP methodology, thus it was integrated into the Information Quality.

E-Cognocracy was given a more positive evaluation than the current system of citizen participation (*current situation*), both in terms of its ability to improve the current system (*future situation*) and to achieve its ultimate goal-the creation of a better society (7.9 and 7.7 compared to 5.5). It is, therefore, clear that the improvement of the current system and the creation of a better society require the introduction of more dynamic and participative mechanisms, such as those employed by e-Cognocracy. All of the above can be defined as effectiveness. The role of the citizen is perceived as being as relevant as that of the associations. The relationships between the indicators were coherent with the constructs.

The confirmatory factor analysis (Table 8) offered sufficient evidence to maintain a first order structure of four correlated factors (Information, Communication, Decision, Expectation) from the eight observed variables (X1, . . . , X8). The structures of the constructs, in theoretical terms, were empirically corroborated at both an exploratory and a confirmatory level. The eight indicators reflected a structure of four interrelated latent variables.

Table 8. Confirmatory factor analysis model.

		INFO.	COM.	DEC.	EXP.	Efficiency	R^2
INFORMATION	X1	0.99 ***					0.98
	X2	0.87 ***					0.76
COMMUNICATION	X3		0.95 ***				0.90
	X4		0.98 ***				0.96
DECISION	X5			0.90 ***			0.81
	X6			0.78 ***			0.61
EXPECTATION	X7				0.70 ***		0.49
	X8				0.97 ***		0.94
EFFICIENCY	X9					0.83 ***	0.69
	X10					0.86 ***	0.74
INFORMATION	INF.	1					
COMMUNICATION	COM.	0.40 *	1				
DECISION	DEC.	0.26*	0.57 **	1			
EXPECTATIONS	EXP.	−0.10	−0.04	−0.26	1		
EFFICIENCY	Efficiency	0.29 *	−0.37 *	−0.28 *	0.27	1	
	C-FL	0.87	0.93	0.71	0.72	0.71	
	Omega	0.93	0.97	0.84	0.84	0.85	

$\chi 2(25)$: 26.97, p-value: 0.52; SRMR: 0.07; GFI: 0.90; CFI: 0.99. * significant to 10%, ** significant to 5% and *** significant to 1%. (Source: the authors).

Table 8 also shows that the relationships between the dimensions of *information, communication* and *decision* were positive and this implies that a higher perception of information signifies a higher perception of *communication* and *decision*, and vice versa. The relationship between *expectations* and *decision* was negative, and this implies that citizen disappointment with regards to the existence of *decision* is greater if they have greater *expectations* of participation. Reliability indices for both these observed variables and their respective latent variables were more than acceptable (R^2, omega and C-FL).

Figure 6 depicts the estimated structural model. Table 9 shows the estimated structural model; there are three determinants of the citizens' perception of the current system: *information, communication* and *efficiency*. Determinants of the *future situation* are *decision* (negative) and *expectations*. The *current and future situations* affect the perception of the creation of a better society, though the effect of the *future situation* (e-Cognocracy) is greater.

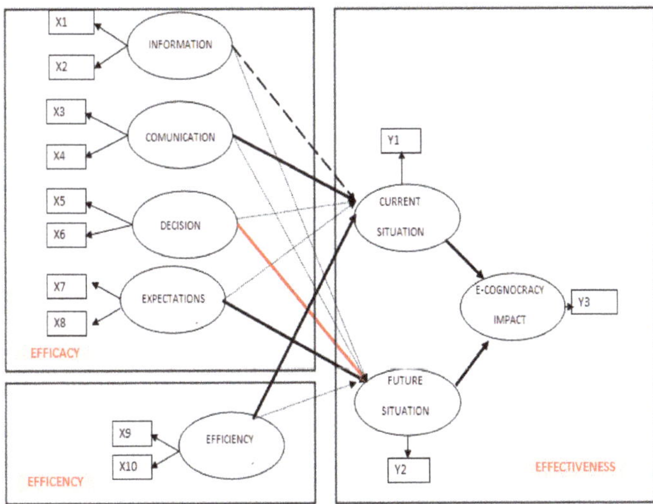

Figure 6. Estimated structural model for an EF3-evaluation of e-Cognocracy. (Source: the authors).

Table 9. Structural model.

	INFO.	COM.	DEC.	EXP.	Efficiency	CUR.S.	FUT.S.	R^2
CURRENT SYSTEM	0.31 *	0.38 **	0.11	0.04	0.50 **			0.58
FUTURE SYSTEM	0.14	0.05	−0.45 *	0.38 *	0.03			0.40
E-COGNOCRACY						0.37 **	0.51 ***	0.41

$\chi 2(46)$: 43.16, *p*-value: 0.71; SRMR: 0.07; GFI: 0.90; CFI: 0.99. * significant to 10%, ** significant to 5% and *** significant to 1%. (Source: the authors)

If citizens perceive that information exists, that is to say, that the administration informs society of the participation mechanisms and the decisions that are taken (top-down unidirectional flow), then they have a positive perception of the current system of representation.

If communication exists, that is to say, information flows in both directions (feedback), it is also a determinant, but this is not the case with *decision* and *expectations*. Moreover, if citizens feel that they have no influence on the taking of political decisions, irrespective of their perceptions of the existence of information and communication, they will favour a change in the participation system. If the citizens have higher expectations of involvement in the design and formulation of public policies, they will also favour change.

Due to the limited number of responses, it was not possible to validate a general framework for the conjoint evaluation of all the aspects outlined in the theoretical EF3-framework. Nevertheless, the results obtained from the 20 valid responses identified a series of relationships that contribute to the formulation of a general framework.

The small sample size means that the evaluation and selection of the models is governed by goodness of fit indicators (SRMR, GFI and CFI) that do not directly depend on the number of observations [49]. For all the measured and/or structural models, the estimated parameters were presented in their completely standardised version, norm 0–1, and, in addition, all the equations were given their corresponding coefficients of explained variance.

The assessment of the construct is based on the methodology proposed by Bagozzi [55] for the validation of multidimensional constructs and the covariance structure analysis of observed variables (McDonald's omega coefficient [56] and Fornell and Larcher's coefficient, C-FL [57]. The stability of the parameters of the models was estimated and evaluated sequentially.

Although the simplified analysis of the theoretical EF3-framework has not allowed significant statistical conclusions, it has meant that, in conjunction with a review of the existing literature, this framework may be extended to evaluate any e-Participation experience, not only e-Cognocracy.

4. Conclusions

The traditional democratic system finds it difficult to efficaciously react in the context of a dynamic, complex and uncertain environment. The democratic legitimacy of public institutions is being questioned by a citizenry that is more and more educated, reflexive, critical and interconnected. Citizens are demanding more open and receptive governments that are prepared to listen, share, and co-decide.

In the search for an appropriate response to the needs of democracy in the epoch of the Knowledge Society, new models of e-participation and systems are being advanced; the validity of these models needs to be analysed by taking into account their effectiveness, efficacy and efficiency (EF3-approach). The integral evaluation of the three perspectives in citizen participation models is still pending; in this work, the EF3-approach has been applied to a real-life experience resolved by means of the cognitive democracy model known as e-Cognocracy. Through the use of structural equations, the relevant factors extracted for the EF3 evaluation of this model will help to analyse the performance of the remaining citizen participation models.

The theoretical model (EF3) was applied to the design of cultural and sporting policies in the Spanish municipality of Cadrete. Due to the limited number of observations, it was not possible to validate a generalised model for the integral evaluation of e-participation experiences. Nevertheless, it was possible to extract a series of relationships and relevant factors that will be considered when dealing with similar cases in the future.

Despite the small number of responses to the evaluation questionnaire, it appears that the experience was seen as very positive by the people of Cadrete. These kinds of projects should be continued; they make a contribution to the conjoint construction of a better society.

Among the relevant aspects derived from the study we can highlight the followings: (i) The efficiency is determined by information and system quality; (ii) The efficacy is explained by information, communication, decision and expectations; and, the effectiveness (creation of a better society) is studied by analysing the change (impact) between the current and future situations.

More specifically, the high value given for the evaluation of expectation is a reflection of the uncertainty that citizens feel with regards to actively participating in the system. The role of the citizen was perceived to be of the same importance as that of the local associations.

The confirmatory factor analysis offered sufficient evidence to maintain a structure of four, interrelated, first order factors based on the eight observed variables. The relationships between the dimensions of information, communication and decision imply that a greater perception of levels of Information will mean that perceptions of communication and decision are higher, and vice versa.

The relationship between participation expectation and decision was negative. This implies that when citizen disappointment regarding the existence of decision increases, there is also an increase in participation expectation. The sample gave a more positive evaluation of e-Cognocracy than the traditional democratic model, both in terms of its ability to improve the present system and to achieve its ultimate aim-the construction of a better society.

As already mentioned, a limitation of the study (that could be resolved with further research) is the small number of responses to the evaluation questionnaire. It is well known that a higher number of responses improves statistical robustness so it is difficult to validate more generalised models. However, one of the strengths of the study is that the analysis was of a real-life situation, and this enhances the internal validity of the research.

Finally, it should be emphasised that the improvement of the system and the creation of a better society require the introduction of more dynamic participation mechanisms such as e-Cognocracy. The ideas extracted in this project for the evaluation of e-Cognocracy could be used for the evaluation of any democracy model that combines, as Michels and De Graff [9] assert, participatory processes and formal decision-making.

Author Contributions: Conceptualization, J.M.M.-J. and P.R.-T.; Formal analysis, J.M.M.-J., C.P.-E. and P.R.-T.; Investigation, J.M.M.-J., C.P.-E. and P.R.-T.; Methodology, J.M.M.-J., C.P.-E. and P.R.-T.; Project administration, J.M.M.-J.; Writing—original draft, J.M.M.-J. and C.P.-E.; Writing—review & editing, J.M.M.-J and C.P.-E. All authors have read and agreed to the published version of the manuscript.

Funding: This research was funded by the Spanish Ministry of Economy and Competitiveness and FEDER funding, project ECO2015-66673-R.

Acknowledgments: The authors would like to acknowledge the work of English translation professional David Jones in preparing the final text.

Conflicts of Interest: The authors declare no conflict of interest.

Appendix A. The Cadrete Questionnaire

After finishing the real-life e-Participation experience based on e-Cognocracy (Cadrete), participants were asked to complete an online questionnaire; 24 residents responded though only 20 replies were valid. The questionnaire was considered as invalid if: (i) less than 80% of the questions were answered; and/or (ii) there was zero variability with regards to the total number of questions. The measurement scale was from 0 to 10 (0 = total disagreement, 10 = total agreement). A total of 51 questions were grouped into seven sections:

(i) The System of Citizen Participation:

1. With the current system of citizen participation, the representatives defend my interests
2. The Citizen has weight in political decision making
3. Associations have weight in political decision making
4. The Citizen should participate in the design of public policies
5. Associations should participate in the design of public policies
6. The Citizen should decide the design of public policies in conjunction with the elected representatives
7. Associations should decide the design of public policies in conjunction with elected representatives
8. Political powers take citizens' opinions into account for the design public policies
9. Political powers take associations' opinions into account for the design public policies
10. The Administration informs society about the existing mechanisms of citizen participation
11. The Administration informs society about the decisions taken
12. E-Cognocracy contributes to the creation of a better society

(ii) The Creation of a Better Society:

1. Participation is limited to Citizen consultation by the Administration
2. Participation includes Debate/Discussion with the Citizen, but the Decision is taken by the Administration
3. Participation allows the joint decision between the Administration and the Citizen

(iii) Motivation:

1. I cannot miss the opportunity to be part of a citizen participation initiative like this one
2. I think it is a very important opportunity to express my opinions
3. I believe that this initiative will allow me to enrich myself as a person
4. I am interested in participating in the planning of cultural/sports activities
5. I do not agree with the current management of cultural and sports activities

(iv) Evaluation of the Technology Support and Applications:

1. The computer equipment was adequate
2. The presentation structure of the program was simple and understandable
3. It was easy and comfortable to move from screen to screen (navigate)
4. There were too many errors/incidents in the computer application *
5. The number of screens was not excessive *
6. The voting system was easy to use
7. The discussion system for the incorporation of arguments was adequate
8. The discussion system allowed me to know and share opinions
9. I consider that my anonymity was assured throughout the entire process
10. In general, I liked the design of the software application
11. In general, I am satisfied with the computer application used

(v) Evaluation of the Information:

1. It was easy to understand
2. It was suitable
3. It was received on time
4. There were virtually no errors
5. In general, I am satisfied with the information that I have received

(vi) Evaluation of the Support Staff

1. They helped my involvement in the citizen participation process
2. They gave me additional information
3. Without the support staff, I would not have been able to participate
4. Overall, I am satisfied with the help of the support staff

(vii) Overall Evaluation:

1. I really enjoyed participating in this initiative
2. I have learned a lot from the experience
3. I feel that my participation has improved my ingenuity and creativity
4. The experience allowed me to feel involved in political decision making
5. My perception of social belonging in my municipality has increased (identity)
6. The discussions in the forum influenced my decisions
7. Participating in this experience was not a waste of time
8. I would participate again in a similar experience
9. Other municipalities should incorporate this type of citizen participation
10. E-Cognocracy improves the current democratic system
11. I feel satisfied with my participation in this initiative

References

1. García-Lizana, A.; Moreno-Jiménez, J.M. Economía y Democracia en la Sociedad del Conocimiento. *Estudios de Economía Aplicada* **2008**, *26*, 181–212.
2. Moreno-Jiménez, J.M.; Pérez Espés, C.; Wimmer, M.A. The Effectiveness of e-Governance Experiences in the Knowledge Society. In *Proceedings 13th European Conference on eGovernment (ECEG)*; Castelnovo, W., Ferrari, E., Eds.; Academic Conferences and Publishing International Ltd.: Como, Italy, 2013; pp. 354–363.
3. Macintosh, A.; Whyte, A. Towards an Evaluation Framework for eParticipation. *Transform. Gov. People Process Policy* **2008**, *2*, 16–30. [CrossRef]
4. Aichholzer, G.; Westholm, H. Evaluating eParticipation Projects: Practical Examples and Outline of an Evaluation Framework. *Eur. J. ePract.* **2009**, *7*, 1–18.
5. Rogers, E.M. *Diffusion of Innovations*, 5th ed.; Free Press: New York, NY, USA, 2003.
6. Phang, C.W.; Kankanhalli, A. A framework of ICT exploitation for e-participation initiatives. *Commun. ACM* **2008**, *51*, 128–132. [CrossRef]
7. Geissel, B. Participatory Governance: Hope or Danger for Democracy? A Case Study of Local Agenda 21. *Local Gov. Stud.* **2009**, *35*, 401–414. [CrossRef]
8. Panopoulou, E.; Tambouris, E.; Konstantinos, T. eParticipation Initiatives in Europe: Learning from Practitioners. In *IFIP WG 8.5 International Conference on Electronic Participation (EPART)*; Tambouris, E., Macintosh, A., Glassey, O., Eds.; Lecture Notes in Computer Science series; Springer: Lausanne, Italy, 2010; Volume 6229, pp. 54–65.
9. Michels, A.; De Graaf, L. Examining Citizen Participation: Local Participatory Policymaking and Democracy Revisited. *Local Gov. Stud.* **2017**, *43*, 875–881. [CrossRef]
10. Moreno-Jiménez, J.M. Las Nuevas Tecnologías y la Representación Democrática del Inmigrante. In *IV Jornadas Jurídicas de Albarracín*; TSJA, Memoria Judicial Anual de Aragón; Arenere, J., Ed.; Consejo General del Poder Judicial: Madrid, Spain, 2003; Volume 66, p. 22.
11. Moreno-Jiménez, J.M. E-cognocracia. Nueva Sociedad, Nueva Democracia. *Estudios de Economía Aplicada* **2006**, *24*, 559–581.
12. Moreno-Jiménez, J.M.; Polasek, W. E-Democracy and Knowledge. A Multicriteria Framework for the New Democratic Era. *J. Multicriteria Decis. Anal.* **2003**, *12*, 163–176. [CrossRef]
13. Moreno-Jiménez, J.M.; Piles, J.P.; Ruiz, J.; Escobar, M.T.; Salazar, J.L.; Turón, A. Securization of policy making social computing. An application to e-cognocracy. *Comput. Hum. Behav.* **2011**, *27*, 1382–1388. [CrossRef]

14. Moreno-Jiménez, J.M.; Aguarón, J.; Cardeñosa, J.; Escobar, M.T.; Salazar, J.L.; Toncovich, A.; Turón, A. A collaborative platform for cognitive decision making in the Knowledge Society. *Comput. Hum. Behav.* **2012**, *28*, 1921–1928. [CrossRef]
15. Moreno-Jiménez, J.M.; Cardeñosa, J.; Gallardo, C.; de la Villa-Moreno, M.A. A new e-learning tool for cognitive democracies in the Knowledge Society. *Comput. Hum. Behav.* **2014**, *30*, 409–418. [CrossRef]
16. Moreno-Jiménez, J.M.; Pérez Espés, C.; Velázquez, M. E-Cognocracy and the Design of Public Policies. *Gov. Inf. Q.* **2014**, *31*, 185–194. [CrossRef]
17. Pérez-Espés, C.; Moreno-Jiménez, J.M. Social-Economic Approach to an eParticipation experience based on eCognocracy. In *Electronic Participation: 7th IFIP WG 8.5 Intern. Conference, Thessaloniki, Greece, 30 August–2 September 2015*; Tambouris, E., Panagiotopoulos, P., Sæbø, Ø., Tarabanis, K., Wimmer, M.A., Milano, M., Pardo, T., Eds.; Lecture Notes in Computer Science series; Springer: Berlin/Heidelberg, Germany, 2015; Volume 9249, pp. 120–131.
18. Davis, F.D. Perceived Usefulness, Perceived Ease of Use, and User Acceptance of Information Technology. *MIS Q.* **1989**, *13*, 319–340. [CrossRef]
19. Delone, W.H.; MacLean, E.R. Information Systems Success: The Quest for the Dependent Variable. *Inf. Syst. Res.* **1992**, *3*, 60–95. [CrossRef]
20. Sartori, A. An Estimator for Some Binary-Outcome Selection Models without Exclusion Restrictions. *Polit. Anal.* **2003**, *11*, 111–138. [CrossRef]
21. Layne, K.; Lee, J. Developing Fully Functional E-Government: A Four Stage Model. *Gov. Inf. Q.* **2001**, *18*, 122–136. [CrossRef]
22. Moreno-Jiménez, J.M.; Pérez Espés, C.; Velázquez, M. E-Participation and E-Cognocracy. Next Stages in e-Participation Evolution. 2015; Private document.
23. Moreno-Jiménez, J.M. E-cognocracia y Representación Democrática del Inmigrante. In *Proceedings of the XVIII Anales de Economía Aplicada*; University of León: León, Spain, 2004; ISBN 84-609-4715-7.
24. Moreno-Jiménez, J.M.; Polasek, W. E-Cognocracy: Combining E-democracy and Knowledge Networks. *Res. Comput. Sci.* **2004**, *8*, 165–175.
25. Moreno-Jiménez, J.M.; Polasek, W. E-Cognocracy and the Participation of Immigrants in E-Governance. In *Electronic Democracy: The Challenge Ahead, TED Conference on e-Government*; Schriftenreihe Informatik; Böhlen, Ed.; University Rudolf Trauner-Verlag: Linz, Austria, 2005; Volume 13, pp. 18–26.
26. Moreno-Jiménez, J.M. *Participación Ciudadana Electrónica en el Diseño de Políticas Públicas Locales*; Informe Proyecto OTRI2009-0410 del Gobierno de Aragón: Zaragoza, Spain, 2009. Available online: http://aragonparticipa.aragon.es/dmdocuments/Resumen%20e-cognocracia.pdf (accessed on 18 September 2018).
27. Dahl, R.A. *Democracy and its Critics*; Yale University Press: New Haven, CT, USA, 1989.
28. Dahl, R.A. *On Democracy*; Yale Nota Bene: New Haven, CT, USA, 2000.
29. Habermas, J. *Between Facts and Norms: Contributions to a Discourse Theory of Law and Democracy*; MIT Press: Cambridge, UK, 1996.
30. Bagozzi, R.P.; Warshaw, P.R. Development and Test of a Theory of Technological Learning and Usage. *Hum. Relat.* **1992**, *45*, 660–686. [CrossRef]
31. Delone, W.H.; McLean, E.R. The DeLone and McLean Model of Information Systems Success: A Ten-Year Update. *J. Manag. Inf. Syst.* **2003**, *19*, 9–30.
32. Rowe, G.; Frewer, L.J. Evaluating Public-Participation Exercises: A Research Agenda. *Sci. Technol. Hum. Values* **2004**, *29*, 512–557. [CrossRef]
33. Henderson, M.; Henderson, P. E-Democracy Evaluation Framework. 2005; Unpublished manuscript.
34. DEMO-net D51. Report on Current ICT's to Enable Participation. DEMO-netDeliverable 5.1. 31 August 2006. Available online: http://www.ifib.de/publikationsdateien/IntroducingeParticipation_DEMO-net_booklet_1.pdf (accessed on 2 February 2015).
35. Mamaqui, X.; Moreno-Jiménez, J. The Effectiveness of E-cognocracy. *Lect. Notes Artif. Intell. (LNAI)* **2009**, *5736*, 417–426.
36. Luna-Reyes, L.F.; Gil-Garcia, J.R.; Romero, G. Towards a Multidimensional Model for Evaluating Electronic Government: Proposing a More Comprehensive and Integrative Perspective. *Gov. Inf. Q.* **2012**, *29*, 324–334. [CrossRef]

37. Wimmer, M.A.; Bicking, M. Method and Lessons from Evaluating the Impact of E-Participation Projects in MOMENTUM. In *E-Government Success Factors and Measures: Theories, Concepts, and Methodologies*; Gil-Garcia, J.R., Ed.; IGI-Global Book: Hershey, PA, USA, 2013; pp. 213–233.
38. Al-Adawi, Z.; Yousafzai, S.; Pallister, J. Conceptual Model of Citizen Adoption of E-government. In Proceedings of the The Second International Conference on Innovations in Information Technology (IIT'05), Dubai, UAE, 26–28 September 2005; pp. 1–10.
39. Davis, F.D.; Bagozzi, R.P.; Warshaw, P.R. User Acceptance of Computer Technology: A Comparison of Two Theoretical Models. *Manag. Sci.* **1989**, *35*, 982–1003. [CrossRef]
40. Swanson, E.B. Management Information System: Appreciation and Involvement. *Manag. Sci.* **1974**, *21*, 178–188. [CrossRef]
41. Schultz, R.L.; Slevin, D.P. Implantation an Organizational Validity: An Empirical Investigation. In *Implementing Operations Research/Management Science*; Schultz, R.L., Slevin, D.P., Eds.; American Elsevier: New York, NY, USA, 1975; pp. 153–182.
42. Zmud, R.W. An Empirical Investigation of Dimensionality of the Concept of Information. *Decis. Sci.* **1978**, *9*, 187–195. [CrossRef]
43. Ajzen, I.; Fisbein, M. *Understanding Attitudes and Predicting Social Behavior*; Prentice-Hall: Englewood Cliffs, NJ, USA, 1980.
44. Bagozzi, R.P. Attitudes, Intentions and Behavior: A Test of Some Key Hypotheses. *J. Personal. Soc. Psychol.* **1981**, *41*, 607–627. [CrossRef]
45. Warshaw, P.R.; Davis, F.D. Self-understanding and the accuracy of behavioral expectations. *Personal. Soc. Psychol. Bull.* **1984**, *10*, 111–118. [CrossRef]
46. Warshaw, P.R. A New Model for Predicting Behaviroal Intentions: An Alternative to Fishbein. *J. Mark. Res.* **1980**, *17*, 153–172. [CrossRef]
47. Bandura, A. Self-Efficacy Mechanism in Human Agency. *Am. Psychol.* **1982**, *37*, 122–147. [CrossRef]
48. Shannon, C.E.; Weaver, W. *The Mathematical Theory of Communication*; University of Illinois Press: Champaign, IL, USA, 1949.
49. Bollen, K.A. *Structural Equations with Latent Variables*; John Wiley & Sons, Inc.: Hoboken, NJ, USA, 1989.
50. Bollen, K.A.; Lennox, R. Conventional Wisdom on Measurement: A Structural Equation Perspective. *Psichol. Bull.* **1991**, *110*, 305–314. [CrossRef]
51. Jöreskog, K.G.; Sörbom, D. *LISREL 8: User's Reference Guide*; SSI Scientific Software International: Chicago, IL, USA, 1996.
52. Bentler, P.M. *EQS Structural Equations Program Manual*; Multivariate Software, Inc.: Encino, CA, USA, 2006.
53. Saaty, T.L. *The Analytic Hierarchy Process*; McGraw-Hill: New York, NY, USA, 1980.
54. Saaty, T.L. *Fundamentals of Decision Making*; RSW Publicat: Pittsburgh, PA, USA, 1994.
55. Bagozzi, R.P. A Prospectus for Theory Construction in Marketing. *J. Mark.* **1984**, *48*, 11–29. [CrossRef]
56. McDonald, R.P. *Factor Analysis and Related Methods*; American Educational Research Association: Hillsdale, NJ, USA, 1985.
57. Fornell, C.; Larcker, D.F. Evaluating Structural Equation Models with Unobservable and Measurement Error. *J. Mark. Res.* **1981**, *XVIII*, 39–50. [CrossRef]

Article

The Basic Algorithm for the Constrained Zero-One Quadratic Programming Problem with k-diagonal Matrix and Its Application in the Power System

Shenshen Gu *,† and Xinyi Chen

School of Mechatronic Engineering and Automation, Shanghai University, Shanghai 200444, China
* Correspondence: gushenshen@shu.edu.cn
† Current address: 99 Shangda Road, Shanghai 200444, China.

Received: 30 December 2019; Accepted: 16 January 2020; Published: 19 January 2020

Abstract: Zero-one quadratic programming is a classical combinatorial optimization problem that has many real-world applications. However, it is well known that zero-one quadratic programming is non-deterministic polynomial-hard (NP-hard) in general. On one hand, the exact solution algorithms that can guarantee the global optimum are very time consuming. And on the other hand, the heuristic algorithms that generate the solution quickly can only provide local optimum. Due to this reason, identifying polynomially solvable subclasses of zero-one quadratic programming problems and their corresponding algorithms is a promising way to not only compromise these two sides but also offer theoretical insight into the complicated nature of the problem. By combining the basic algorithm and dynamic programming method, we propose an effective algorithm in this paper to solve the general linearly constrained zero-one quadratic programming problem with a k-diagonal matrix. In our algorithm, the value of k is changeable that covers different subclasses of the problem. The theoretical analysis and experimental results reveal that our proposed algorithm is reasonably effective and efficient. In addition, the placement of the phasor measurement units problem in the power system is adopted as an example to illustrate the potential real-world applications of this algorithm.

Keywords: zero-one quadratic problem; combinatorial optimization; k-diagonal matrix; power system

---

1. Introduction

Optimization problems normally fall into two categories, one for continuous variables and the other for discrete variables. The latter one is called combinatorial optimization, which is an active field in applied mathematics [1]. Common problems are the maximum flow problem, traveling salesman problem, matching problem, knapsack problem, etc. Among those famous combinatorial optimizations, zero-one quadratic programming (01QP), whose variables can only be either 0 or 1 [2], is very important and attracts a lot of attention.

Zero-one quadratic programming, which can be divided into constrained (01CQP) and unconstrained (01UQP), is a combinatorial optimization and has practical significance. For example, it can be applied in circuit design [3], pattern recognition [4], capital budgeting [5], portfolio optimization [6], etc. Except for these well known applications, 01QP also has potential applications in the nonlinear control related fields [7–11]. Among these applications, the phasor measurement unit (PMU) placement has been widely studied. Phasor measurement units can be used in dynamic monitoring, system protection, and system analysis and prediction. Therefore, the placement of PMU has become an important issue. As the scale of the electric grid grows, the PMU placement problem becomes more difficult and must be addressed considering certain requirements. In [12], a modified bisecting search combined with simulated annealing method was proposed, and the latter randomly selected arrays were used to test the observability of the system. Considering the incomplete observability of the system, a calculation method based on graph theory was proposed in [13]. This method is time-consuming, and with the increase of dimensions, the calculation load is too heavy. An integer linear programming method [14] and an improved algorithm [15] were put forward, considering system redundancy, and full and incomplete observability. Researchers in [16] proposed a binary programming method considering the joint placement of the conventional measurement units and phasor measurement units.

Zero-one quadratic programming also has some theoretical significance. Many classical problems, such as the max-cut problem [17,18] and max-bisection problem [19] can be converted to zero-one quadratic programming. Therefore, designing the algorithm that can solve 01QP effectively and efficiently is very meaningful not only in practical fields but also in theoretical fields.

However, zero-one quadratic programming is a well known NP-hard problem in general. The common solutions are exact solution and heuristic algorithms. Exact solution algorithms can guarantee the global optimum. In [20], the branch-and-bound method was used to solve zero-one quadratic programming problems without constraints. In [21,22], some exact solutions were proposed by means of geometric properties. Penalty parameters were introduced to solve zero-one quadratic programming problems with constraints in [23]. But exact solution algorithm are very time consuming and suitable for small scale problems only. Oppositely, heuristic algorithms, such as simulated annealing [24], genetic algorithms [25], neural networks [26], and ant colony algorithms [27], can solve the medium and large scale problems quickly in general. But most of them can only find the local optimum.

Therefore, identifying polynomially solvable subclasses of zero-one quadratic programming problems and their corresponding algorithms is a promising way to not only compromise these two sides but also offer theoretical insight into the complicated nature of the problem. In our past studies [28,29], the problem of five-diagonal matrix quadratic programming with linear constraints has been solved effectively. In [30], an algorithm for solving 01QP with a seven-diagonal matrix Q was presented. However, for these algorithms, the applicable problem is very specific. This narrows their applications. Then, we proposed an algorithm for the unconstrained problems with a k-diagonal matrix in [31]. In this paper, based on our previous results, we further propose an algorithm for the general linearly constrained zero-one quadratic programming problem with a k-diagonal matrix by combining the basic algorithm and dynamic programming method. For the algorithm we proposed, the value of k is changeable. That means the algorithm can cover different subclasses of the problem. The theoretical analysis and experimental results reveal that our proposed algorithm is reasonably effective and efficient. We also apply the algorithm to real-world applications and then verify its feasibility.

The main contributions of this paper are reflected in the following aspects: (1) The previous algorithm targeted a fixed k value in Q matrix. While the algorithm in this paper targeted a general problem with changeable k values. (2) We analyze the time complexity and give the proof process of the rationality of the algorithm. (3) We apply the algorithm to the phasor measurement units placement in real-world application.

This paper is organized as follows: in Section 2, we review the algorithm of solving unconstrained zero-one k-diagonal matrix quadratic programming. In Section 3, a constrained zero-one k-diagonal matrix quadratic programming algorithm with the proof of the algorithm is proposed. Application of zero-one quadratic programming in the phasor measurement units placement is put forward in Section 4. Experimental results and discussion are given in Section 5. We draw our conclusions and put forward the prospects in Section 6.

2. Basic Algorithm to 01UQP

The following Equation (1) shows the form of the k-diagonal matrix zero-one quadratic programming problem. The special point is the form of the matrix Q called the k-diagonal matrix, where $k = 2m + 1$ $(m = 0, 1, 2, \dots, n-1)$.

$$\min_{x \in \{0,1\}^n} f(x) = \frac{1}{2} x^T Q x + c^T x \tag{1}$$

where $Q = (q_{ij})_{n \times n}$, $q_{ij} = q_{ji} (i, j = 1, 2, \dots, n)$ indicates that it is a symmetric matrix. Note that all the numbers in this matrix are zero except q_{ij} and $q_{ji} (i = 1, 2, \dots, n-1, j = i+1, i+2, \dots, i+m)$.

$$\begin{pmatrix} 0 & q_{1,2} & q_{1,3} & \cdots & 0 & 0 & 0 \\ q_{2,1} & 0 & q_{2,3} & \cdots & 0 & 0 & 0 \\ q_{3,1} & q_{3,2} & 0 & \cdots & 0 & 0 & 0 \\ \vdots & \vdots & \vdots & \ddots & \vdots & \vdots & \vdots \\ 0 & 0 & 0 & \cdots & 0 & q_{n-2,n-1} & q_{n-2,n} \\ 0 & 0 & 0 & \cdots & q_{n-1,n-2} & 0 & q_{n-1,n} \\ 0 & 0 & 0 & \cdots & q_{n,n-2} & q_{n,n-1} & 0 \end{pmatrix}$$

Based on past works [30,31], we can have the algorithm, as follows, to solve 01UQP. The feasibility and effect of the algorithm can be seen in [31]. Figure 1 shows the Algorithm 1 process intuitively.

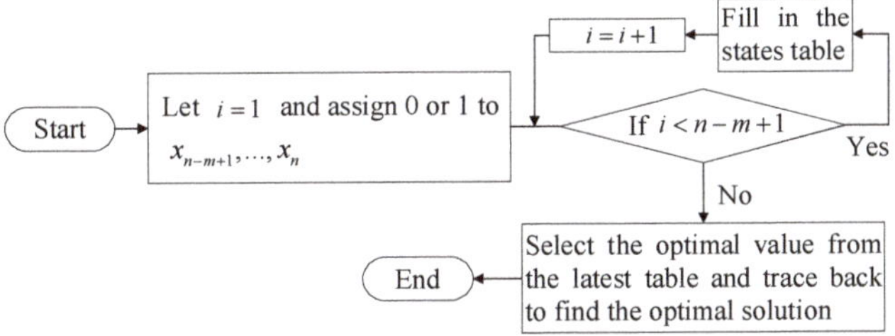

Figure 1. Flow chart of the zero-one unconstrained quadratic programming (01UQP) algorithm.

Algorithm 1 Process of solving 01UQP

Step 1: Assign $x_{n-m+1}, \dots, x_n$ to 0 or 1.

(1) Adjacent terms are the combination of $x_{n-m+1}, \dots, x_n$, whose value is the same except for the value of x_n.
(2) Get the corresponding $f(x)$.
(3) Label these states $state0, state1, \dots, state(2^m - 1)$.
(4) Compare $f(x)$ in the two adjacent terms. (only the coefficient of x_{n-m} and the constant term are different.)

Step 2: Change the value of x_{n-m-i} to 0 and $1 (i = 0, 1, \dots, n-m-1)$.

(1) Compare every two adjacent states and take the result with the smaller constant term as the new state.
(2) Update all the 2^m states.

Step 3: Get the optimal solution x.

(1) Update the states based on **Step 2** until only the constant term is in $f(x)$.
(2) The optimal value is the minimal one.
(3) Trace back and get the optimal solution x.

---

3. Basic Algorithm to 01CQP

3.1. 01CQP Algorithm Description

Consider the constrained k-diagonal matrix zero-one quadratic programming problem:

$$\begin{array}{ll} \min\limits_{x \in \{0,1\}^n} & \frac{1}{2} x^T Q x + c^T x \\ \text{s.t.} & a^T x \leq b \end{array} \tag{2}$$

where Q has the same meaning as that of the 01UQP formula in Section 2, $a \in \mathbb{Z}_+^n$, $b \in \mathbb{Z}_+$.

In this section, we utilize the dynamic programming method to solve the 01CQP problem. To apply the dynamic programming method, we introduce a state variable $s_k(s_k \in \mathbb{Z})$ and a stage variable $k(0 < k \leq n)$, which should satisfy the following iteration. s_{k+1} can be expressed as:

$$s_{k+1} = s_k + a_{k+1} x_{k+1} \quad (k = 1, \ldots, n-1).$$

We only need to consider the integer point of the state space since $a \in \mathbb{Z}_+^n$ and $b \in \mathbb{Z}_+$. Since s_k satisfies $0 \leq s_k \leq b$, we define a set $s_k = \{s_k | 0 \leq s_k \leq b, s_k \in \mathbb{Z}\}$.

Algorithms 2 and 3 show the detailed calculation process.

Algorithm 2 Calculation Method of $f(s_k)$

Case 1: When $k = 1$, there are two cases.

(1a) $s_1 < a_1$
$x_1 = 0, x_1^* = 0, f(s_1) = f(0, \ldots, x_n)$
(1b) $s_1 \geq a_1$
$x_1 = 1, x_1^* = 1, f(s_1) = f(1, \ldots, x_n)$

We will get a series of functions $f(x)$ after executing Case 1.

Case 2: When $k \geq 2$, there are also two cases.

(2a) $s_k < a_k$
x_k must be 0 to satisfy $s_{k-1} = s_k - a_k x_k$. At this time, $s_k = s_{k-1}$, by which we can obtain the function $f(s_k) = f(s_{k-1})|_{x_k=0}$.
(2b) $s_k \geq a_k$
In this case, x_k can be both 0 or 1, which generates two more situations:
1) If $x_k{}^* = 0$ and $s_k = s_{k-1}$, we can get $f(s_k) = f(s_{k-1})|_{x_k=0}$.
2) If $x_k{}^* = 1$ and $s_{k-1} = s_k - a_k x_k$, we can get $f(s_k) = f(s_{k-1})|_{x_k=1}$.

We can see that there is only one function $f(s_k)$ corresponding to each state s_k when $k = 1$, and there are several $f(s_k)$ to each state s_k when $k > 1$. To save storage and computational time, $f(s_k)$ should be selected satisfactorily in the next step. We need to find the optimal $f(s_k) = f(s_{k-1})|_{x_k=0}$ and $f(s_k) = f(s_{k-1})|_{x_k=1}$, that the optimizing process refers to for Algorithm 3.

Algorithm 3 $state_0$ and $state_1$

$state_0(x_k = 0)$
Case 1: There is only one $f(s_k) = f(s_{k-1})|_{x_k=0}$:
Set $x_k{}^* = 0, f(s_k) = f(s_{k-1})|_{x_k=0}$
Case 2: There are more than one $f(s_k) = f(s_{k-1})|_{x_k=0}$:
1) Compare $f(s_{k-1})|_{x_k=0}$, which are almost the same except for the constant term and pick up the smallest one.
2) Set $x_k{}^* = 0$ and $f(s_k) = f(s_{k-1})|_{x_k=0}$.
$state_1(x_k = 1)$
Case 1: There is only one $f(s_k) = f(s_{k-1})|_{x_k=1}$:
Set $x_k{}^* = 1, f(s_k) = f(s_{k-1})|_{x_k=1}$
Case 2: There are more than one $f(s_k) = f(s_{k-1})|_{x_k=1}$:
1) Find the $f(x)$ using a similar approach to that in $state_0$ Case 2.
2) Set $x_k{}^* = 1$ and $f(s_k) = f(s_{k-1})|_{x_k=1}$.

According to the above algorithm, the maximum number of functions per *State* is shown in Table 1. The primary time is spent generating the state table. The number of times that the core steps calculated is focused on the state number in Table 1. In total, we need to update the state table n times and the number of state is b, so the time complexity is $O(2^{m-1} \times n \times b)$.

Table 1. Maximum number of functions per *State*.

Situation	**The Maximum Number of Functions**
Algorithm 2 Case 1	1
Algorithm 3 Case 1	2^{m-1}
Algorithm 3 Case 2	2^{m-1}

3.2. Analysis on the Effectiveness and Rationality of 01CQP Algorithm

In this section, we analyze the properties of the polynomial time algorithm for 01CQP. To analyze the algorithm, we need to demonstrate the rationality of Algorithms 2 and 3. Suppose $f(x)$ has the form:

$$f(x) = q_{12}x_1x_2 + q_{13}x_1x_3 + \cdots + q_{1,1+m}x_1x_{1+m} + \cdots + q_{n-m,n-m+1}x_{n-m} x_{n-m+1} + \cdots + q_{n-m,n}x_{n-m}x_n + \cdots + q_{n-1,n}x_{n-1}x_n + c_1x_1 + \cdots + c_nx_n.$$

Step 1: Set $k = 1$
(1) $s_1 < a_1, x_1 = 0$

$$\begin{aligned} f_1(s_1) = f(x)|_{x_1=0} &= q_{i,i+1}x_ix_{i+1} + q_{i,i+2}x_ix_{i+2} + \cdots + q_{i,i+m}x_ix_{i+m} \\ &+ \cdots + q_{n-m,n-m+1}x_{n-m}x_{n-m+1} + \cdots + q_{n-m,n}x_{n-m}x_n \\ &+ \cdots + q_{n-1,n}x_{n-1}x_n + c_2x_2 + \cdots + c_nx_n (i = 2, \ldots, n-m-1). \end{aligned} \tag{3}$$

(2) $s_1 \geq a_1, x_1 = 1$

$$\begin{aligned} f_1(s_1) = f(x)|_{x_1=1} &= q_{i,i+1}x_ix_{i+1} + q_{i,i+2}x_ix_{i+2} + \cdots + q_{i,i+m}x_ix_{i+m} \\ &+ \cdots + q_{n-m,n-m+1}x_{n-m}x_{n-m+1} + \cdots + q_{n-m,n}x_{n-m}x_n \\ &+ \cdots + q_{n-1,n}x_{n-1}x_n + \widehat{c_2}x_2 + \cdots + \widehat{c_{1+m}}x_{1+m} + c_{m+2}x_{m+2} \\ &+ \cdots + c_nx_n + c_1. \end{aligned} \tag{4}$$

Clearly, when $k = 1$, each state s_1 has only one function $f_1(s_1)$. The coefficient of $x_2, \ldots, x_{m+1}$ and the constants are the only different terms in the function.

Step 2: Set $k = 2, 3, ..., m$

(1) If $s_k < a_k$, execute $state_0$.

If $s_{k-1} < a_{k-1}$, $f_k(s_k)$ has the same form as function (3):

$$\begin{aligned}
&f(s_k) = q_{k,k+1}x_k x_{k+1} + q_{k,k+2}x_k x_{k+2} + \cdots + q_{k,k+m}x_k x_{k+m} \\
&+ \cdots + q_{i,i+1}x_i x_{i+1} + q_{i,i+2}x_i x_{i+2} + \cdots + q_{i,i+m}x_i x_{i+m} + \cdots + \\
&q_{n-m,n-m+1}x_{n-m}x_{n-m+1} + \cdots + q_{n-m,n}x_{n-m}x_n + \cdots + \\
&q_{n-1,n}x_{n-1}x_n + c_{k+1}x_{k+1} + \cdots + c_n x_n + const.
\end{aligned} \tag{5}$$

($const$ represents the constant term of the $f(s_{k-1})$.)

If $s_k \geq a_{k-1}$, $f_k(s_k)$ has the same form as function (4):

$$\begin{aligned}
&f(s_{k-1}) = q_{k,k+1}x_k x_{k+1} + q_{k,k+2}x_k x_{k+2} + \cdots + q_{k,k+m}x_k x_{k+m} \\
&+ \cdots + q_{i,i+1}x_i x_{i+1} + q_{i,i+2}x_i x_{i+2} + \cdots + q_{i,i+m}x_i x_{i+m} + \cdots + \\
&q_{n-m,n-m+1}x_{n-m}x_{n-m+1} + \cdots + q_{n-m,n}x_{n-m}x_n + \cdots + \\
&q_{n-1,n}x_{n-1}x_n + \widehat{c_{k+1}}x_{k+1} + \cdots + \widehat{c_{k+m}}x_{k+m} + c_{k+m}x_{k+m} \\
&+ \ldots + c_n x_n + const + c_k.
\end{aligned} \tag{6}$$

Then, $f_k(s_k)$ can be shown as:

$$\begin{aligned}
&f_k(s_k) = f_{k-1}(s_{k-1})|_{x_k=0} = q_{k+1,k+2}x_{k+1}x_{k+2} + q_{k+1,k+3}x_{k+1}x_{k+3} + \cdots \\
&+ q_{k+1,k+m+1}x_{k+1}x_{k+m+1} + \cdots + q_{i,i+1}x_i x_{i+1} + q_{i,i+2}x_i x_{i+2} + \cdots + \\
&q_{i,i+m}x_i x_{i+m} + \cdots + q_{n-m,n-m+1}x_{n-m}x_{n-m+1} + \cdots + q_{n-m,n}x_{n-m}x_n \\
&+ \cdots + q_{n-1,n}x_{n-1}x_n + \widehat{c_{k+1}}x_{k+1} + \cdots + \widehat{c_{k+m}}x_{k+m} + c_{k+m+1}x_{k+m+1} \\
&+ \ldots + c_n x_n + \widehat{const}.
\end{aligned} \tag{7}$$

At this point, the maximum number of $f_k(s_k)$ corresponding to each s_k is 2^{k-2}.

(2) If $s_k \geq a_k$, execute $state_0$ and $state_1$.

Here, the $state_0$ is the same as the one in Case 1, so we do not need to repeat it.

Consider $state_1$, $x_k = 1$, $f_k(s_k) = f_{k-1}(s_k - a_k x_k)$, $s_{k-1} = s_k - a_k x_k = s_k - a_k$.

If $s_{k-1} < a_{k-1}$, $f(s_{k-1})$ has the form of function (5). If $s_{k-1} \geq a_{k-1}$, $f(s_{k-1})$ has the form of function (6), then, $f_k(s_k)$ can be expressed as:

$$\begin{aligned}
&f_k(s_k) = f_{k-1}(s_{k-1})|_{x_k=1} = q_{k+1,k+2}x_{k+1}x_{k+2} + q_{k+1,k+3}x_{k+1}x_{k+3} + \cdots \\
&+ q_{k+1,k+m+1}x_{k+1}x_{k+m+1} + \cdots + q_{i,i+1}x_i x_{i+1} + q_{i,i+2}x_i x_{i+2} + \cdots + \\
&q_{i,i+m}x_i x_{i+m} + \cdots + q_{n-m,n-m+1}x_{n-m}x_{n-m+1} + \cdots + q_{n-m,n}x_{n-m}x_n \\
&+ \cdots + q_{n-1,n}x_{n-1}x_n + \widehat{c_{k+1}}'x_{k+1} + \cdots + \widehat{c_{k+m}}'x_{k+m} + c_{k+m+1}x_{k+m+1} \\
&+ \ldots + c_n x_n + \widehat{const}'.
\end{aligned} \tag{8}$$

All $f_k(s_k)$ are only different in the constant terms and the coefficient of $x_{k+1}, ..., x_{m+2}$.

Step 3: $k = n - m - 1$

Consider the most complex case, $s_k \geq a_k$ and the state variable s_k corresponds to 2^{m-1} $f_k(s_k)$ and 2^{m-1} $f_k(s_k) = f_{k-1}(s_k - a_k)$ respectively (2^{m-1} is the number of both). Therefore, we need to execute $state_0$ first, and then execute $state_1$.

(1) $state_0$:

Since $f_k(s_k)$ has the form as function (8), the $state_0$ is executed for the $f_k(s_k)$ respectively, and the following expression is obtained:

$$\begin{aligned}
&f_k(s_k) = f_{k-1}(s_k)|_{x_k=0} = q_{n-m,n-m+1}x_{n-m}x_{n-m+1} + \cdots + \\
&q_{n-m,n}x_{n-m}x_n + \cdots + q_{n-1,n}x_{n-1}x_n + \widehat{c_{n-m}}'x_{n-m} + c_{n-m+1}x_{n-m+1} \\
&+ \ldots + c_n x_n + \widehat{const}'.
\end{aligned} \tag{9}$$

Clearly, these $f_k(s_k)$ are only different in the constants and the coefficients of x_{n-m}.

(2) $state_1$:

It is possible to obtain $f_k(s_k)$, which has the following form:

$$\begin{aligned} & f_k(s_k) = f_{k-1}(s_{k-1})|_{x_k=1} = q_{n-m,n-m+1}x_{n-m}x_{n-m+1} + \cdots + \\ & q_{n-m,n}x_{n-m}x_n + \cdots + q_{n-1,n}x_{n-1}x_n + \widehat{c_{n-m}}'' x_{n-m} + c_{n-m+1}x_{n-m+1} \\ & + \ldots + c_n x_n + \widehat{const}' + \widehat{c_{n-m-1}}. \end{aligned} \tag{10}$$

For both case (1) and (2), they are only different in the constants and the coefficients of x_{k+1}.

Step 4: $k = m+2, ..., n$

Based on Algorithm 2 Case 2, we calculate the function $f_k(s_k)$, and the core is the implementation of $state_0$ and $state_1$. When executing $state_0$ and $state_1$, the difference between the two $f_k(s_k)$ is the constant and the coefficients of x_k. Therefore, when executing $state_0$, we only need to compare the constant terms. When executing $state_1$, we only need to compare the sum of the constants and the coefficients of x_k. In this way, we can avoid solving the excess $f_k(s_k)$.

The form of the function $f_k(s_k)$ and the maximum number of $f_k(s_k)$ in each stage k are similar to the one in stage $k = n - m$, which also ensures the repeatability of the algorithm. The algorithm is supplemented by concrete examples and numerical simulations.

3.3. Calculation Example of 01CQP

For example, the parameters Q, c, a, and b are:

$$Q = \begin{pmatrix} 0 & 23 & -37 & -56 & 0 & 0 & 0 & 0 \\ 23 & 0 & 41 & 16 & -34 & 0 & 0 & 0 \\ -37 & 41 & 0 & -62 & -27 & 76 & 0 & 0 \\ -56 & 16 & -62 & 0 & -81 & 14 & -58 & 0 \\ 0 & -34 & -27 & 81 & 0 & 90 & 25 & -42 \\ 0 & 0 & 76 & 14 & 90 & 0 & -12 & 31 \\ 0 & 0 & 0 & -58 & 25 & -12 & 0 & -94 \\ 0 & 0 & 0 & 0 & -42 & 31 & -94 & 0 \end{pmatrix}$$

$$c = \begin{pmatrix} 24, & -54, & -17, & 36, & -72, & 63, & 46, & -18 \end{pmatrix}^T$$

$$a = \begin{pmatrix} 1, & 2, & 3, & 2, & 4, & 2, & 3, & 2 \end{pmatrix}^T$$

$$b = 6.$$

In this example, we will omit some of the states.

(1) Set $k = 1, s_1 = a_1x_1, a_1 = 1$

We can set $s_1 = 0, 1, \ldots, 6$. According to Algorithm 2 Case 1, we can obtain Table 2.
When $s_1 = 0$, we can calculate $x_1{}^*$ and $f(s_1)$ through Algorithm 2 Case (1a) for $s_1 < a_1$.
When $s_1 = 1, \ldots, 6$, we can calculate $x_1{}^*$ and $f(s_1)$ through Algorithm 2 Case (1b) for $s_1 \geq a_1$.
We omit $state3, \ldots, state6$, which is the same as $state1, state2$.

(2) Set $k = 2, 3, s_2 = a_1x_1 + a_2x_2 = s_1 + a_2x_2, s_3 = s_2 + a_3x_3$

We can set $s_3 = 0, 1, \ldots, 6$. According to Algorithm 2 Case 2, we obtain the Table 3. The case of $s_2 = 0, 1, \ldots, 6$ is omitted.

(3) Set $k = 4, 5, 6, 7$

According to Algorithm 2 Case 2 and Algorithm 3, we can gain $f_k(s_k)$ with $k = 3, ..., 7$. Tables 4 and 5 show part of the calculation process. From the table, we can see that only the coefficient of s_{k+1} and constant terms are different in the adjacent items.

(4) Set $k = 8$

Finally, we have Table 6, which only has the constant terms. Through it, we can see that $x_8{}^* = 0$, $f(x^*) = -160$, and by the backtracking method, we find the optimum solution $x^* = (0, 1, 0, 0, 1, 0, 0, 0)$.

Table 2. Functions $f_1(s_1)$.

s_1	x_1	$f_1(s_1)(a_1 = 1)$
0	0	$41x_2x_3 + 16x_2x_4 - 34x_2x_5 - 62x_3x_4 - 27x_3x_5 + 76x_3x_6 - 81x_4x_5$ $+14x_4x_6 - 58x_4x_7 + 90x_5x_6 + 25x_5x_7 - 42x_5x_8 - 12x_6x_7 + 31x_6x_8$ $-94x_7x_8 - 54x_2 - 17x_3 + 36x_4 - 72x_5 + 63x_6 + 46x_7 - 18x_8$
1	1	$41x_2x_3 + 16x_2x_4 - 34x_2x_5 - 62x_3x_4 - 27x_3x_5 + 76x_3x_6 - 81x_4x_5$ $+14x_4x_6 - 58x_4x_7 + 90x_5x_6 + 25x_5x_7 - 42x_5x_8 - 12x_6x_7 + 31x_6x_8$ $-94x_7x_8 - 31x_2 - 54x_3 - 20x_4 - 72x_5 + 63x_6 + 46x_7 - 18x_8 + 24$
2	1	$41x_2x_3 + 16x_2x_4 - 34x_2x_5 - 62x_3x_4 - 27x_3x_5 + 76x_3x_6 - 81x_4x_5$ $+14x_4x_6 - 58x_4x_7 + 90x_5x_6 + 25x_5x_7 - 42x_5x_8 - 12x_6x_7 + 31x_6x_8$ $-94x_7x_8 - 31x_2 - 54x_3 - 20x_4 - 72x_5 + 63x_6 + 46x_7 - 18x_8 + 24$

Table 3. Functions $f_3(s_3)$.

s_3	x_3	$f_3(s_3)(a_3 = 3)$
0	0	$f_3(s_3) = f_2(0)_{\mid x_3=0} = -81x_4x_5 + 14x_4x_6 - 58x_4x_7 + 90x_5x_6 +$ $25x_5x_7 - 42x_5x_8 - 12x_6x_7 + 31x_6x_8 - 94x_7x_8 + 36x_4 - 72x_5 +$ $63x_6 + 46x_7 - 18x_8$
1	0	$f_3(s_3) = f_2(1)_{\mid x_3=0} = -81x_4x_5 + 14x_4x_6 - 58x_4x_7 + 90x_5x_6 +$ $25x_5x_7 - 42x_5x_8 - 12x_6x_7 + 31x_6x_8 - 94x_7x_8 - 20x_4 - 72x_5 +$ $63x_6 + 46x_7 - 18x_8 + 24$
3	0	$f_3(s_3) = f_2(3)_{\mid x_3=0} = -81x_4x_5 + 14x_4x_6 - 58x_4x_7 + 90x_5x_6 +$ $25x_5x_7 - 42x_5x_8 - 12x_6x_7 + 31x_6x_8 - 94x_7x_8 - 20x_4 - 72x_5 +$ $63x_6 + 46x_7 - 18x_8 + 24$
3	0	$f_3(s_3) = f_2(3)_{\mid x_3=0} = -81x_4x_5 + 14x_4x_6 - 58x_4x_7 + 90x_5x_6 +$ $25x_5x_7 - 42x_5x_8 - 12x_6x_7 + 31x_6x_8 - 94x_7x_8 - 4x_4 - 106x_5 +$ $63x_6 + 46x_7 - 18x_8 - 7$
3	1	$f_3(s_3) = f_2(3)_{\mid x_3=1} = -81x_4x_5 + 14x_4x_6 - 58x_4x_7 + 90x_5x_6 +$ $25x_5x_7 - 42x_5x_8 - 12x_6x_7 + 31x_6x_8 - 94x_7x_8 - 26x_4 - 99x_5 +$ $139x_6 + 46x_7 - 18x_8 - 17$
5	0	$f_3(s_3) = f_2(5)_{\mid x_3=0} = -81x_4x_5 + 14x_4x_6 - 58x_4x_7 + 90x_5x_6 +$ $25x_5x_7 - 42x_5x_8 - 12x_6x_7 + 31x_6x_8 - 94x_7x_8 - 20x_4 - 72x_5 +$ $63x_6 + 46x_7 - 18x_8 + 24$
5	0	$f_3(s_3) = f_2(5)_{\mid x_3=0} = -81x_4x_5 + 14x_4x_6 - 58x_4x_7 + 90x_5x_6 +$ $25x_5x_7 - 42x_5x_8 - 12x_6x_7 + 31x_6x_8 - 94x_7x_8 - 4x_4 - 106x_5$ $+63x_6 + 46x_7 - 18x_8 - 7$
5	1	$f_3(s_3) = f_2(2)_{\mid x_3=1} = -81x_4x_5 + 14x_4x_6 - 58x_4x_7 + 90x_5x_6 +$ $25x_5x_7 - 42x_5x_8 - 12x_6x_7 + 31x_6x_8 - 94x_7x_8 - 84x_4 - 99x_5 +$ $139x_6 + 46x_7 - 18x_8 - 30$
5	1	$f_3(s_3) = f_2(2)_{\mid x_3=1} = -81x_4x_5 + 14x_4x_6 - 58x_4x_7 + 90x_5x_6 +$ $25x_5x_7 - 42x_5x_8 - 12x_6x_7 + 31x_6x_8 - 94x_7x_8 - 10x_4 - 133x_5 +$ $139x_6 + 46x_7 - 18x_8 - 30$

Table 4. Functions $f_4(s_4)$.

s_4	x_4	$f_4(s_4)(a_4 = 2)$
1	0	$f_4(s_4) = f_3(1)_{\mid x_4=0} = 90x_5x_6 + 25x_5x_7 - 42x_5x_8 - 12x_6x_7 + 31x_6x_8$ $-94x_7x_8 - 72x_5 + 63x_6 + 46x_7 - 18x_8 + 24$
5	0	$f_4(s_4) = f_3(5)_{\mid x_4=0} = 90x_5x_6 + 25x_5x_7 - 42x_5x_8 - 12x_6x_7 + 31x_6x_8$ $-94x_7x_8 - 72x_5 + 63x_6 + 46x_7 - 18x_8 + 24$
5	0	$f_4(s_4) = f_3(5)_{\mid x_4=0} = 90x_5x_6 + 25x_5x_7 - 42x_5x_8 - 12x_6x_7 + 31x_6x_8$ $-94x_7x_8 - 106x_5 + 63x_6 + 46x_7 - 18x_8 - 7$
5	0	$f_4(s_4) = f_3(5)_{\mid x_4=0} = 90x_5x_6 + 25x_5x_7 - 42x_5x_8 - 12x_6x_7 + 31x_6x_8$ $-94x_7x_8 - 99x_5 + 139x_6 + 46x_7 - 18x_8 - 30$
5	0	$f_4(s_4) = f_3(5)_{\mid x_4=0} = 90x_5x_6 + 25x_5x_7 - 42x_5x_8 - 12x_6x_7 + 31x_6x_8$ $-94x_7x_8 - 133x_5 + 139x_6 + 46x_7 - 18x_8 - 30$
5	1	$f_4(s_4) = f_3(3)_{\mid x_4=1} = 90x_5x_6 + 25x_5x_7 - 42x_5x_8 - 12x_6x_7 + 31x_6x_8$ $-94x_7x_8 - 153x_5 + 77x_6 - 12x_7 - 18x_8 + 4$
5	1	$f_4(s_4) = f_3(3)_{\mid x_4=1} = 90x_5x_6 + 25x_5x_7 - 42x_5x_8 - 12x_6x_7 + 31x_6x_8$ $-94x_7x_8 - 187x_5 + 77x_6 - 12x_7 - 18x_8 - 11$
5	1	$f_4(s_4) = f_3(3)_{\mid x_4=1} = 90x_5x_6 + 25x_5x_7 - 42x_5x_8 - 12x_6x_7 + 31x_6x_8$ $-94x_7x_8 - 180x_5 + 153x_6 - 12x_7 - 18x_8 - 43$

Table 5. Functions $f_5(s_5)$.

s_5	x_5	$f_5(s_5)(a_5 = 4)$
5	0	$f_5(s_5) = f_4(5)_{\mid x_5=0} = -12x_6x_7 + 31x_6x_8$ $-94x_7x_8 - 72x_6 + 46x_7 - 18x_8 - 7$
5	0	$f_5(s_5) = f_4(5)_{\mid x_5=0} = -12x_6x_7 + 31x_6x_8$ $-94x_7x_8 - 72x_6 + 46x_7 - 18x_8 - 30$
5	0	$f_5(s_5) = f_4(5)_{\mid x_5=0} = -12x_6x_7 + 31x_6x_8$ $-94x_7x_8 + 77x_6 - 12x_7 - 18x_8 - 11$
5	0	$f_5(s_5) = f_4(5)_{\mid x_5=0} = -12x_6x_7 + 31x_6x_8$ $-94x_7x_8 + 153x_6 - 12x_7 - 18x_8 - 43$
5	1	$f_5(s_5) = f_4(1)_{\mid x_5=1} = -12x_6x_7 + 31x_6x_8$ $-94x_7x_8 + 153x_6 + 71x_7 - 60x_8 - 48$

Table 6. Functions $f_8(s_8)$.

s_8	x_8	$f_8(s_8)$
0	0	0
1	0	24
2	0	−54
2	1	−18
3	0	−17
3	1	6
4	0	−72
4	1	−72
5	0	−48
5	1	−66
6	0	−160
6	1	−132

4. Application of Zero-One Quadratic Programming in Phasor Measurement Units Placement

How to find the installation location and the number of installations is the focus of the phasor measurement units placement problem. The matrix in the PMU placement problem differs from the previous k-diagonal matrix in the elements of the main diagonal. However, since x is either 0 or 1, we can convert it into a k-diagonal matrix. Then, we can take advantage of the algorithms designed above to work out PMU placement without considering too many constraints and realistic requirements.

4.1. Modelling

For the power system composed of n nodes, the placement of PMU is represented by n-dimensional vector $X = (x_1, x_2, \ldots, x_n)$. Here, $i = 1, 2, \ldots, n$,

$$x_i = \begin{cases} 0 & \text{a PMU is installed at bus } i \\ 1 & \text{otherwise.} \end{cases}$$

The matrix H represents the network graph structure,

$$h_{ij} = \begin{cases} 1 & i = j \\ 1 & i \text{ is connected to } j \\ 0 & \text{otherwise,} \end{cases}$$

and the objective function is

$$V(x) = \lambda(N - HX)^T R(N - HX) + X^T QX \tag{11}$$

where λ is the weight. $N \in \mathbb{R}^n$ represents the upper bound of the maximum redundancy of each bus. $R \in \mathbb{R}^{n\times n}$ and $Q \in \mathbb{R}^{n\times n}$ are diagonal arrays, representing the importance of each bus and the cost of placing the PMU on the bus respectively.

We expand Equation (11),

$$\begin{aligned} V(x) &= \lambda(N - HX)^T R(N - HX) + X^T QX \\ &= \lambda[N^T RN - N^T RHX - X^T H^T RN + X^T H^T RHX] + X^T QX \\ &= \lambda N^T RN - 2\lambda N^T RHX + \lambda X^T H^T RHX + X^T QX \\ &= \tfrac{1}{2} X^T (2\lambda H^T RH + 2Q)X + (-2\lambda N^T RH)X + \lambda N^T RN. \end{aligned} \tag{12}$$

Then, Equation (12) can be expressed as integer quadratic programming as,

$$\begin{aligned} \min \quad & \tfrac{1}{2} x^T Gx + f^T x \\ \text{s.t.} \quad & M(x) = 0 \\ & x_i \in \{0, 1\}^n \end{aligned} \tag{13}$$

where $G = 2\lambda H^T RH + 2Q$, $f = (-2\lambda N^T RH)^T$. $M(x)$ is the column vector consisting of $m_i(x) = (1 - x_i)\prod_{j\in A_i}(1 - x_j)(i \in \Omega)$. A_i and Ω represent a set of nodes adjacent to the bus i and the bus set respectively. The above constraints require at least one PMU unit to be placed between the bus and its adjacent nodes to ensure that each adjacent node of the bus can be observed. The above problem can be expressed as unconstrained problems by the weighted least square method:

$$\begin{aligned} \min \quad & \tfrac{1}{2} x^T Gx + f^T x + M(x)^T VM(x) \\ & = \tfrac{1}{2} x^T Gx + f^T x + \textstyle\sum_{i=1}^{n} (v_i (1 - x_i)^2 (\prod_{j\in A_i} (1 - x_j))^2) \\ \text{s.t.} \quad & x_i \in \{0, 1\}^n \end{aligned} \tag{14}$$

where $V = diag(v_i)$.

4.2. Example and Experimental Result

Suppose the coefficient matrix H of a power system is

$$H = \begin{pmatrix} 1 & 1 & 1 & 0 & 0 & 0 \\ 1 & 1 & 0 & 1 & 0 & 0 \\ 1 & 0 & 1 & 1 & 0 & 0 \\ 0 & 1 & 1 & 1 & 1 & 0 \\ 0 & 0 & 0 & 1 & 1 & 1 \\ 0 & 0 & 0 & 0 & 1 & 1 \end{pmatrix}.$$

Set Q as the unit matrix, $\lambda = 0.5$,

$$R = \begin{pmatrix} 12 & 0 & 0 & 0 & 0 & 0 \\ 0 & 128 & 0 & 0 & 0 & 0 \\ 0 & 0 & 50 & 0 & 0 & 0 \\ 0 & 0 & 0 & 140 & 0 & 0 \\ 0 & 0 & 0 & 0 & 72 & 0 \\ 0 & 0 & 0 & 0 & 0 & 10 \end{pmatrix}, N = \begin{pmatrix} 2, & 2, & 2, & 3, & 2, & 1 \end{pmatrix}^T.$$

Then, we can have

$$G = \begin{pmatrix} 192 & 140 & 62 & 178 & 0 & 0 \\ 140 & 282 & 152 & 268 & 140 & 0 \\ 62 & 152 & 204 & 190 & 140 & 0 \\ 178 & 268 & 190 & 392 & 212 & 72 \\ 0 & 140 & 140 & 212 & 224 & 82 \\ 0 & 0 & 0 & 72 & 82 & 84 \end{pmatrix},$$

$$f = \begin{pmatrix} -380, & -700, & -544, & -920, & -574, & -154 \end{pmatrix}^T.$$

The main diagonal of the matrix for this problem is 1, while the definition of k-diagonal matrix Q requires the main diagonal elements to be zero. Since $x \in \{0,1\}^n$, the diagonal of G can be converted into a linear term and can be considered as a seven-diagonal matrix. This indicates that the diagonal of G becomes zero and f is updated as seen in Equation (15). Using the algorithm in Section 2, we can find the optimal solution $(0,1,1,1,0,1)$ without considering the constraints. This is to install four PMUs in bus 2, 3, 4, and 6. Through observation, it can be seen that the configuration results meet the system observability and the constraint is reached.

$$f = \begin{pmatrix} -284, & -559, & -442, & -724, & -462, & -112 \end{pmatrix}^T. \tag{15}$$

The above problem considers the placement of PMU under the condition that the system is completely observable. When considering the PMU placement problem with the $N-1$ principle, constraint conditions $2x_i + 2\sum_{j\in P_i^2} x_j + \sum_{j\in P_i^1} x_j \geq 2 (i \in N)$ are added based on the definition of a single node observable. N is a set that includes the nodes required to have objectivity when a fault occurs in the system. P_i^2 and P_i^1 represents a node set connected to node i with two and one line respectively.

If the bus between nodes 5 and 6 breaks down, we set $H(5,6)$ and $H(6,5)$ as zero. Then, the constraint is added.

$$\begin{pmatrix} 0 & 0 & 0 & 1 & 1 & 0 \\ 0 & 0 & 0 & 0 & 0 & 1 \end{pmatrix} x \geq \begin{pmatrix} 1 \\ 1 \end{pmatrix}.$$

We can get the optimal solution is the same as that in the former example. As the connection between nodes 5 and 6 fails, the system after placing PMU will not lose its observability. We only consider the algorithm to solve the constrained k-diagonal zero-one matrix quadratic programming and consider that it can be applied to some form of PMU placement problems. Some of the conditions to the placement of PMU have not been taken into account here, which has yet to be studied by specialized engineers.

5. 01CQP Algorithm Simulation and Discussion

5.1. Algorithm Experimental Simulation

Now, the experimental results are provided for this algorithm to illustrate its performance. We implement the algorithm with C++ and run it on an Intel(R) Core(TM) i7-8550U CPU. For the problems we tested, the dimension of matrix Q ranges from 10 to 100 and k takes $5, 7, \ldots, 25$. All simulation data (Q, c, a, b) are generated randomly. Q, c is set up as in Section 2 with numbers ranging from -100 to 100, and a, b range between 0 and 20. Then, we obtain Table 7, which shows the detailed computation time for different dimensional problems with different diagonal numbers.

Table 7. Corresponding computation time in different diagonal numbers with dimension n = 10 to 100 of 01CQP.

n	10	20	30	40	50	60	70	80	90	100
$k = 5$	0.0356	0.0678	0.1404	0.2039	0.3289	0.4102	0.4589	0.6942	0.7391	1.0060
$k = 7$	0.0336	0.0703	0.1419	0.2030	0.3097	0.4024	0.4839	0.6908	0.7475	1.0519
$k = 9$	0.0320	0.0638	0.1424	0.1851	0.3129	0.3640	0.4896	0.6831	0.7835	0.9771
$k = 11$	0.0335	0.0619	0.1470	0.2025	0.2749	0.3357	0.4155	0.5482	0.6621	0.8812
$k = 13$	0.0246	0.0482	0.1093	0.1577	0.2591	0.3500	0.4176	0.6029	0.7109	0.8872
$k = 15$	0.0185	0.0544	0.1138	0.1734	0.2561	0.3624	0.4469	0.5782	0.7356	0.8873
$k = 17$	0.0234	0.0549	0.1185	0.1749	0.2838	0.3766	0.4732	0.7220	0.8908	1.1064
$k = 19$	0.0397	0.0744	0.1692	0.2210	0.3944	0.4888	0.6059	0.9147	0.9886	1.2304
$k = 21$	--	0.0872	0.1797	0.2836	0.4656	0.5714	0.6995	1.1097	1.2544	1.7283
$k = 23$	--	0.1283	0.2640	0.4409	0.5499	0.9615	1.0942	1.3158	1.8381	1.8951
$k = 25$	--	0.1961	0.3673	0.6169	0.8295	1.3556	1.3235	2.0418	2.3717	4.2689

5.2. Experimental Results Discussion

Here, based on the experimental results, we investigate the influence of n and k on the algorithm and discuss the importance and implications of our study results. Firstly, we fix the values of k but increase n. Figure 2 shows the experimental situations when the matrix dimension n changes from 10 to 100 with k is fixed at 9, 13, 19 and 23 respectively. As can be seen, the calculation time increases slightly with the increase of dimension. Secondly, we fix the values of n but increase k. Figure 3 shows the experimental situations when k changes from 5 to 25 with n is fixed at 20, 40, 60 and 80 respectively. It can be observed clearly that time changes significantly with the increase of diagonal number. Finally, we increase n and k simultaneously. Figure 4 illustrates this situation. It is obvious that k has an obvious influence on the calculation speed. All these observations coincide with the time complexity we derived in Section 3. It means that if the diagonal number is within an appropriate range, our algorithm can perform effectively and efficiently even for very large scale problems. And these kind of problems cover a large portion of the whole problem set. Therefore, the algorithm we proposed will have great potential in many real-world applications.

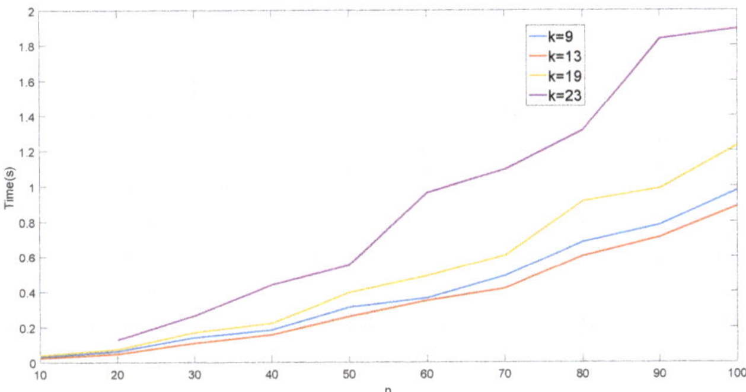

Figure 2. Calculation time with different dimensions when the diagonal number $k = 9, 13, 19, 23$.

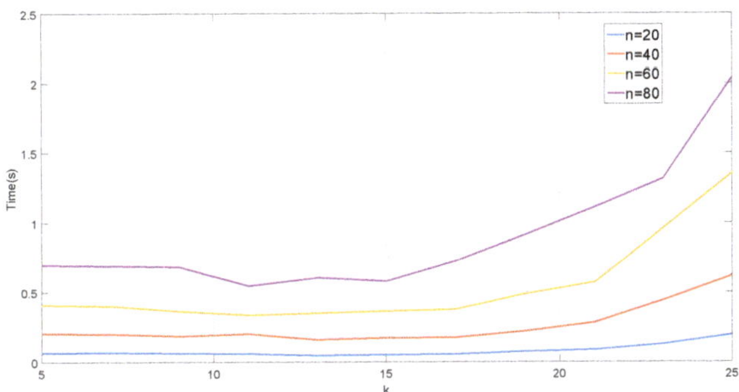

Figure 3. Calculation time with different diagonal numbers when the dimension $n = 20, 40, 60, 80$.

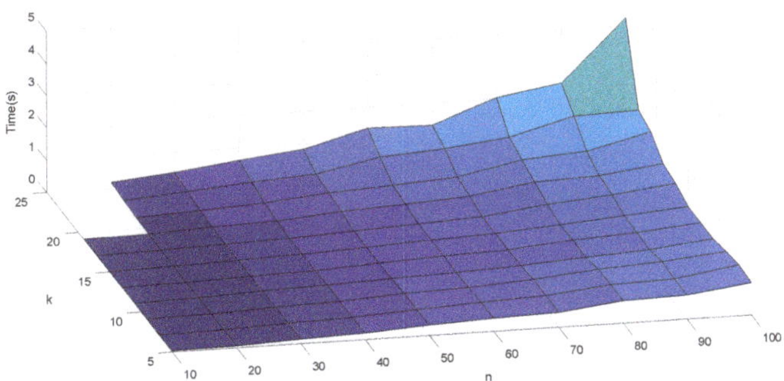

Figure 4. The variation of calculation time with different dimensions n and diagonal number k.

6. Conclusions

In this paper, a novel exact algorithm to the general linearly constrained zero-one quadratic programming problem with k-diagonal matrix is proposed. The algorithm is designed by analyzing the property of matrix Q and then combining the famous basic algorithm and dynamic programming method. The complexity of the algorithm is analyzed and shows that it is polynomially solvable when m is fixed. The experimental results also illustrate the feasibility and efficiency of the algorithm. Designing efficient algorithm to this special class of problem $01CQP$ not only provides useful information for designing efficient algorithms for other special classes but also can provide hints and facilitate the derivation of efficient relaxations for the general problems. And finally, the phasor measurement units placement problem is used to demonstrate that the algorithm has wide potential applications in decision-making real-life problems.

Author Contributions: S.G. put forward ideas and algorithms. S.G. and X.C simulated the results and wrote important parts of the article. X.C. formatted the whole paper, and summarized the results in tables. All authors have read and agreed to the published version of the manuscript.

Funding: The work described in the paper was supported by the National Science Foundation of China under Grants 61876105 and 61503233.

Conflicts of Interest: The authors declare no conflict of interest.

Abbreviations

The following abbreviations are used in this manuscript:

01QP	Zero-One Quadratic Programming
01UQP	Unconstrained Zero-One Quadratic Programming
01CQP	Constrained Zero-One Quadratic Programming
PMU	Phasor Measurement Units
NP-hard	Non-deterministic polynomial-hard

Variables

The following variables are used in this manuscript:

Q	$Q = (q_{ij})_{n\times n}$ and only q_{ij}=$q_{ji}(i = 1, 2, \ldots, n-1, j = i+1, i+2, \ldots, i+m)$ are not zero
c	$c = (c_{ij})_{n\times 1}$
a	$a = (a_{ij})_{n\times 1}$
b	$b \in \mathbb{Z}_+$
n	Dimensions of matrix Q
m	$m \in [0, n-1]$ and $m \in \mathbb{Z}$
k	$k = 2m+1$ $(m = 0, 1, 2, \ldots, n-1)$
s_k	State variable $s_k \in \mathbb{Z}$, stage variable $k(0 < k \leq n)$
$f(x)$	$f(x) = \frac{1}{2}x^T Qx + c^T x$
H	$H = (h_{ij})_{n\times n}$ is a network graph structure
N	$N \in \mathbb{R}^n$ represents the upper bound of the maximum redundancy of each bus
R	$R \in \mathbb{R}^{n\times n}$ is the importance of each bus
Q	$Q \in \mathbb{R}^{n\times n}$ in the application is the cost of placing the PMU on the bus
$M(x)$	Column vector consisted of $m_i(x) = (1 - x_i)\prod_{j\in A_i}(1 - x_j)(i \in \Omega)$
Ω	Bus set
λ	Weight
A_i	Nodes adjacent to the bus i

References

1. Cook, W.J.; Cunningham, W.H.; Pulleyblank, W.R.; Schrijver, A. *Combinatorial Optimization;* Wiley-Interscience: Hoboken, NJ, USA, 1997.
2. Hammer, P.L.; Rudeanu, S. *Boolean Methods in Operations Research and Related Areas;* Spring: Berlin/Heidelberg, Germany; New York, NY, USA, 1968.
3. Chang, K.C.; Du, H.C. Efficient algorithms for layer assignment problem. *IEEE Trans. Comput.-Aided Des. Integr. Circuits Syst.* **1987**, *6*, 67–78. [CrossRef]
4. Dehghan, A.; Mubarak, S. Binary quadratic programing for online tracking of hundreds of people in extremely crowded scenes. *IEEE Trans. Pattern Anal. Mach. Intell.* **2016**, *40*, 568–581. [CrossRef] [PubMed]
5. Laughhunn, D.J. Quadratic binary programming with application to capital-budgeting problems. *Oper. Res.* **1970**, *18*, 454–461. [CrossRef]

6. Kizys, R.; Juan, A.; Sawik, B.; Calvet, L. A biased-randomized iterated local search algorithm for rich portfolio optimization. *Appl. Sci.* **2019**, *9*, 3509. [CrossRef]
7. Huang, L.; Zhang, Q.; Sun, L.; Sheng, Z. Robustness analysis of iterative learning control for a class of mobile robot systems with channel noise. *IEEE Access* **2019**, *7*, 34711–34718. [CrossRef]
8. Manoli, G.; Rossi, M.; Pasetto, D.; Deiana, R.; Ferraris, S.; Cassiani, G.; Putti, M. An iterative particle filter approach for coupled hydro-geophysical inversion of a controlled infiltration experiment. *J. Comput. Phys.* **2015**, *283*, 37–51. [CrossRef]
9. Mesbah, A. Stochastic model predictive control with active uncertainty learning: A Survey on dual control. *Ann. Rev. Control* **2018**, *45*, 107–117. [CrossRef]
10. Imani, M.; Braga-Neto, U. Maximum-Likelihood adaptive filter for partially-observed boolean dynamical systems. *IEEE Trans. Signal Process.* **2017**, *65*, 359–371. [CrossRef]
11. Imani, M.R.; Dougherty, E.; Braga-Neto, U. Boolean Kalman filter and smoother under model uncertainty. *Automatica* **2020**, *111*, 108609. [CrossRef]
12. Baldwin, T.L.; Mili, L.; Boisen, M.B. Power system observability with minimal phasor measurement placement. *IEEE Trans. Power Syst.* **1993**, *8*, 707–715. [CrossRef]
13. Nuqui, R.F.; Phadke, A.G. Phasor measurement unit placement techniques for complete and incomplete observability. *IEEE Trans. Power Deliv.* **2005**, *20*, 2381–2388. [CrossRef]
14. Xu, B.; Abur, A. Observability analysis and measurement placement for system with PMUs. *IEEE PES Power Syst. Conf. Expos.* **2004**, *2*, 943–946.
15. Gou, B. Generalized integer linear programming formulation for optimal PMU placement. *IEEE Trans. Power Syst.* **2008**, *23*, 1099–1104. [CrossRef]
16. Kavasseri, R.; Srinivasan, S.K. Joint placement of phasor and power flow measurements for observability of power systems. *IEEE Trans. Power Syst.* **2011**, *26*, 1929–1936. [CrossRef]
17. Poljak, S.; Rendl, F. Solving the max-cut problem using eigenvalues. *Discret. Appl. Math.* **1995**, *62*, 249–278. [CrossRef]
18. Rendl, F.; Rinaldi, G.; Wiegele, A. Solving Max-Cut to optimality by intersecting semidefinite and polyhedral relaxations. *Lect. Notes Comput.* **2007**, *4513*, 295–309. [CrossRef]
19. Ye, Y. A .699-approximation algorithm for Max-Bisection. *Math. Program.* **2001**, *90*, 101–111. [CrossRef]
20. Pardalos, P.M.; Rodgers, G.P. Computational aspects of a branch and bound algorithm for quadratic zero-one programming. *Computing* **1990**, *45*, 131–144. [CrossRef]
21. Li, D.; Sun, X.L.; Liu, C.L. An exact solution method for unconstrained quadratic 0-1 programming: A geometric approach. *J. Glob. Optim.* **2012**, *52*, 797–829. [CrossRef]
22. Gu, S.; Chen, X.Y.; Wang L. Global optimization of binary quadratic programming: A neural network based algorithm and its FPGA Implementation. *Neural Process. Lett.* **2019**, 1–20. [CrossRef]
23. Zhu, W.X. Penalty parameter for linearly constrained 0-1 quadratic programming. *J. Optim. Theory Appl.* **2003**, *116*, 229–239. [CrossRef]
24. Katayama, K.; Narihisa, H. Performance of simulated annealing-based heuristic for the unconstrained binary quadratic programming problem. *Eur. J. Oper. Res.* **2001**, *134*, 103–119. [CrossRef]
25. Jiang, D.; Zhu, S. A method to solve nonintersection constraint 0-1 quadratic programming model. *J. Southwest Jiaotong Univ.* **1997**, *32*, 667–671.
26. Ranjbar, M.; Effati, S.; Miri, S.M. An artificial neural network for solving quadratic zero-one programming problems. *Neurocomputing* **2017**, *235*, 192–198. [CrossRef]
27. Ping, W.; Weiqing, X. Binary ant colony algorithm with controllable search bias for unconstrained binary quadratic problem. In Proceedings of the 2012 International Conference on Electronics, Communications and Control, Zhoushan, China, 16–18 October 2012.
28. Gu, S.; Cui, R. Polynomial time solvable algorithm to linearly constrained binary quadratic programming problems with Q being a five-diagonal matrix. In Proceedings of the Fiveth International Conference on Intelligent Control and Information Processing, Dalian, China, 18–20 August 2014; pp. 366–372.
29. Gu, S.; Cui, R.; Peng, J. Polynomial time solvable algorithms to a class of unconstrained and linearly constrained binary quadratic programming problems. *Neurocomputing* **2016**, *198*, 171–179. [CrossRef]

30. Gu, S.; Peng, J.; Cui, R. A polynomial time solvable algorithm to binary quadratic programming problems with Q being a seven-diagonal matrix and its neural network implementation. In Proceedings of the ISNN 2014—Advances in Neural Networks, Macao, China, 28 November–1 December 2014; pp. 338–346.
31. Gu, S.; Chen, X.Y. The basic algorithm for zero-one unconstrained quadratic programming problem with k-diagonal matrix. In Proceedings of the Twelfth International Conference on Advanced Computational Intelligence International, Dali, China, 14–16 March 2020.

Article

Evaluating Personal Default Risk in P2P Lending Platform: Based on Dual Hesitant Pythagorean Fuzzy TODIM Approach

Xiaonan Ji, Lixia Yu * and Jiapei Fu

Business School, Sichuan Normal University, Chengdu 610101, Sichuan, China; 20181901002@stu.sicnu.edu.cn (X.J.); 20181901005@stu.sicnu.edu.cn (J.F.)

* Correspondence: yulixia@sicnu.edu.cn

Received: 26 November 2019; Accepted: 11 December 2019; Published: 18 December 2019

Abstract: An extended approach proposed in this paper is to make a more reasonable assessment of personal default risk in peer to peer (P2P) online lending platform, which reduces uncertainty while taking into account the psychological characteristics of lenders to avoid risk. The TODIM (an acronym in Portuguese of interactive and multi-criteria decision making) approach, which can describe the psychological behaviors of decision maker, has been proved to be effective to solve multi-attribute decision making (MADM) problems. The definitions of dual hesitant Pythagorean fuzzy set (DHPFS) and the processes of traditional TODIM approach are firstly introduced in this paper. Then, the TODIM approach is extended to solve the MADM problems with a dual hesitant Pythagorean fuzzy number (DHPFN). Finally, a case study evaluating the personal default risk in P2P online lending is conducted to demonstrate that the proposed approach is applicable to solve MADM problems. In addition, some comparative analyses are performed to compare the dual hesitant Pythagorean fuzzy TODIM method with the other two integrated operators of DHPFS. Through the comparisons, we conclude that the advantage of the proposed method over other methods is that it reduces uncertainty while taking into account the psychological characteristics of lenders to avoid risk. Today's credit environment is fraught with risks, and the psychological behaviors of decision makers are important factors that cannot be ignored. For these reasons, the dual Pythagorean hesitant fuzzy TODIM method is applicable for evaluating personal default risk.

Keywords: dual hesitant Pythagorean fuzzy set; multi-attribute decision making; TODIM approach; psychological behaviors; peer to peer; evaluating personal credit

---

1. Introduction

P2P network lending is a new type of lending model. As a website that provides intermediary services, it connects capital applicants and providers together. Due to its relatively high rate of return while low threshold, requirement and cost, P2P online lending provides an important channel for borrowers who have difficulty in obtaining financing from the traditional financial industry. This lending model has been popularized nationwide. In 2007, the first P2P platform in China was established in Shanghai, then there was a blowout in 2015. In fact, there was little interference in P2P by regulators before 2018. But with the continuous upgrading of financial regulation after that, P2P continued to suffer from explosion, suspension, losing contact, withdrawal difficulties, platform fraud and other problems, and those phenomena disrupted the normal financial order. By November 2019, the total loan balance of normal operating platforms of P2P was 540.828 billion yuan, and the transaction volume was 50.623 billion yuan. Compared with the November of last year, the total loan balance declined by 33.33 percent, and the transaction volume dropped by 54.58 percent. Recently, some provinces, such as Hunan, Shandong, Chongqing, and Sichuan have announced a crackdown on non-compliant P2P business. At present, the biggest risk of P2P lending is default risk, which has become the main bottleneck of the development of P2P industry. How to evaluate the borrower's default risk is particularly important for the normal operation of P2P. Therefore, this article attempts to propose a new method to evaluate the borrower's default risk in P2P online lending.

In recent years, a lot of researchers studied on the personal default risk in the P2P online lending. Malekipirbazari [1] established a credit risk assessment model based on random forest classification method. Serrano-cinca [2] exceeded the traditional credit model, established the profit scoring model, and proved that the profit scoring model based on multiple regression was superior to the credit scoring system based on the logistic model. Oreski S. [3] proposed a personal credit assessment method combining genetic algorithm and neural network. Guo et al. [4] verified that the instance-based credit risk assessment model was superior to the score-based model.

Wei and Lu [5] proposed the dual hesitant Pythagorean fuzzy set (DHPFS), the definitions and basic algorithms of DHPFS were given in their paper. In DHPFS, both membership and non-membership are represented by sets. It is clear that DHPFS can get more valuable decision information from decision makers. Pythagorean fuzzy sets and dual hesitant fuzzy sets have their own outstanding advantages and application prospects. The combination of the two will make great contribution to the extension of fuzzy sets and make a difference in decision-making process [6].

MADM is an important part of decision science and expected utility theory is the theoretical basis of traditional MADM research. Expected utility theory believes that complete rationality is the characteristic of decision makers, and those persons' goal is to maximize benefits. In real life, the practical option of decision makers and best option based on rational decision theory have a certain deviation, owing to cognitive ability, emotion, psychology and other factors [7]. At present, some scholars have studied the MADM method which takes the psychological behavior of decision makers into account. Knhneman and Tversky put forward prospect theory [8], which is regarded to be the most influential theory in the field of behavioral decision making and has been widely used in MADM problems [9]. The theory mainly draws the following conclusions. First, most people have a risk-averse attitude to gains and a risk-appetite attitude to losses. Second, people are more sensitive to loss than gain. The pain of loss greatly exceeds the pleasure of gain, which is called "loss aversion". Third, most people tend to rely on reference points for their judgment of gains and losses, which is called "reference dependence". Gomes and Lima [10] proposed a TODIM decision making method based on prospect theory, which is a typical multi-attribute decision making method that takes the psychological behavior of decision makers into account. The TODIM method's main concepts are as follows. First, calculating the advantages of alternatives over other alternatives in various attributes based on the decision maker's reference dependence and the psychological behavior characteristics of loss avoidance. Then, the overall advantage of each scheme over other schemes is calculated, and finally, the objective function which can maximize the overall advantage and the optimal model is established respectively. The TODIM method has been deeply studied and applied to various decision-making problems by many scholars, such as choosing the best destination for natural gas storage [11] and evaluating green supply chain [12]. In order to describe both the uncertainty of MADM problems and the attitude of lenders to avoid risks, some scholars have successively proposed fuzzy TODIM method [13], intuitionistic fuzzy TODIM [14], Pythagoras fuzzy TODIM [15] and hesitant fuzzy TODIM [16]. Wei [17] proposed picture fuzzy TODIM method for MADM problems. Huang and Wei [18] proposed the Pythagoras binary semantic TODIM method. Ji et al. [19] applied the project-based multi-valued neutrophil set TODIM method to personnel selection. Tian et al. [20] proposed an extended TODIM based on cumulative prospect theory, and then applied it in venture capital. Yin et al. [21] proposed an extended TODIM to evaluate the competency of project manager, which combines lambda-fuzzy measure with Choquet integral. However, the dual hesitant Pythagoras fuzzy TODIM method has not been studied.

Although there are a large number of methods for evaluating personal default risk, they do not take people's attitude to risk into account. At the same time, the decision environment contains too much uncertain information. So, risk and uncertainty are two important factors that should be taken into consideration when evaluating personal credit default risk. Previous research has found that TODIM is an effective method that can reflect people's attitude to risk, and dual hesitant Pythagoras fuzzy set can get more information from decision maker. Therefore, we are trying to apply TODIM method with dual hesitant Pythagoras fuzzy number to the personal default risk evaluation in P2P online lending. Our goal is to demonstrate the applicability of this approach in individual default risk evaluation and, more broadly, to apply it to more similar MADM problems.

In the next section, we will introduce the process of traditional TODIM approach, the concepts of Pythagorean fuzzy set, dual hesitant fuzzy set and dual hesitant Pythagorean fuzzy set. In Section 3, we shall propose the dual hesitant Pythagorean TODIM approach. In Section 4, a numerical example for evaluating personal credit risk in the P2P online lending platform with dual hesitant Pythagorean fuzzy information is used to demonstrate the approach's applicability. In Section 5, we will draw the conclusions about this paper.

2. Preliminaries

In this part, the process of traditional TODIM approach in decision making will be introduced briefly. Then, we retrospect the concepts and theorics of dual hesitant Pythagorean fuzzy set (DHPFS).

2.1. The TODIM Approach

Let $A = \{A_1, A_2, \ldots, A_m\}$ be a discrete set with m alternatives, and A_i represents the ith alternative in the set. Let $C = \{C_1, C_2, \ldots, C_n\}$ be a set with n attributes, and C_j represents the jth attribute in the set. Let $\omega = \{\omega_1, \omega_2, \ldots, \omega_n\}^T$ be the attributes set's weight vector, with $\omega_j \in [0,1]$ and $\sum\limits_{j=1}^{n} \omega_j = 1$. Suppose that $X = \left(x_{ij}\right)_{m \times n}$ be a decision matrix, where x_{ij} is the attribute value according to a decision maker over the alternative A_i with respect to attribute C_j. We define $\omega_{jr} = \omega_j / \omega_r (j, r = 1, 2, \ldots, n)$ as the relative weight of the attribute C_j to C_r, $\omega_r = \max\left\{\omega_j \middle| j = 1, 2, \ldots, n\right\}$, and $\omega_{jr} \in [0,1]$. Then, the steps of the traditional TODIM approach are as follows:

Step 1. Standardize the decision matrix $X = \left(x_{ij}\right)_{m \times n}$ into the matrix $Y = \left(y_{ij}\right)_{m \times n}$.

Step 2. Compute the dominance degree of Ai over each alternative Ak for attribute C_j:

$$\delta(A_i, A_k) = \sum_{j=1}^{n} \phi_j(A_i, A_k), (i, k = 1, 2, \ldots, m) \tag{1}$$

where

$$\phi_j(A_i, A_k) = \begin{cases} \sqrt{\left(y_{ij} - y_{kj}\right)\omega_{jr} / \left(\sum_{j=1}^{n} \omega_{jr}\right)}, & y_{ij} > y_{kj} \\ 0, & y_{ij} = y_{kj} \\ -\frac{1}{\theta}\sqrt{\left(y_{kj} - y_{ij}\right)\left(\sum_{j=1}^{n} \omega_{jr}\right) / \omega_{jr}}, & y_{ij} < y_{kj} \end{cases} \tag{2}$$

and the attenuation factor of the losses is expressed by the parameter value θ.

Step 3. Compute the overall value of the alternative A_i by:

$$\xi(A_i) = \frac{\sum_{k=1}^{m} \delta(A_i, A_k) - \min\limits_{i \in M}\left\{\sum_{k=1}^{m} \delta(A_i, A_k)\right\}}{\max\limits_{i \in M}\left\{\sum_{k=1}^{m} \delta(A_i, A_k)\right\} - \min\limits_{i \in M}\left\{\sum_{k=1}^{m} \delta(A_i, A_k)\right\}}, (i = 1, 2, \ldots, m) \tag{3}$$

Step 4. Rank all the alternatives according to the overall values $\xi(A_i), (i = 1, 2, \ldots, m)$. Based on this result, we can get the conclusion that the most ideal alternative has largest overall value, and conversely, the worst alternative has the smallest value.

2.2. Dual Pythagorean Hesitant Fuzzy Set

In this part, some definitions of PFS, DHFS and DHPFS are introduced. The score function and accuracy function of DHPFN are also provided.

Definition 1 [22]). *A Pythagorean fuzzy set (PFS) P with an object X is defined as follows:*

$$P = \{\langle x, u_P(x), v_P(x)\rangle | x \in X\} \tag{4}$$

where $u_P(x) \in [0,1]$ is defined as the membership degree function, $v_P(x) \in [0,1]$ is defined as the non-membership degree function. For all $x \in X$, $u_P(x)$ and $v_P(x)$ meet the following requirements: $u_P(x)^2 + v_P(x)^2 \le 1$. The hesitant degree of $x \in X$ is presented as:

$$\pi_P(x) = \sqrt{1 - u_P(x)^2 - v_P(x)^2}, \forall x \in X \tag{5}$$

For convenience, $(u_P(x), v_P(x))$ is called a PFN, which can be denoted by $P = (u_P, v_P)$.

Definition 2 ([23]). *Let X be a fixed set, then a dual hesitant fuzzy set (DHFS) D with an object X is defined as follows:*

$$D = \{\langle x, \Gamma_D(x), \Psi_D(x)\rangle | x \in X\} \tag{6}$$

where $\Gamma_D(x) \in [0,1]$ is defined as a set of all the possible membership degrees, $\Psi_D(x) \in [0,1]$ is defined as a set of all the possible non-membership degrees.

Definition 3 ([5]). *Let X be a fixed set, then a dual Pythagorean hesitant fuzzy set (DHPFS) A with an object X is defined as follows:*

$$A = \{\langle x, \Gamma_A(x), \Psi_A(x)\rangle | x \in X\} \tag{7}$$

where $\Gamma_A(x) \in [0,1]$ is defined as a set of all the possible membership degrees, $\Psi_A(x) \in [0,1]$ is defined as a set of all the possible non-membership degrees. Furthermore, for all $u_A(x) \in \Gamma_A(x)$ and $v_A(x) \in \Psi_A(x)$ meet the following requirements: $u_A(x)^2 + v_A(x)^2 \le 1$. The set of hesitant degree $x \in X$ is presented as:

$$\Pi_A(x) = \bigcup_{\substack{u_A(x) \in \Gamma_A(x) \\ v_A(x) \in \Psi_A(x)}} \left\{\pi_A(x) = \sqrt{1 - u_A(x)^2 - v_A(x)^2} \ge 0\right\}, \forall_x \in X \tag{8}$$

For convenience, $(\Gamma_A(x), \Psi_A(x))$ is called a DHPFN, which can be denoted by $A = (\Gamma_A, \Psi_A)$.

Wei and Lu [5] have proposed some operational rules for DHPFN.

Definition 4 ([5]). *Let $\beta_1 = (\Gamma_{\beta 1}, \Psi_{\beta 1})$, $\beta_2 = (\Gamma_{\beta 2}, \Psi_{\beta 2})$ and $\beta = (\Gamma_\beta, \Psi_\beta)$ be three DHPFNs, then*

(1) $\beta_1 \oplus \beta_2 = < \bigcup_{\substack{u_{\beta 1(x)} \in \Gamma_{\beta 1(x)} \\ u_{\beta 2(x)} \in \Gamma_{\beta 2(x)}}} \left\{\sqrt{u_{\beta 1}{}^2 + u_{\beta 2}{}^2 - u_{\beta 1}{}^2 u_{\beta 1}{}^2}\right\}, \bigcup_{\substack{v_{\beta 1(x)} \in \Psi_{\beta 1(x)} \\ v_{\beta 2(x)} \in \Psi_{\beta 2(x)}}} \{v_{\beta 1} v_{\beta 2}\} >;$

(2) $\beta_1 \otimes \beta_2 = < \bigcup_{\substack{u_{\beta 1(x)} \in \Gamma_{\beta 1(x)} \\ u_{\beta 2(x)} \in \Gamma_{\beta 2(x)}}} \{v_{\beta 1} v_{\beta 2}\}, \bigcup_{\substack{v_{\beta 1(x)} \in \Psi_{\beta 1(x)} \\ v_{\beta 2(x)} \in \Psi_{\beta 2(x)}}} \left\{\sqrt{u_{\beta 1}{}^2 + u_{\beta 2}{}^2 - u_{\beta 1}{}^2 u_{\beta 1}{}^2}\right\} >;$

(3) $\lambda\beta = < \bigcup_{u_\beta \in \Gamma_\beta} \left\{\sqrt{1 - (1 - u_\beta{}^2)^\lambda}\right\}, \bigcup_{v_\beta \in \Psi_\beta} \{v_\beta{}^\lambda\} >;$

(4) $\beta^\lambda = < \bigcup_{u_\beta \in \Gamma_\beta} \{u_\beta{}^\lambda\}, \bigcup_{v_\beta \in \Psi_\beta} \left\{\sqrt{1 - (1 - v_\beta{}^2)^\lambda}\right\} >;$

(5) $\beta^c = <\Psi_\beta, \Gamma_\beta>$;

Wei and Lu [5] gave a score function of DHPFNs to rank two different numbers.

Definition 5 ([5]). *Let $\beta = (\Gamma_\beta, \Psi_\beta)$ be a DHPFN, then the score function of β is defined as:*

$$S_\beta = \frac{1}{2}\left(1 + \frac{1}{|\Gamma_\beta|}\sum_{u_\beta \in \Gamma_\beta} {u_\beta}^2 - \frac{1}{|\Psi_\beta|}\sum_{v_\beta \in \Psi_\beta} {v_\beta}^2\right) \tag{9}$$

the accuracy function of β is defined as:

$$p_\beta = \frac{1}{|\Gamma_\beta|}\sum_{u_\beta \in \Gamma_\beta} {u_\beta}^2 + \frac{1}{|\Psi_\beta|}\sum_{v_\beta \in \Psi_\beta} {v_\beta}^2 \tag{10}$$

$|\Gamma_\beta|$ and $|\Psi_\beta|$ represents the length of set Γ_β and Ψ_β respectively. The higher the score, the better the number.

3. Dual Pythagorean Hesitant Fuzzy TODIM Approach

In order to describe the dual Pythagorean hesitant fuzzy TODIM approach in detail, first of all, we put forward some fundamental concepts that are essential to this paper.

3.1. Fundamental Knowledge

According to Definition 5, Wei and Lu [5] gave a solution to compare two DHPFNs.

Definition 6 ([5]). ***Let*** *$\beta_1 = (\Gamma_{\beta 1}, \Psi_{\beta 1})$ and $\beta_2 = (\Gamma_{\beta 2}, \Psi_{\beta 2})$ be two DHPFNs, S_{β_1} and S_{β_2} be the scores value of β_1 and β_2 respectively, p_{β_1} and p_{β_2} be the accuracy values of β_1 and β_2 respectively.*

Then, if $S_{\beta_1} > S_{\beta_2}$, then $\beta_1 > \beta_2$; if $S_{\beta_1} = S_{\beta_2}$, then

(1) *if $p_{\beta_1} = p_{\beta_2}$, then $\beta_1 = \beta_2$;*

(2) *if $p_{\beta_1} > p_{\beta_2}$, then $\beta_1 > \beta_2$;*

Definition 7 ([24]). *Let $\beta_1 = (\Gamma_{\beta 1}, \Psi_{\beta 1})$ and $\beta_2 = (\Gamma_{\beta 2}, \Psi_{\beta 2})$ be two DHPFNs, then the Hamming distance between β_1 and β_2 is:*

$$d(\beta_1, \beta_2) = \frac{1}{2}\left(\frac{1}{l}\sum\nolimits_{j=1}^{l}\left|{\Gamma_{\beta 1}}^{\sigma(j)} - {\Gamma_{\beta 2}}^{\sigma(j)}\right| + \frac{1}{m}\sum\nolimits_{j=1}^{m}\left|{\Psi_{\beta 1}}^{\sigma(j)} - {\Psi_{\beta 2}}^{\sigma(j)}\right|\right) \tag{11}$$

where l, m are the lengths of the set Γ and *Ψ, respectively. ${\Gamma_{\beta 1}}^{\sigma(j)}, {\Gamma_{\beta 2}}^{\sigma(j)}$ represents the jth largest element in set $\Gamma_{\beta 1}$ and $\Gamma_{\beta 2}$, respectively. ${\Psi_{\beta 1}}^{\sigma(j)}, {\Psi_{\beta 2}}^{\sigma(j)}$ represents the jth largest element in set $\Psi_{\beta 1}$ and $\Psi_{\beta 2}$, respectively.*

In general, the number of elements in two DHPFS is different. Therefore, we use the extension principle to compare two sets, that is, adding elements to a set with fewer elements, so that the number of elements in the two sets is the same. There are two types of extension principles. The first is to add the largest element in the set, while the other is to add the smallest element in the set. In this article, we use the first principle.

An example is shown to illustrate the Hamming distance between β_1 and β_2: If $\beta_1 = (\{0.4, 0.2\}, \{0.7, 0.3, 0.1\}$ and $\beta_1 = (\{0.7, 0.3, 0.1\}, \{0.4, 0.1\}$ are two DHPFNs. Ar first, we extend set $\Gamma_{\beta 1}(x) = \{0.4, 0.2\}$ to $\{0.4, 0.4, 0.2\}$ and set $\Psi_{\beta 2}(x) = \{0.4, 0.1\}$ to $\{0.4, 0.4, 0.1\}$. Then the distance between β_1 and β_2 is: $d(\beta_1, \beta_2) = 0.1500$.

3.2. TODIM Method for Dual Hesitant Pythagorean Fuzzy MADM Problems

Now, we give procedures to use TODIM approach to deal with MADM Problems with dual hesitant Pythagorean fuzzy set. In order to understand it easily, we describe the problems briefly as follows:

Let $A = \{A_1, A_2, \ldots, A_m\}$ be a discrete set of alternatives and $C = \{C_1, C_2, \ldots, C_n\}$ be the set of attributes; Let $= \{\omega_1, \omega_2, \ldots, \omega_n\}$ be the attributes set's weight vector, with $\omega_j \in [0,1]$ and $\sum_{j=1}^{n} \omega_j = 1$. Suppose $R = (r_{ij})_{m \times n} = (\Gamma_{ij}, \Psi_{ij})_{m \times n}$ be a dual hesitant Pythagorean fuzzy decision matrix, where r_{ij} is a criterion value according to a decision maker over the alternative. A_i with respect to attribute C_j, where $\Gamma_{A(x)} \in [0,1]$ is defined as a set of all the possible membership degrees, $\Psi_{A(x)} \in [0,1]$ is defined as a set of all the possible non-membership degrees. Furthermore, for all $u_{A(x)} \in \Gamma_{A(x)}$ and $v_{A(x)} \in \Psi_{A(x)}$ meet the following requirements: $u_A(x)^2 + v_A(x)^2 \leq 1$.

Then, we extend the TODIM model to deal with the MADM problems with dual hesitant Pythagorean fuzzy set. At first, it is necessary to standardize the decision matrix, because there may be some benefit attribute and cost attribute in C, as they are two opposite DHPFNs:

$$l_{ij} = \begin{cases} r_{ij}, forbenefitattributeC_j \\ r_{ij}^c, for \cos tattributeC_j \end{cases} \tag{12}$$

where $r_{ij}^c = (\Psi_{ij}, \Gamma_{ij})$ is the complement of r_{ij}^c. After that, we can get the normalized dual hesitant Pythagorean fuzzy decision matrix $L = (l_{ij})_{m \times n}$.

Next, we calculate the relative weight of the attribute C_j to C_r as

$$\omega_{jr} = \omega_j / \omega_r (j, r = 1, 2, \ldots, n) \tag{13}$$

where $\omega_r = \max\{\omega_j | j = 1, 2, \ldots, n\}, \omega_{jr} \in [0,1]$.

Then we can get the dominance degree of A_i over attribute C_j for each alternative A_k:

$$\phi j(Ai, Ak) = \begin{cases} \sqrt{d(l_{ij}.l_{kj})\omega_{jr} / (\sum_{j=1}^{n} \omega_{jr})}, & l_{ij} > l_{kj} \\ 0, & l_{ij} = l_{kj} \\ -\frac{1}{\theta}\sqrt{d(l_{ij}.l_{kj})(\sum_{j=1}^{n} \omega_{jr}) / \omega_{jr}}, & l_{ij} < l_{kj} \end{cases} \tag{14}$$

where the parameter θ represents the attenuation factor of the losses, $d(l_{ij}.l_{kj})$ describe the distance between the DHPFNs l_{ij} and l_{kj}. In order to show the function $\phi_j(A_i, A_k)$ visually, we express it in a matrix way:

$$\phi_j(A_i, A_k) = \begin{array}{c} \\ \mathrm{A}_1 \\ \mathrm{A}_2 \\ \vdots \\ \mathrm{Am} \end{array} \begin{array}{c} \begin{array}{cccc} \mathrm{A}_1 & \mathrm{A}_2 & \cdots & A_m \end{array} \\ \begin{bmatrix} 0 & \phi_j(A_1, A_2) & \cdots & \phi_j(A_1, A_m) \\ \phi_j(A_2, A_1) & 0 & \cdots & \phi_j(A_2, A_m) \\ \vdots & \vdots & \cdots & \vdots \\ \phi_j(A_m, A_1) & \phi_j(A_m, A_2) & \cdots & 0 \end{bmatrix} \end{array} \tag{15}$$

based on which, we can calculate the overall dominance degree of the A_i over each alternative A_k by:

$$\delta(A_i, A_k) = \sum_{j=1}^{n} \phi_j(A_i, A_k), (i, k = 1, 2, \ldots, m) \tag{16}$$

and we also show the function $\delta(A_i, A_k)$ visually in a matrix:

$$\delta = \delta(A_i, A_k)_{m\times n} = \begin{matrix} & \mathrm{A}_1 & \mathrm{A}_2 & \cdots & A_m \\ \mathrm{A}_1 \\ \mathrm{A}_2 \\ \vdots \\ A_m \end{matrix} \begin{bmatrix} 0 & \phi_j(A_1, A_2) & \cdots & \phi_j(A_1, A_m) \\ \phi_j(A_2, A_1) & 0 & \cdots & \phi_j(A_2, A_m) \\ \vdots & \vdots & \cdots & \vdots \\ \phi_j(A_m, A_1) & \phi_j(A_m, A_2) & \cdots & 0 \end{bmatrix} \tag{17}$$

Finally, the overall value of each alternative A_i can be computed by the following formula:

$$\xi(A_i) = \frac{\sum_{k=1}^{m} \delta(A_i, A_k) - \min\limits_{i\in M}\left\{\sum_{k=1}^{m} \delta(A_i, A_k)\right\}}{\max\limits_{i\in M}\left\{\sum_{k=1}^{m} \delta(A_i, A_k)\right\} - \min\limits_{i\in M}\left\{\sum_{k=1}^{m} \delta(A_i, A_k)\right\}}, (i = 1, 2, \ldots, m) \tag{18}$$

In order to rank all alternatives, we construct a principle whereby that the greater the overall value of $\xi(A_i), (i = 1, 2, \ldots, m)$, the better the alternative A_i.

4. Numerical Example

4.1. The Application of Dual Hesitant Pythagorean Fuzzy TODIM Approach

In order to evaluate the borrower's default risk accurately, many models have been constructed. The first step in building the model is to select some reasonable credit evaluation indicators. Lin et al. [25] explored the factors that determine the default risk based on the demographic characteristics of borrowers. The empirical results show that the following ten factors played an important role in loan default: (1) gender, (2) age, (3) marital status, (4) educational level, (5) working years, (6) company size, (7) monthly payment, (8) loan amount, (9) debt to income ratio, and (10) delinquency history.

We denote these ten attributes as $C_j(j = 1, 2, \ldots, 10)$ respectively, and their weight vector is $\omega = (0.0550, 0.0466, 0.2005, 0.1692, 0.0766, 0.0804, 0.0821, 0.1530, 0.0655, 0.0711)^T$, which is given by the decision maker. Suppose that there are four applicants to be ranked, which denoted as $A_j(j = 1, 2, 3, 4)$, and dual hesitant Pythagorean numbers $r_{ij}(\Gamma_{ij}, \Psi_{ij})$ appear in the MADM process. $r_{ij}(\Gamma_{ij}, \Psi_{ij})$(i = 1,2,3,4; $j = 1, 2, \ldots, 10$) represents the value of alternative A_i relative to attribute C_j. All these values are contained in the decision matrix $R = \left(r_{ij}\right)_{m\times n} = \left(\Gamma_{ij}, \Psi_{ij}\right)_{m\times n}$, as shown in Table 1.

Table 1. The decision matrix R.

	C_1	C_2	C_3	C_4	C_5
A_1	({0.7,0.6},{0.7})	({0.4},{0.7,0.6})	({0.3},{0.8})	({0.4,0.3},{0.7,0.6})	({0.4},{0.7,0.6})
A_2	({0.5,0.4},{0.6,0.5})	({0.5},{0.5,0.4})	({0.3,0.2},{0.8})	({0.5},{0.7})	({0.6,0.5},{0.7})
A_3	({0.6,0.5},{0.5})	({0.5},{0.6})	({0.4},{0.7,0.6})	({0.4,0.3},{0.6})	({0.6,0.5},{0.8})
A_4	({0.6,0.5},{0.5})	({0.5,0.4},{0.7})	({0.6},{0.8,0.7})	({0.7},{0.4,0.3})	({0.4,0.3},{0.7})
	C_6	C_7	C_8	C_9	C_{10}
A_1	({0.6,0.5},{0.6})	({0.5},{0.6,0.5})	({0.6},{0.7,0.6})	({0.5,0.4},{0.7})	({0.6},{0.7,0.6})
A_2	({0.7},{0.6,0.5})	({0.5},{0.6,0.5})	({0.5,0.4},{0.8})	({0.5},{0.7})	({0.6,0.5},{0.7})
A_3	({0.6,0.5},{0.5,0.4})	({0.7},{0.7})	({0.4},{0.6,0.5})	({0.7,0.6},{0.4})	({0.6},{0.8,0.7})
A_4	({0.6},{0.6,0.5})	({0.6,0.5},{0.5})	({0.5},{0.7,0.6})	({0.6},{0.5,0.4})	({0.5,0.4},{0.6})

Then, we can apply dual hesitant Pythagorean fuzzy TODIM approach to rank the personal default risk. Before that, we use extension principle which is claimed in this paper to extend the decision matrix, as shown in Table 2.

Table 2. The extended decision matrix R.

	C_1	C_2	C_3	C_4	C_5
A_1	({0.7,0.6},{0.7,0.7})	({0.4,0.4},{0.7,0.6})	({0.3,0.3},{0.8,0.8})	({0.4,0.3},{0.7,0.6})	({0.4,0.4},{0.7,0.6})
A_2	({0.5,0.4},{0.6,0.5})	({0.5,0.5},{0.5,0.4})	({0.3,0.2},{0.8,0.8})	({0.5,0.5},{0.7,0.7})	({0.6,0.5},{0.7,0.7})
A_3	({0.6,0.5},{0.5,0.5})	({0.5,0.5},{0.6,0.6})	({0.4,0.4},{0.7,0.6})	({0.4,0.3},{0.6,0.6})	({0.6,0.5},{0.8,0.8})
A_4	({0.6,0.5},{0.5,0.5})	({0.5,0.4},{0.7,0.7})	({0.6,0.6},{0.8,0.7})	({0.7,0.7},{0.4,0.3})	({0.4,0.3},{0.7,0.7})
	C_6	C_7	C_8	C_9	C_{10}
A_1	({0.6,0.5},{0.6,0.6})	({0.5,0.5},{0.6,0.5})	({0.6,0.6},{0.7,0.6})	({0.5,0.4},{0.7,0.7})	({0.6,0.6},{0.7,0.6})
A_2	({0.7,0.7},{0.6,0.5})	({0.5,0.5},{0.6,0.5})	({0.5,0.4},{0.8,0.8})	({0.5,0.5},{0.7,0.7})	({0.6,0.5},{0.7,0.7})
A_3	({0.6,0.5},{0.5,0.4})	({0.7,0.7},{0.7,0.7})	({0.4,0.4},{0.6,0.5})	({0.7,0.6},{0.4,0.4})	({0.6,0.6},{0.8,0.7})
A_4	({0.6,0.6},{0.6,0.5})	({0.6,0.5},{0.5,0.5})	({0.5,0.5},{0.7,0.6})	({0.6,0.6},{0.5,0.4})	({0.5,0.4},{0.6,0.6})

Firstly, $\omega_3 = \max\{\omega_j | j = 1, 2, \ldots, 10\} = 0.2005$, so we calculate the relative weights of the attribute are:

$$\omega_{13} = 0.2743, \omega_{23} = 0.2324, \omega_{33} = 1.0000, \omega_{43} = 0.8439, \omega_{53} = 0.3820, \omega_{63} = 0.4010, \omega_{73} = 0.4095, \omega_{83} = 0.7631, \omega_{93} = 0.3267, \omega_{10\,3} = 0.3546.$$

Then, we need to calculate the dominance degree of A_i over each alternative A_k with respect to the attribute $C_j (i, k = 1, 2, 3, 4; j = 1, 2, \ldots, 10)$. Let $\theta = 2.5$:

$$\phi_1 = \begin{array}{c} \\ \mathrm{A}_1 \\ \mathrm{A}_2 \\ \mathrm{A}_3 \\ \mathrm{A}_4 \end{array} \begin{array}{c} \begin{array}{cccc} \mathrm{A}_1 & \mathrm{A}_2 & \mathrm{A}_3 & \mathrm{A}_4 \end{array} \\ \left[\begin{array}{cccc} 0 & 0.0981 & -0.7340 & -0.7340 \\ -0.7928 & 0 & -0.5190 & -0.5190 \\ 0.0908 & 0.0642 & 0 & 0 \\ 0.0908 & 0.0642 & 0 & 0 \end{array}\right] \end{array}$$

$$\phi_2 = \begin{array}{c} \\ \mathrm{A}_1 \\ \mathrm{A}_2 \\ \mathrm{A}_3 \\ \mathrm{A}_4 \end{array} \begin{array}{c} \begin{array}{cccc} \mathrm{A}_1 & \mathrm{A}_2 & \mathrm{A}_3 & \mathrm{A}_4 \end{array} \\ \left[\begin{array}{cccc} 0 & -0.7974 & -0.5638 & 0.0483 \\ 0.0836 & 0 & 0.0591 & 0.0836 \\ 0.0591 & -0.5638 & 0 & 0.0591 \\ -0.4604 & -0.7974 & -0.5638 & 0 \end{array}\right] \end{array}$$

$$\phi_3 = \begin{array}{c} \\ \mathrm{A}_1 \\ \mathrm{A}_2 \\ \mathrm{A}_3 \\ \mathrm{A}_4 \end{array} \begin{array}{c} \begin{array}{cccc} \mathrm{A}_1 & \mathrm{A}_2 & \mathrm{A}_3 & \mathrm{A}_4 \end{array} \\ \left[\begin{array}{cccc} 0 & 0.0708 & -0.3509 & -0.4152 \\ -0.1569 & 0 & -0.3844 & -0.4439 \\ 0.1583 & 0.1734 & 0 & -0.3844 \\ 0.1873 & 0.2002 & 0.1734 & 0 \end{array}\right] \end{array}$$

$$\phi_4 = \begin{array}{c} \\ \mathrm{A}_1 \\ \mathrm{A}_2 \\ \mathrm{A}_3 \\ \mathrm{A}_4 \end{array} \begin{array}{c} \begin{array}{cccc} \mathrm{A}_1 & \mathrm{A}_2 & \mathrm{A}_3 & \mathrm{A}_4 \end{array} \\ \left[\begin{array}{cccc} 0 & -0.3417 & -0.1708 & -0.6160 \\ 0.1301 & 0 & -0.3820 & -0.5666 \\ 0.0650 & 0.1454 & 0 & -0.5918 \\ 0.2345 & 0.2157 & 0.2253 & 0 \end{array}\right] \end{array}$$

$$\phi_5 = \begin{array}{c} \\ \mathrm{A}_1 \\ \mathrm{A}_2 \\ \mathrm{A}_3 \\ \mathrm{A}_4 \end{array} \begin{array}{c} \begin{array}{cccc} \mathrm{A}_1 & \mathrm{A}_2 & \mathrm{A}_3 & \mathrm{A}_4 \end{array} \\ \left[\begin{array}{cccc} 0 & -0.5078 & 0.1072 & 0.0619 \\ 0.0875 & 0 & 0.0619 & 0.0875 \\ -0.6219 & -0.3591 & 0 & 0.1072 \\ -0.3591 & -0.5078 & -0.6219 & 0 \end{array}\right] \end{array}$$

$$\phi_6 = \begin{array}{c} \\ A_1 \\ A_2 \\ A_3 \\ A_4 \end{array} \begin{array}{c} \begin{array}{cccc} A_1 & A_2 & A_3 & A_4 \end{array} \\ \begin{bmatrix} 0 & -0.4957 & -0.4293 & -0.3505 \\ 0.0897 & 0 & 0.1002 & 0.0634 \\ 0.0777 & -0.5542 & 0 & 0.0777 \\ 0.0634 & -0.3505 & -0.4293 & 0 \end{bmatrix} \end{array}$$

$$\phi_7 = \begin{array}{c} \\ A_1 \\ A_2 \\ A_3 \\ A_4 \end{array} \begin{array}{c} \begin{array}{cccc} A_1 & A_2 & A_3 & A_4 \end{array} \\ \begin{bmatrix} 0 & 0 & -0.6489 & -0.3468 \\ 0 & 0 & -0.6489 & -0.3468 \\ 0.1199 & 0.1199 & 0 & -0.6489 \\ 0.0641 & 0.0641 & 0.1199 & 0 \end{bmatrix} \end{array}$$

$$\phi_8 = \begin{array}{c} \\ A_1 \\ A_2 \\ A_3 \\ A_4 \end{array} \begin{array}{c} \begin{array}{cccc} A_1 & A_2 & A_3 & A_4 \end{array} \\ \begin{bmatrix} 0 & 0.1515 & 0.1515 & 0.0875 \\ -0.4401 & 0 & -0.4401 & -0.3593 \\ -0.4401 & 0.1515 & 0 & 0.1237 \\ -0.2541 & 0.1237 & -0.3593 & 0 \end{bmatrix} \end{array}$$

$$\phi_9 = \begin{array}{c} \\ A_1 \\ A_2 \\ A_3 \\ A_4 \end{array} \begin{array}{c} \begin{array}{cccc} A_1 & A_2 & A_3 & A_4 \end{array} \\ \begin{bmatrix} 0 & -0.2746 & -0.8683 & -0.7766 \\ 0.0405 & 0 & -0.8237 & -0.7265 \\ 0.1280 & 0.1214 & 0 & 0.0572 \\ 0.1145 & 0.1071 & -0.3883 & 0 \end{bmatrix} \end{array}$$

$$\phi_{10} = \begin{array}{c} \\ A_1 \\ A_2 \\ A_3 \\ A_4 \end{array} \begin{array}{c} \begin{array}{cccc} A_1 & A_2 & A_3 & A_4 \end{array} \\ \begin{bmatrix} 0 & 0.0596 & 0.0596 & 0.0843 \\ -0.3727 & 0 & 0.0596 & -0.5271 \\ -0.3727 & -0.3727 & 0 & -0.6455 \\ -0.5271 & 0.0843 & 0.1033 & 0 \end{bmatrix} \end{array}$$

Secondly, we calculate the overall dominance degree of the A_i over each alternative A_k:

$$\delta = \begin{array}{c} \\ A_1 \\ A_2 \\ A_3 \\ A_4 \end{array} \begin{array}{c} \begin{array}{cccc} A_1 & A_2 & A_3 & A_4 \end{array} \\ \begin{bmatrix} 0 & -2.0371 & -3.4477 & -2.9572 \\ -1.3312 & 0 & -2.9172 & -3.2547 \\ -0.7359 & -1.0740 & 0 & -1.8458 \\ -0.8460 & -0.7964 & -1.7408 & 0 \end{bmatrix} \end{array}$$

Finally, we can get the overall value of each alternative A_i by the follows:

$$\xi(A_1) = 0; \xi(A_2) = 0.1856; \xi(A_3) = 0.9461; \xi(A_4) = 1;$$

Thus, the ranking of alternatives is $A_4 > A_3 > A_2 > A_1$, and A_4 has best credit.

4.2. Comparative Analysis

In this section, some comparative analysis will be performed to compare our proposed method with the other two operators defined by Wei and Lu [5].

4.2.1. Comparison with the Dual Hesitant Pythagorean Fuzzy Weighted Average Operator

Then, we contrast this method with a dual hesitant Pythagorean fuzzy weighted average (DHPFWA) operator proposed by Wei and Lu [5] as follows:

Definition 8 ([5]). Let $\alpha_i = (\Gamma_{\alpha_i}(x), \Psi_{\alpha_i}(x)) \in DHPFNs(i = 1, 2, \ldots, n)$ and $\omega = \{\omega_1, \omega_2, \ldots, \omega_n\}^T$ be the attributes set's weight vector, with $\omega_j \in [0,1]$ and $\sum_{j=1}^{n} \omega j = 1$, then

$$DHPFWA_{\omega}(\alpha_1, \alpha_2, \ldots, \alpha_n) = \omega_1\alpha_1 \oplus \omega_2\alpha_2 \oplus \ldots \oplus \omega_n\alpha_n$$
$$= < \bigcup_{u_{\alpha_i} \in \Gamma_{\alpha_i}} \left\{ \sqrt{1 - \prod_{i=1}^{n} (1 - u_{\alpha_i}^2)^{\omega_i}} \right\}, \bigcup_{v_{\alpha_i} \in \Psi_{\alpha_i}} \left\{ \prod_{i=1}^{n} v_{\alpha_i}^{\omega_i} \right\} > \tag{19}$$

We can get:

$$\begin{aligned}
A_1 = \; & (\{0.4982, 0.4925, 0.4884, 0.4826, 0.4878, 0.4819, 0.4777, 0.4716,\\
& 0.4886, 0.4828, 0.4785, 0.4725, 0.4779, 0.4718, 0.4674, 0.4611\},\\
& \{0.7012, 0.6936, 0.6849, 0.6774, 0.6908, 0.6833, 0.6747, 0.6673,\\
& 0.6930, 0.6854, 0.6768, 0.6694, 0.6827, 0.6752, 0.6668, 0.6595,\\
& 0.6832, 0.6757, 0.6672, 0.6600, 0.6730, 0.6657, 0.6573, 0.6502,\\
& 0.6751, 0.6678, 0.6594, 0.6522, 0.6651, 0.6579, 0.6496, 0.6425,\\
& 0.6962, 0.6886, 0.6800, 0.6725, 0.6858, 0.6784, 0.6699, 0.6626,\\
& 0.6880, 0.6805, 0.6720, 0.6647, 0.6778, 0.6704, 0.6620, 0.6548,\\
& 0.6783, 0.6709, 0.6625, 0.6552, 0.6682, 0.6609, 0.6526, 0.6455,\\
& 0.6703, 0.6630, 0.6547, 0.6475, 0.6603, 0.6531, 0.6450, 0.6379\})\\
A_2 = \; & (\{0.5115, 0.5033, 0.4987, 0.4901, 0.5026, 0.4941, 0.4894, 0.4806,\\
& 0.5037, 0.4952, 0.4906, 0.4817, 0.4945, 0.4858, 0.4810, 0.4718,\\
& 0.5070, 0.4986, 0.4940, 0.4853, 0.4980, 0.4893, 0.4846, 0.4755,\\
& 0.4990, 0.4904, 0.4857, 0.4767, 0.4897, 0.4809, 0.4760, 0.4667\},\\
& \{0.6986, 0.6882, 0.6884, 0.6782, 0.6914, 0.6811, 0.6813, 0.6712,\\
& 0.6916, 0.6814, 0.6816, 0.6714, 0.6845, 0.6743, 0.6745, 0.6645\})\\
A_3 = \; & (\{0.5303, 0.5201, 0.5216, 0.5110, 0.5220, 0.5115, 0.5130, 0.5021,\\
& 0.5210, 0.5104, 0.5120, 0.5011, 0.5124, 0.5015, 0.5031, 0.4918,\\
& 0.5244, 0.5139, 0.5154, 0.5046, 0.5159, 0.5051, 0.5066, 0.4955,\\
& 0.5149, 0.5041, 0.5056, 0.4945, 0.5061, 0.4949, 0.4965, 0.4850\},\\
& \{0.6213, 0.6154, 0.6042, 0.5985, 0.6102, 0.6044, 0.5934, 0.5878,\\
& 0.6023, 0.5967, 0.5858, 0.5802, 0.5916, 0.5860, 0.5754, 0.5699\})\\
A_4 = \; & (\{0.5861, 0.5816, 0.5788, 0.5741, 0.5827, 0.5781, 0.5752, 0.5705,\\
& 0.5832, 0.5786, 0.5757, 0.5710, 0.5797, 0.5750, 0.5721, 0.5674,\\
& 0.5812, 0.5766, 0.5737, 0.5690, 0.5777, 0.5730, 0.5701, 0.5653,\\
& 0.5782, 0.5735, 0.5706, 0.5658, 0.5746, 0.5699, 0.5670, 0.5621\},\\
& \{0.5968, 0.5882, 0.5829, 0.5745, 0.5882, 0.5796, 0.5744, 0.5661,\\
& 0.5685, 0.5602, 0.5552, 0.5472, 0.5602, 0.5521, 0.5472, 0.5392,\\
& 0.5811, 0.5726, 0.5675, 0.5593, 0.5726, 0.5643, 0.5593, 0.5512,\\
& 0.5535, 0.5454, 0.5406, 0.5327, 0.5454, 0.5375, 0.5327, 0.5250\})
\end{aligned}$$

then, the score values of each alternative are as follows: $s_{A_1} = 0.3914$, $s_{A_2} = 0.3878$, $s_{A_3} = 0.4521$, $s_{A_4} = 0.5080$. We can get the ranking of alternatives according to the score value: $s_{A_4} > s_{A_3} > s_{A_1} > s_{A_2}$, and A_4 has best credit.

4.2.2. Comparison with the Dual Hesitant Pythagorean Fuzzy Weighted Geometric Operator

Then, we contrast this method with dual hesitant Pythagorean fuzzy weighted geometric (DHPFWG) operator proposed by Wei and Lu [5] as follows:

Definition 9 ([5]). Let $\alpha_i = (\Gamma_{\alpha_i}(x), \Psi_{\alpha_i}(x)) \in DHPFNs(i = 1, 2, \ldots, n)$ and $\omega = \{\omega_1, \omega_2, \ldots, \omega_n\}^T$ be the attributes set's weight vector, with $\omega_j \in [0, 1]$ and $\sum_{j=1}^{n} \omega_j = 1$,then

$$DHPFWG_{\omega}(\alpha_1, \alpha_2, \ldots, \alpha_n) = \omega_1\alpha_1 \otimes \omega_2\alpha_2 \otimes \ldots \otimes \omega_n\alpha_n$$
$$= < \bigcup_{u_{\alpha_i}\in\Gamma_{\alpha_i}} \left\{ \prod_{i=1}^{n} u_{\alpha_i}^{\omega_i} \right\}, \bigcup_{v_{\alpha_i}\in\Psi_{\alpha_i}} \left\{ \sqrt{1 - \prod_{i=1}^{n} (1 - v_{\alpha_i}^2)^{\omega_i}} \right\} > \tag{20}$$

We can get:

A_1 = ({0.4553, 0.4487, 0.4487, 0.4422, 0.4337, 0.4274, 0.4274, 0.4212, 0.4515, 0.4449, 0.4449, 0.4384, 0.4300, 0.4238, 0.4238, 0.4176}, {0.7117, 0.7060, 0.6993, 0.6934, 0.7071, 0.7014, 0.6945, 0.6884, 0.7056, 0.6998, 0.6929, 0.6868, 0.7009, 0.6950, 0.6880, 0.6817, 0.6980, 0.6920, 0.6849, 0.6785, 0.6932, 0.6870, 0.6798, 0.6733, 0.6915, 0.6854, 0.6781, 0.6715, 0.6866, 0.6803, 0.6728, 0.6662, 0.7080, 0.7023, 0.6954, 0.6894, 0.7034, 0.6975, 0.6906, 0.6844, 0.7018, 0.6959, 0.6889, 0.6827, 0.6970, 0.6910, 0.6839, 0.6775, 0.6941, 0.6880, 0.6808, 0.6743, 0.6892, 0.6829, 0.6756, 0.6690, 0.6875, 0.6812, 0.6738, 0.6672, 0.6825, 0.6761, 0.6685, 0.6617})

A_2 = ({0.4764, 0.4702, 0.4604, 0.4544, 0.4698, 0.4637, 0.4540, 0.4481, 0.4392, 0.4335, 0.4244, 0.4190, 0.4331, 0.4275, 0.4185, 0.4131, 0.4705, 0.4645, 0.4548, 0.4489, 0.4640, 0.4580, 0.4484, 0.4427, 0.4338, 0.4282, 0.4192, 0.4138, 0.4278, 0.4223, 0.4134, 0.4081}, {0.7195, 0.7151, 0.7152, 0.7107, 0.7177, 0.7133, 0.7134, 0.7088, 0.7165, 0.7121, 0.7122, 0.7076, 0.7147, 0.7102, 0.7103, 0.7058})

A_3 = ({0.4924, 0.4874, 0.4852, 0.4803, 0.4856, 0.4807, 0.4785, 0.4737, 0.4690, 0.4643, 0.4622, 0.4575, 0.4625, 0.4578, 0.4558, 0.4512, 0.4875, 0.4826, 0.4804, 0.4755, 0.4807, 0.4759, 0.4737, 0.4690, 0.4643, 0.4596, 0.4576, 0.4530, 0.4579, 0.4533, 0.4512, 0.4467}, {0.6531, 0.6420, 0.6423, 0.6307, 0.6491, 0.6378, 0.6381, 0.6263, 0.6324, 0.6203, 0.6206, 0.6080, 0.6280, 0.6158, 0.6160, 0.6033})

A_4 = ({0.5683, 0.5593, 0.5598, 0.5510, 0.5559, 0.5471, 0.5476, 0.5390, 0.5624, 0.5535, 0.5540, 0.5453, 0.5501, 0.5415, 0.5420, 0.5334, 0.5626, 0.5537, 0.5542, 0.5455, 0.5503, 0.5417, 0.5422, 0.5336, 0.5568, 0.5480, 0.5485, 0.5399, 0.5446, 0.5361, 0.5366, 0.5281}, {0.6487, 0.6454, 0.6328, 0.6292, 0.6430, 0.6396, 0.6267, 0.6230, 0.6426, 0.6392, 0.6263, 0.6226, 0.6367, 0.6332, 0.6200, 0.6163, 0.6156, 0.6118, 0.5975, 0.5935, 0.6091, 0.6052, 0.5906, 0.5864, 0.6087, 0.6048, 0.5901, 0.5860, 0.6020, 0.5981, 0.5830, 0.5788})

then, the score values of each alternative are as follows: $s_{A_1} = 0.3587, s_{A_2} = 0.3436, s_{A_3} = 0.4122, s_{A_3} = 0.4607$. We can get the rank of all alternatives according to the score values: $s_{A_4} > s_{A_3} > s_{A_1} > s_{A_2}$, and A_4 has best credit.

The results of comparative analyses by using different methods are as follows:

As you can see from Table 3, the three methods have the same best choice A_4 and their results are slightly different. Obviously, the proposed method considers the attitude of decision makers towards risk, that is, people are more sensitive to loss than gain, which makes the results of decisions consistent with the expectations of decision makers. Each of these methods has its own advantages: (1) DHPFWA operator is affected by groups, (2) DHPFWG operator is affected by individuals, and (3) the dual hesitant Pythagoras fuzzy TODIM method reduces uncertainty while taking into account the psychological characteristics of lenders to avoid risk. Today's credit environment is full of risks, and the psychological behaviors of decision makers are important factors that cannot be ignored. So, the dual Pythagorean hesitant fuzzy TODIM method is applicable for evaluating personal default risk.

Table 3. Rank of alternatives by using different methods.

	Ordering
Dual Hesitant Pythagoras Fuzzy TODIM	$A_4 > A_3 > A_2 > A_1$
DHPFWA	$A_4 > A_3 > A_1 > A_2$
DHPFWG	$A_4 > A_3 > A_1 > A_2$

5. Concluding Remarks and Suggestions

In the decision-making environment full of risks, the TODIM method enables decision-makers' psychological behaviors to be described. But it doesn't work when it's used directly to solve the MADM problems with fuzzy information. Through existing research, we can find that the membership and non-membership of the dual hesitant Pythagoras fuzzy set are respectively represented by a set containing all possible values, this fuzzy set is a good tool for people to express hesitancy in daily life. Therefore, we have proposed a method named dual hesitant Pythagoras fuzzy TODIM to solve MADM problem with dual hesitant Pythagoras fuzzy information. One of the most important advantages of the proposed approach is that it can reduce uncertainty while taking the psychological characteristics of lenders into account to avoid risk. At the same time, this method can be further extended to other similar MADM problems with interdependent attribute under dual hesitant Pythagoras fuzzy environments, such as performance evaluation, risk investment and so on. At last, the dual hesitant Pythagoras fuzzy has made a great contribution to the expansion of fuzzy set and also plays a significant role in practical decision process.

In this paper, the dual hesitant Pythagoras fuzzy TODIM method is applied to the case of personal default risk evaluation in the P2P lending platforms, and some comparative analyses are performed to compare the dual hesitant Pythagorean fuzzy TODIM method with the other two integrated operators DHPFWA and DHPFWG. The advantage of the dual hesitant Pythagoras fuzzy TODIM method over these two operators is that it takes the decision maker's reference dependence and the psychological behavior characteristics of loss aversion into account. We have demonstrated that the proposed approach is applicable for evaluating personal default risk through the comparisons. More broadly, it can also be applied to solve more similar MADM problems. In the future, we shall continue studying the MADM problems with the application and extension of the developed operators to other domains [26–33].

Finally, in view of the macroeconomic downturn and strong financial supervision, we would like to put forward the following suggestions to prevent default risk of P2P online platforms: (1) Improve the information disclosure system to reduce information asymmetry. On the one hand, online loan platforms should actively participate in relevant regulatory associations and fully disclose the information of the borrowers; on the other hand, it is necessary and even desirable to examine the authenticity and legitimacy of the information and issue a reliable risk assessment report. (2) Establish a third-party depository system. On the one hand, the funds of the loan project should be managed by a special third-party financial institution like a bank, and the online loan platforms are only used as a medium for information interaction. On the other hand, it avoids short-term goals and extracts a certain percentage of risk reserves to protect the investment. This can improve the anti-risk ability of online loan platforms, and also enhance investors' confidence in the platform.

Author Contributions: Data curation, L.Y.; writing—review and editing, L.Y.; investigation, X.J.; writing—original draft preparation, X.J.; software, J.F. All authors have read and agreed to the published version of the manuscript.

Funding: This work was supported by the National Social Science Foundation the People's Republic of China (No. 19BJL025) and the Humanities and Social Sciences Foundation of the Ministry of Education of the People's Republic of China (No.17YJA790088)

Conflicts of Interest: The authors declare no conflicts of interest.

References

1. Malekipirbazari, M.; Aksakalli, V. Risk assessment in social lending via random forests. *Expert Syst. Appl.* **2015**, *42*, 4621–4631. [CrossRef]
2. Serrano-Cinca, C.; Gutiérrez-Nieto, B. The use of profit scoring as an alternative to credit scoring systems in peer-to-peer (P2P) lending. *Decis. Support Syst.* **2016**, *89*, 113–122. [CrossRef]
3. Oreski, S.; Oreski, G. Genetic algorithm-based heuristic for feature selection in credit risk assessment. *Expert Syst. Appl.* **2014**, *41*, 2052–2064. [CrossRef]
4. Guo, Y.H.; Zhou, W.J.; Luo, C.Y.; Liu, C.R.; Hui, X. Instance-based credit risk assessment for investment decisions in P2P lending. *Eur. J. Oper. Res.* **2016**, *249*, 417–426. [CrossRef]
5. Wei, G.W.; Lu, M. Dual hesitant pythagorean fuzzy Hamacher aggregation operators in multiple attribute decision making. *Arch. Control Sci.* **2017**, *27*, 365–395. [CrossRef]
6. Lu, J.P.; Tang, X.Y.; Wei, G.W.; Wei, C.; Wei, Y. Bidirectional project method for dual hesitant Pythagorean fuzzy multiple attribute decision-making and their application to performance assessment of new rural construction. *Int. J. Intell. Syst.* **2019**, *34*, 1920–1934. [CrossRef]
7. Simon, H.A. Administrative Behavior. *Am. J. Nurs.* **1950**, *50*, 46–47. [CrossRef]
8. Tversky, K.A. Prospect Theory: An Analysis of Decision under Risk. *Econometrica* **1979**, *47*, 263–292.
9. Fan, Z.P.; Zhang, X.; Chen, F.D.; Liu, Y. Multiple attribute decision making considering aspiration-levels: A method based on prospect theory. *Comput. Ind. Eng.* **2013**, *65*, 341–350. [CrossRef]
10. Gomes, L.F.A.M.; Lima, M.M.P.P. TODIM: Basics and application to multicriteria ranking of projects with environmental impacts. *Found. Comput. Decis. Sci.* **1992**, *16*, 113–127.
11. Gomes, L.F.A.M.; Rangel, L.A.D.; Maranh?o, F.J.C. Multicriteria analysis of natural gas destination in Brazil: An application of the TODIM method. *Math. Comput. Model.* **2009**, *50*, 92–100. [CrossRef]
12. Tseng, M.L.; Lin, Y.H.; Tan, K.; Chen, R.H.; Chen, Y.H. Using TODIM to evaluate green supply chain practices under uncertainty. *Appl. Math. Model.* **2014**, *38*, 2983–2995. [CrossRef]
13. Fan, Z.P.; Zhang, X.; Chen, F.D.; Liu, Y. Extended TODIM method for hybrid multiple attribute decision making problems. *Knowl.-Based Syst.* **2013**, *42*, 40–48. [CrossRef]
14. Lourenzutti, R.; Krohling, R.A. A study of TODIM in a intuitionistic fuzzy and random environment. *Expert Syst. Appl.* **2013**, *40*, 6459–6468. [CrossRef]
15. Ren, P.J.; Xu, Z.S.; Gou, X.J. Pythagorean fuzzy TODIM approach to multi-criteria decision making. *Appl. Soft. Comput.* **2016**, *42*, 246–259. [CrossRef]
16. Liu, N.Y. Hesitant Fuzzy TODIM Multiple Attribute Decision-making Method Based on Attribute Association. *J. Chongqing Technol. Bus. Univ.* **2018**, *35*, 32–39.
17. Wei, G.W. TODIM Method for Picture Fuzzy Multiple Attribute Decision Making. *Informatica* **2018**, *29*, 555–566. [CrossRef]
18. Huang, Y.H.; Wei, G.W. TODIM method for Pythagorean 2-tuple linguistic multiple attribute decision making. *J. Intell. Fuzzy Syst.* **2018**, *35*, 901–915. [CrossRef]
19. Ji, P.; Zhang, H.Y.; Wang, J.Q. A projection-based TODIM method under multi-valued neutrosophic environments and its application in personnel selection. *Neural Comput. Appl.* **2016**, *29*, 221–234. [CrossRef]
20. Tian, X.L.; Xu, Z.S.; Gu, J. An Extended TODIM Based on Cumulative Prospect Theory and Its Application in Venture Capital. *Informatica* **2019**, *30*, 413–429. [CrossRef]
21. Yin, J.L.; Guo, J.; Ji, T.M. An Extended TODIM Method for Project Manager's Competency Evaluation. *J. Civ. Eng. Manag.* **2019**, *25*, 673–686. [CrossRef]
22. Yager, R.R. Pythagorean Membership Grades in Multicriteria Decision Making. *IEEE Trans. Fuzzy Syst.* **2014**, *22*, 958–965. [CrossRef]
23. Zhu, B.; Xu, Z.S.; Xia, M.M. Dual Hesitant Fuzzy Sets. *J. Appl. Math.* **2012**, *2012*, 2607–2645. [CrossRef]
24. Ren, Z.L.; Xu, Z.S.; Wang, H. Dual hesitant fuzzy VIKOR method for multi-criteria group decision making based on fuzzy measure and new comparison method. *Inf. Sci.* **2017**, *388–389*, 1–16. [CrossRef]
25. Lin, X.C.; Li, X.L.; Zheng, Z. Evaluating borrower's default risk in peer-to-peer lending: Evidence from a lending platform in China. *Appl. Econ. Lett.* **2016**, 1–8. [CrossRef]

26. Lu, J.P.; Wei, C.; Wu, J.; Wei, G.W. TOPSIS method for probabilistic linguistic magdm with entropy weight and its application to supplier selection of new agricultural machinery products. *Entropy* **2019**, *21*, 953. [CrossRef]
27. Zhang, S.Q.; Wei, G.W.; Gao, H.; Wei, C.; Wei, Y. EDAS method for multiple criteria group decision making with picture fuzzy information and its application to green suppliers selections. *Technol. Econ. Dev. Econ.* **2019**, *25*, 1123–1138. [CrossRef]
28. Gao, H.; Wu, J.; Wei, C.; Wei, G.W. MADM method with Interval-valued Bipolar Uncertain Linguistic Information for Evaluating the Computer Network Security. *IEEE Access* **2019**, *7*, 151506–151524. [CrossRef]
29. Gao, H.; Lu, M.; Wei, Y. Dual hesitant bipolar fuzzy hamacher aggregation operators and their applications to multiple attribute decision making. *J. Intell. Fuzzy Syst.* **2019**, *37*, 5755–5766. [CrossRef]
30. Wu, L.P.; Gao, H.; Wei, C. VIKOR method for financing risk assessment of rural tourism projects under interval-valued intuitionistic fuzzy environment. *J. Intell. Fuzzy Syst.* **2019**, *37*, 2001–2008. [CrossRef]
31. Li, Z.X.; Lu, M. Some novel similarity and distance and measures of Pythagorean fuzzy sets and their applications. *J. Intell. Fuzzy Syst.* **2019**, *37*, 1781–1799. [CrossRef]
32. Deng, X.M.; Gao, H. TODIM method for multiple attribute decision making with 2-tuple linguistic Pythagorean fuzzy information. *J. Intell. Fuzzy Syst.* **2019**, *37*, 1769–1780. [CrossRef]
33. Wang, R. Research on the Application of the Financial Investment Risk Appraisal Models with Some Interval Number Muirhead Mean Operators. *J.Intell.Fuzzy Syst.* **2019**, *37*, 1741–1752. [CrossRef]

Article

Pythagorean 2-Tuple Linguistic VIKOR Method for Evaluating Human Factors in Construction Project Management

Tingting He [1], Guiwu Wei [1], Jianping Lu [1], Cun Wei [2] and Rui Lin [3,*]

1 School of Business, Sichuan Normal University, Chengdu 610101, China; m_hetingting@163.com (T.H.); weiguiwu1973@sicnu.edu.cn (G.W.); lujp2002@163.com (J.L.)
2 School of Statistics, Southwestern University of Finance and Economics, Chengdu 611130, China; weicun1990@163.com
3 School of Economics and Management, Chongqing University of Arts and Sciences, Chongqing 402160, China
* Correspondence: linrui@cqwu.edu.cn

Received: 28 October 2019; Accepted: 18 November 2019; Published: 24 November 2019

Abstract: Since the reform and opening up, Chinese economic and social development has undergone great changes, and the people's living standards have improved markedly. For the national economy, the engineering construction is not only a carrier for specific economic tasks, but also a driving force for rapid and sustained economic development. With the continuous expansion of the scale of construction projects, safety management problems of construction projects are constantly exposed. How to effectively avoid accidents has become an important issue to be solved urgently in the construction industry. This paper mainly evaluates human factors in the process of construction project management, such as workers' proficiency, workers' safety awareness, technical workers' quality, and workers' emergency capacity, with the purpose of helping China's construction projects proceed smoothly. In this research, we provide a multiple attribute group decision-making (MAGDM) technique based on Pythagorean 2-tuple linguistic numbers (P2TLNs) and the VIseKriterijumska Optimizacija I KOmpromisno Resenje (VIKOR) method for evaluating the human factors of construction projects. P2TLNs are used to represent the performance assessments of decision makers. Relying on a P2TLWA operator, P2TLWG operator, and the essential VIKOR method, a general framework is established. An application is presented to test the validity of the new method, and a comparative analysis with two algorithms and the P2TL-TODIM method is illustrated with detail.

Keywords: multiple attribute group decision making (MAGDM); Pythagorean 2-tuple linguistic numbers (P2TLNs); P2TLWA operator; P2TLWG operator; VIKOR method

1. Introduction

In the related projects of engineering construction, the management problem mainly involves the safety management problem, and is closely related to human, material, and environmental factors. Safety management problems involved in project construction management should be eliminated or avoided in time to ensure the timely and smooth completion of construction and protect the safety of construction personnel and property. However, in project management, the human factor plays a very important role, as it can effectively guide the construction project construction management results. Ning and Wang [1] promoted the TOPSIS method into an intuitionistic fuzzy environment to evaluate and select the optimal site selection scheme according to the specific attributes of the construction project, so as to effectively improve the construction operation and improve the safety of the working environment. Lu [2] studied how to utilize the 2-tuple linguistic model to deal with assessment information in the information management process of construction projects, and then followed the steps of TOPSIS model for evaluation. Gu et al. [3] introduced the IVIFECA (interval-valued intuitionistic fuzzy Einstein correlated averaging) operator, which was applied to choose construction projects. Wu et al. [4] expanded the HM (Hamy Mean) operator to 2TLNNs and then introduced some operators. Finally, they utilized these operators to assess the risk of construction projects.

The fuzzy set theory [5] was first introduced to describe the uncertainty and fuzziness of things. In order to reflect the objective world as faithfully as possible, many people offered some extended forms of the fuzzy set, such as IVHFS (interval-valued hesitant fuzzy set), T2FS (Type-2 fuzzy sets), IFS (intuitionistic fuzzy set) [6–11], etc. The IFS theory was proposed by Atanassov [6] in 1986 as an important extension of the classical fuzzy set theory [5]. The research on its theory and application has achieved extensive research results in the field of fuzzy set theory and produced far-reaching influence. In an intuitionistic fuzzy set, the membership degree was defined as the degree of affirmation about the same concept, and non-membership degree was defined as the degree of negation. However, when using intuitionistic fuzzy to make decisions, the following situation may occur: the membership degree plus non-membership degree of the scheme satisfying attributes given by the decision makers is greater than 1. Based on this, in 2013, the American scholar Yager [12] proposed the Pythagorean fuzzy set, which satisfies conditions where the membership degree plus non-membership degree is greater than 1, but the sum of squares does not exceed 1. Therefore, the decision maker does not need to modify the values of membership and non-membership, and thus the model can be a more accurate and detailed description of the reality [13–16].

After the Pythagorean fuzzy set was proposed, a large number of researchers combined the Pythagorean fuzzy set [12] with various methods and applied these proposed methods to MADM. Zhang and Xu [17] first put forward the mathematical expression of a Pythagorean fuzzy set, and then they tied the Pythagorean fuzzy set (PFS) and TOPSIS method together. Finally, they gave a practical example to illustrate the developed method and made a comparative analysis of different methods. Zhang [18] presented a Pythagorean fuzzy QUALIFLEX method with the closeness index to address the layered multi-criteria decision-making issue under a Pythagorean fuzzy environment on the basis of PFNs (Pythagorean fuzzy numbers) and IVPFNs (interval-valued Pythagorean fuzzy numbers). Ren et al. [19] provided a case of choosing the governor of the Asian Infrastructure Investment Bank by using the PF-TODIM (Pythagorean fuzzy TODIM) method to observe the feasibility of the model. Bolturk [20] expanded the CODAS (COmbinative Distance-based Assessment) model to Pythagorean fuzzy environment to propose a novel method that is PF-CODAS (Pythagorean fuzzy CODAS). They addressed a MADM problem of supplier selection utilizing the new method to show its validity and effectiveness. In the end, they concluded that the presented model has better results than the general fuzzy conclusion, because it takes into account the decision makers of dithering and expands the scope of membership and non-membership degree. Chen [21] defined a new VIseKriterijumska Optimizacija I KOmpromisno Resenje (VIKOR)-based method for MADM analysis containing Pythagorean fuzzy information. A Pythagorean fuzzy set has certain advantages over un-normalized fuzzy sets such as IFS (intuitionistic fuzzy set) in dealing with fuzziness and complex uncertainty. Based on this, a Pythagorean fuzzy VIKOR method based on a distance index is proposed, which is quite different from the existing VIKOR method. It is unique because that the model considers the uncertain information expressed by the PFNs and introduced some new concepts of measuring distance. Through the practical application and comparative analysis of a certain standard satisfaction problem, the validity and superiority of the used method in practice are verified. Huang and Wei [22] briefly introduced the definition of Pythagorean 2-tuple linguistic numbers (P2TLNs), which calculate the distance between two P2TLNs and the classic TODIM (an acronym in Portuguese for Interactive Multi-criteria Decision Making). On this basis, a new extended TODIM is put forward to deal with the MADM problem. The important feature of this method is to fully take the bounded rationality of the each decision maker into account, which is a practical behavior in the decision-making process. Finally, they also gave an example. Ilbahar et al. [23] proposed three methods, respectively:

Fine Kinney, Pythagorean fuzzy analytic hierarchy process. Meanwhile, these methods are used to assess the excavation risk of a construction site. Based on the procedure of the classic TOPSIS method, Khan et al. [24] presented an extension of TOPSIS under the interval value Pythagorean fuzzy context, using the IVPFCIG (interval-valued Pythagorean fuzzy Choquet integral geometric) operator and distance formula based on the Choquet integral to aggregate all the fuzzy decision matrixes. Finally, it is proved by an example that the technique is practical and effective. Perez-Dominguez et al. [25] combined ratio analysis-based multiple objective optimization with a Pythagorean theorem fuzzy set to select an appropriate alternative. In the end, two decision problems illustrated that the method is valid and practical. A novel LINMAP (linear programming technique for multidimensional analysis of preference) method was expanded by Xue et al. [26] to the fuzzy environment of Pythagorean. Then, they defined PFE (Pythagorean fuzzy evalues) and IVPFE (interval-valued Pythagorean fuzzy evalues) based on similarity and hesitations. Based on the above, the PF-LINMAP method is constructed. According to a numerical example, this method can solve the decision problem related to railway project investment.

The VIseKriterijumska Optimizacija I KOmpromisno Resenje (Hereinafter referred to as VIKOR) model explored by Opricovic [27] initially is a practical tool to deal with the MADM problems and has a wide range of industrial, commercial economy, and science of management applications. Compared with previous methods, such as ELECTRE [28], PROMETHEE [29], GRA [30], TOPSIS [31], and TODIM [32–34], the advantage of the VIKOR model is that it takes into account the contradictory criteria such as the objectivity of decision makers and the complexity of the decision environment, so as to obtain more useful and scientific evaluation information.

Devi [35] explored a new expansion of VIKOR into an intuitionistic fuzzy context to solve the problem of a robot selection material handling task. Du and Liu [36] applied the VIKOR method in the fuzzy context of an intuitionistic trapezoidal to rank the pros and cons of each scheme and select the best one. Park et al. [37] proposed a new method to promote the VIKOR method to the interval value intuitionistic fuzzy environment to choose the right outsourcing partner for multinational organizations. Chatterjee et al. [38] gave an expanding VIKOR model based on IFS, which uses five criteria and four decision makers to evaluate five potential outsourcing partners. Finally, we choose the best outsourcer. Liao and Xu [39] contributed a new way of thinking that integrated the classical VIKOR method with hesitant fuzzy circumstances. An example is given to prove its validity. In order to solve the MADM problem of material selection in engineering design, an interval binary semantic VIKOR method is established to avoid information distortion and loss by Liu et al. [40]. Motivated by the traditional VIKOR method, Wei and Zhang [41] firstly defined the multiple criteria hesitant fuzzy decision making with shapley value-based VIKOR method. Using SLp, mu-metric, they developed an extended VIKOR method to handle related multi-criterion decision problems. Bausys and Zavadskas [42] solved a problem of selecting the location for a logistic terminal in a way where they tied the VIKOR method with interval-valued neutrosophic sets, namely VIKOR-IVNS. Dammak et al. [43] compared three methods based on intuitionistic fuzzy set respectively to TOPSIS, AHP (Analytical Hierarchy Process), and VIKOR, and analyzed the differences in the use of the three methods. Liao and Xu [44] constructed the cosine-distance-based HFL-TOPSIS model and the cosine-distance-based HFL-VIKOR model. Afterwards, they offered a case in point. You et al. [45] proposed the interval 2-tuple language VIKOR method for choosing the perfect supplier among three suppliers. It's worth noting that the method mentioned in this paper is more appropriate to treat with the problem of supplier selection in the context of fuzzy with uncertain and incomplete information. Buyukozkan et al. [46] integrated IF-AHP and IF-VIKOR to form an overall framework, and then used appropriate evaluation criteria to rank the web services of medical institutions according to the steps of the framework to measure the performance of 10 medical institutions in Turkey. Zhang et al. [47] proposed the hesitating fuzzy language VIKOR (HFL-VIKOR) method, and then took a West China hospital as an example to apply this method in the process of inpatient admission evaluation, so as to solve the problem of inpatient admission, which can be used for classified diagnosis and treatment. Hu et al. [48] adopted INSs (interval neutrosophic sets) to cope with evaluation information, and made use of a project-based difference measure VIKOR to solve the issue of online selection of doctors in mobile medical services. Wang et al. [49] explored a VIKOR method into a picture fuzzy context with normalized projection for the risk assessment of engineering construction projects.

Although previous studies deal with the selection of construction projects, they do not mention the evaluation of human factors in the process of construction project management. So, it is very essential to take appropriate measures to evaluate human factors by using relevant assessment criteria. In this paper, we extend the VIKOR method with Pythagorean 2-tuple linguistic numbers for the evaluation of human factors. It is our goal in this article to combine the original VIKOR method with P2TLNs to address MADM problems. The innovativeness of the paper can be summarized as follows: (1) the VIKOR method is extended by P2TLSs; (2) the Pythagorean 2-tuple linguistic VIKOR (P2TL-VIKOR) method is proposed to solve the Pythagorean 2-tuple linguistic multiple attribute group decision-making (MAGDM) problems; (3) a case study for evaluating human factors in construction project management is supplied to show the developed approach; and (4) some comparative studies are provided with the existing methods to give effect to the rationality of P2TL-VIKOR.

The remainder of this article is mainly as follows. Section 2 contains some basic definitions of P2TLNs; Section 3 contains the extending VIKOR method with P2TLNs; Section 4 provides a case study of evaluating human factors in the process of construction project management and contrastive analysis; and Section 5 presents the conclusions.

2. Preliminaries

2.1. Pythagorean 2-Tuple Linguistic Sets

Wei et al. [50] proposed the Pythagorean 2-tuple linguistic sets (P2TLSs) based on the PFSs [51] and 2-tuple linguistic information [52].

Definition 1 ([50]). *A P2TLS A in X is given*

$$A = \left\{ \left(s_{\varphi(x)}, \rho\right), (u_A(x), v_A(x)), x \in X \right\}, \tag{1}$$

where $s_{\varphi(x)} \in S$, $\rho \in [-0.5, 0.5)$, $u_A(x) \in [0,1]$, *and* $v_A(x) \in [0,1]$, $u_A(x)$, *and* $v_A(x)$ *satisfy the following condition:* $0 \le (u_A(x))^2 + (v_A(x))^2 \le 1$, $\forall x \in X$. *The numbers* $u_A(x), v_A(x)$ *represent the degree of membership and degree of non-membership of the element x to linguistic variable* $\left(s_{\varphi(x)}, \rho\right)$.

$A = \left\langle \left(s_\varphi, \rho\right), (u_A, v_A) \right\rangle$ *can be called a Pythagorean 2-tuple linguistic number (P2TLN).*

Definition 2 ([50]). *Suppose that* $a = \left\langle \left(s_\varphi, \rho\right), (u_a, v_a) \right\rangle$ *is a P2TLN; then, the score function of P2TLN can be depicted as follows:*

$$\begin{aligned} &S(a) = \Delta\left(\Delta^{-1}\left(s_{\varphi(a)}, \rho\right)\frac{1+(u_a)^2-(v_a)^2}{2}\right), \\ &\Delta^{-1}(S(a)) \in [0, L]. \end{aligned} \tag{2}$$

Definition 3 ([50]). *Suppose that* $a = \left\langle \left(s_\varphi, \rho\right), (u_a, v_a) \right\rangle$ *is a P2TLN; then, the accuracy function of P2TLN can be depicted as follows:*

$$\begin{aligned} &H(a) = \Delta\left(\Delta^{-1}\left(s_{\varphi(a)}, \rho\right)\frac{(u_a)^2+(v_a)^2}{2}\right), \\ &\Delta^{-1}(H(a)) \in [0, L]. \end{aligned} \tag{3}$$

Definition 4 ([50]). *Suppose that* $a_1 = \left\langle \left(s_{\varphi_1}, \rho_1\right), (u_{a_1}, v_{a_1}) \right\rangle$ *and* $a_2 = \left\langle \left(s_{\varphi_2}, \rho_2\right), (u_{a_2}, v_{a_2}) \right\rangle$ *are two P2TLNs. Respectively, the scores of* a_1 *and* a_2 *are* $S(a_1) = \Delta\left(\Delta^{-1}\left(s_{\varphi(a_1)}, \rho_1\right) \cdot \frac{1+\left(u_{a_1}\right)^2-\left(v_{a_1}\right)^2}{2}\right)$

and $S(a_2) = \Delta\left(\Delta^{-1}\left(s_{\varphi(a_2)}, \rho_2\right) \cdot \frac{1+\left(u_{a_2}\right)^2-\left(v_{a_2}\right)^2}{2}\right)$, *and let* $H(a_1) = \Delta\left(\Delta^{-1}\left(s_{\varphi(a_1)}, \rho_1\right) \cdot \frac{\left(u_{a_1}\right)^2+\left(v_{a_1}\right)^2}{2}\right)$ *and* $H(a_2) = \Delta\left(\Delta^{-1}\left(s_{\varphi(a_2)}, \rho_2\right) \cdot \frac{\left(u_{a_2}\right)^2+\left(v_{a_2}\right)^2}{2}\right)$ *be the accuracy degrees of* a_1 *and* a_2*; then, some operational laws of P2TLNs can be defined as follows:*

$$\begin{aligned}
&(1) if\ S(a_1) < S(a_2), a_1 < a_2;\\
&(2) if\ S(a_1) > S(a_2), a_1 > a_2;\\
&(3)\ if\ S(a_1) = S(a_2), H(a_1) < H(a_2),\ then\ a_1 < a_2;\\
&(4)\ if\ S(a_1) = S(a_2), H(a_1) > H(a_2),\ then\ a_1 > a_2;\\
&(5)\ if\ S(a_1) = S(a_2), H(a_1) = H(a_2),\ then\ a_1 = a_2.
\end{aligned}$$

Definition 5 ([50]). *Suppose that* $a_1 = \left\langle\left(s_{\varphi_1}, \rho_1\right), \left(u_{a_1}, v_{a_1}\right)\right\rangle$ *and* $a_2 = \left\langle\left(s_{\varphi_2}, \rho_2\right), \left(u_{a_2}, v_{a_2}\right)\right\rangle$ *are two P2TLNs, the normalized Hamming distance* (H_d) *between* a_1 *and* a_2 *can be depicted below:*

$$H_d(a_1, a_2) = \frac{1}{2L}\left\|\begin{array}{c}\left(1+\left(u_{a_1}\right)^2-\left(v_{a_1}\right)^2\right)\cdot\Delta^{-1}\left(s_{\varphi_1}, \rho_1\right)-\\ \left(1+\left(u_{a_2}\right)^2-\left(v_{a_2}\right)^2\right)\cdot\Delta^{-1}\left(s_{\varphi_2}, \rho_2\right)\end{array}\right\|, \tag{4}$$

where L represents the length of the language scale. It is a numerical value.

Definition 6 ([50]). *Suppose that* $a_1 = \left\langle\left(s_{\varphi_1}, \rho_1\right), \left(u_{a_1}, v_{a_1}\right)\right\rangle$ *and* $a_2 = \left\langle\left(s_{\varphi_2}, \rho_2\right), \left(u_{a_2}, v_{a_2}\right)\right\rangle$ *are two P2TLNs, then*

$$\begin{aligned}
a_1 \oplus a_2 &= \left\langle\Delta\left(\Delta^{-1}\left(s_{\varphi_1}, \rho_1\right)+\Delta^{-1}\left(s_{\varphi_2}, \rho_2\right)\right), \left(\sqrt{\left(u_{a_1}\right)^2+\left(u_{a_2}\right)^2-\left(u_{a_1}\right)^2\left(u_{a_2}\right)^2}, v_{a_1}v_{a_2}\right)\right\rangle;\\
a_1 \otimes a_2 &= \left\langle\Delta\left(\Delta^{-1}\left(s_{\varphi_1}, \rho_1\right)\cdot\Delta^{-1}\left(s_{\varphi_2}, \rho_2\right)\right), \left(u_{a_1}u_{a_2}, \sqrt{\left(v_{a_1}\right)^2+\left(v_{a_2}\right)^2-\left(v_{a_1}\right)^2\left(v_{a_2}\right)^2}\right)\right\rangle;\\
\lambda a_1 &= \left\langle\Delta\left(\lambda\Delta^{-1}\left(s_{\varphi_1}, \rho_1\right)\right), \left(\sqrt{1-\left(1-\left(u_{a_1}\right)^2\right)^\lambda}, \left(v_{a_1}\right)^\lambda\right)\right\rangle;\\
\left(a_1\right)^\lambda &= \left\langle\Delta\left(\left(\Delta^{-1}\left(s_{\varphi_1}, \rho_1\right)\right)^\lambda\right), \left(\left(u_{a_1}\right)^\lambda, \sqrt{1-\left(1-\left(v_{a_1}\right)^2\right)^\lambda}\right)\right\rangle.
\end{aligned}$$

Theorem 1 ([50]). *For any two P2TLNs* $a_1 = \left\langle\left(s_{\varphi_1}, \rho_1\right), \left(u_{a_1}, v_{a_1}\right)\right\rangle$ *and* $a_2 = \left\langle\left(s_{\varphi_2}, \rho_2\right), \left(u_{a_2}, v_{a_2}\right)\right\rangle$*; according to Definition 6, naturally, we can get the following properties of the operation laws:*

$$\begin{aligned}
&(1)\ a_1 \oplus a_2 = a_2 \oplus a_1\\
&(2)\ a_1 \otimes a_2 = a_2 \otimes a_1\\
&(3)\ k(a_1 \oplus a_2) = ka_1 \oplus ka_2, 0 \le k \le 1\\
&(4)\ k_1a_1 \oplus k_2a_1 = (k_1 \oplus k_2)a_1, 0 \le k_1, k_2, k_1+k_2 \le 1\\
&(5)\ {a_1}^{k_1} \otimes {a_1}^{k_2} = (a_1)^{k_1+k_2}, 0 \le k_1, k_2, k_1+k_2 \le 1\\
&(6)\ {a_1}^{k_1} \otimes {a_2}^{k_1} = (a_1 \otimes a_2)^{k_1}, k_1 \ge 0\\
&(7)\ \left((a_1)^{k_1}\right)^{k_2} = (a_1)^{k_1k_2}.
\end{aligned}$$

2.2. Pythagorean 2-Tuple Linguistic Arithmetic Aggregation Operators

In this section, some arithmetic aggregation operators with Pythagorean 2-tuple linguistic information will be introduced, such as the P2TLWA operator and P2TLWG operator.

Definition 7 ([50]). *Assume that* $a_j = \langle (s_{\varphi_j}, \rho_j), (u_{a_j}, v_{a_j}) \rangle (j = 1,2,\ldots,n)$ *is a collection of P2TLNs. The P2TLWA operator can be depicted as follows:*

$$\begin{aligned} \text{P2TLWA}_\omega(a_1, a_2, \ldots, a_n) &= \overset{n}{\underset{j=1}{\oplus}} (\omega_j a_j) \\ &= \left\langle \Delta\left(\sum_{j=1}^{n} \omega_j \Delta^{-1}(s_{\varphi_j}, \rho_j) \right), \left(\sqrt{1 - \prod_{j=1}^{n} \left(1 - (u_{a_j})^2\right)^{\omega_j}}, \prod_{j=1}^{n} (v_{a_j})^{\omega_j} \right) \right\rangle, \end{aligned} \tag{5}$$

where $\omega = (\omega_1, \omega_2, \ldots, \omega_n)^T$ *is the weight vector of* $a_j (j = 1,2,\ldots,n)$ *and* $\omega_j > 0$, $\sum_{j=1}^{n} \omega_j = 1$.

Definition 8 ([50]). *Assume that* $a = \langle (s_{\varphi_j}, \rho_j), (u_{a_j}, v_{a_j}) \rangle (j = 1,2,\ldots,n)$ *is a collection of P2TLNs. The P2TLWG operator can be depicted as follows:*

$$\begin{aligned} \text{P2TLWG}_\omega(a_1, a_2, \ldots, a_n) &= \overset{n}{\underset{j=1}{\otimes}} (\omega_j a_j) \\ &= \left\langle \Delta\left(\prod_{j=1}^{n} \Delta^{-1}(s_{\varphi_j}, \rho_j)^{\omega_j} \right), \left(\prod_{j=1}^{n} (u_{a_j})^{\omega_j}, \sqrt{1 - \prod_{j=1}^{n} \left(1 - (v_{a_j})^2\right)^{\omega_j}} \right) \right\rangle, \end{aligned} \tag{6}$$

where $\omega = (\omega_1, \omega_2, \ldots, \omega_n)^T$ *is the weight vector of* $a_j (j = 1,2,\ldots,n)$ *and* $\omega_j > 0$, $\sum_{j=1}^{n} \omega_j = 1$.

3. VIKOR Method for P2TL MADM Problems

Suppose that $\Phi_i = \{\Phi_1, \Phi_2, \ldots \Phi_m\}$ and $\tau_j = \{\tau_1, \tau_2, \ldots \tau_n\}$ are respectively m alternatives and n criteria. Let $\omega = \{\omega_1, \omega_2, \ldots \omega_n\}$ be the criteria's weighting vector, which satisfies $\omega_j \in [0,1]$ and $\sum_{j=1}^{n} \omega_j = 1$. Let $E = \{E_1, E_2, \ldots E_k\}$ be the group of DMs, $w = \{w_1, w_2, \ldots w_k\}$ be the weight of DMs, with $w_t \in [0,1]$ and $\sum_{t=1}^{k} w_t = 1$. Construct a decision matrix $R^{(t)} = \left(r_{ij}^{(t)}\right)_{m \times n}$, where $R^{(t)} = \left(r_{ij}^{(t)}\right)_{m \times n} = \left\langle \left(s_{\varphi_{ij}}^{(t)}, \rho_{ij}^{(t)}\right), \left(u_{r_{ij}}^{(t)}, v_{r_{ij}}^{(t)}\right) \right\rangle_{m \times n}$ means the performance of the alternative $\Phi_i\{i = 1,2\cdots,m\}$ with respect to criteria $\tau_j\{j = 1,2\cdots,n\}$ by expert $E^{(t)}\{t = 1,2,\ldots k\}$ using a P2TLN, $0 \le u_{r_{ij}}^{(t)} \le 1, 0 \le v_{r_{ij}}^{(t)} \le 1$ and $0 \le \left(u_{r_{ij}}^{(t)}\right)^2 + \left(v_{r_{ij}}^{(t)}\right)^2 \le 1, i = 1,2\cdots,m, j = 1,2\cdots,n, t = 1,2\cdots,k$.

In view of both the P2TLN's theories and procedures from the VIKOR method, we put forward a P2TL-VIKOR method to deal with the problem of MADM effectively. The new model is shown below.

Step 1. Set up a decision-making group composed of several experts, choose the best attributes to measure alternatives, and finally get a series of P2TL fuzzy decision matrices $R^{(t)} = \left(r_{ij}^{(t)}\right)_{m \times n}$ from each decision maker.

$$R^{(t)} = \left[r_{ij}^{(t)}\right]_{m \times n} = \begin{bmatrix} r_{11}^{(t)} & r_{12}^{(t)} & \cdots & r_{1n}^{(t)} \\ r_{21}^{(t)} & r_{22}^{(t)} & \cdots & r_{2n}^{(t)} \\ \vdots & \vdots & \vdots & \vdots \\ r_{m1}^{(t)} & r_{m2}^{(t)} & \cdots & r_{mn}^{(t)} \end{bmatrix}, \tag{7}$$

where $r_{ij}^{(t)}$ denotes the fuzzy performance value of the ith alternative $(i = 1,2,\ldots,m)$ with respect to the jth criterion $(j = 1,2,\ldots,n)$ and tth decision-maker $(t = 1,2,\ldots,k)$.

Step 2. Utilize the P2TLWA operator or P2TLWG operator to the fuse assessment information; then, the group P2TL fuzzy decision matrix $R = (r_{ij})_{m \times n}$ can be obtained by the calculation.

$$R = \left[r_{ij}\right]_{m\times n} = \begin{bmatrix} r_{11} & r_{12} & \cdots & r_{1n} \\ r_{21} & r_{22} & \cdots & r_{2n} \\ \vdots & \vdots & \vdots & \vdots \\ r_{m1} & r_{m2} & \cdots & r_{mn} \end{bmatrix}, \tag{8}$$

$$\begin{aligned} r_{ij} &= \overset{k}{\underset{t=1}{\oplus}} r_{ij}^{(k)} = \text{P2TLWA}(r_{ij}^{(1)}, r_{ij}^{(2)}, \ldots, r_{ij}^{(k)}) \\ &= \left\langle \Delta\left(\sum_{t=1}^{k} w_t \Delta^{-1}\left(s_{\varphi_{ij}}^{(t)}, \rho_{ij}^{(t)}\right)\right), \left(\sqrt{1-\prod_{t=1}^{k}\left(1-\left(u_{r_{ij}}^{(t)}\right)^2\right)^{w_t}}, \prod_{t=1}^{k}\left(v_{r_{ij}}^{(t)}\right)^{w_t}\right)\right\rangle. \end{aligned} \tag{9}$$

Or

$$\begin{aligned} r_{ij} &= \overset{k}{\underset{t=1}{\otimes}} r_{ij}^{(k)} = \text{P2TLWG}(r_{ij}^{(1)}, r_{ij}^{(2)}, \ldots, r_{ij}^{(k)}) \\ &= \left\langle \Delta\left(\prod_{t=1}^{k} \Delta^{-1}\left(s_{\varphi_{ij}}^{(t)}, \rho_{ij}^{(t)}\right)^{w_t}\right), \left(\prod_{t=1}^{k}\left(u_{r_{ij}}^{(t)}\right)^{w_t}, \sqrt{1-\prod_{t=1}^{k}\left(1-\left(v_{r_{ij}}^{(t)}\right)^2\right)^{w_t}}\right)\right\rangle \end{aligned} \tag{10}$$

Here, r_{ij} means the average fuzzy performance value of the ith alternative relative to the jth criterion.

Step 3. Determine the positive ideal solutions R_j^+ and negative ideal solutions R_j^-.

$$R_j^+ = \left\{\left(\Delta^{-1}\left(s_{\varphi_j}, \rho_j\right)\right)^+, \left(u_{a_j}^+, v_{a_j}^+\right)\right\}, \tag{11}$$

$$R_j^- = \left\{\left(\Delta^{-1}\left(s_{\varphi_j}, \rho_j\right)\right)^-, \left(u_{a_j}^-, v_{a_j}^-\right)\right\}. \tag{12}$$

For all the benefit criteria:

$$\left\{\left(\Delta^{-1}\left(s_{\varphi_j}, \rho_j\right)\right)^+, \left(u_{a_j}^+, v_{a_j}^+\right)\right\} = \left\{\max\left(\Delta^{-1}\left(s_{\varphi_j}, \rho_j\right)\right), \left(\max\left(u_{a_j}\right), \min\left(v_{a_j}\right)\right)\right\}, \tag{13}$$

$$\left\{\left(\Delta^{-1}\left(s_{\varphi_j}, \rho_j\right)\right)^-, \left(u_{a_j}^-, v_{a_j}^-\right)\right\} = \left\{\min\left(\Delta^{-1}\left(s_{\varphi_j}, \rho_j\right)\right), \left(\min\left(u_{a_j}\right), \max\left(v_{a_j}\right)\right)\right\}. \tag{14}$$

For all the cost criteria:

$$\left\{\left(\Delta^{-1}\left(s_{\varphi_j}, \rho_j\right)\right)^+, \left(u_{a_j}^+, v_{a_j}^+\right)\right\} = \left\{\min\left(\Delta^{-1}\left(s_{\varphi_j}, \rho_j\right)\right), \left(\min\left(u_{a_j}\right), \max\left(v_{a_j}\right)\right)\right\}, \tag{15}$$

$$\left\{\left(\Delta^{-1}\left(s_{\varphi_j}, \rho_j\right)\right)^-, \left(u_{a_j}^-, v_{a_j}^-\right)\right\} = \left\{\max\left(\Delta^{-1}\left(s_{\varphi_j}, \rho_j\right)\right), \left(\max\left(u_{a_j}\right), \min\left(v_{a_j}\right)\right)\right\}. \tag{16}$$

Step 4. Calculate S_i and P_i values using the following equations:

$$S_i = \sum_{j=1}^{n} \omega_j \frac{d\left(\left(\left(\Delta^{-1}\left(s_{\varphi_j}, \rho_j\right)\right)^+, \left(u_{a_j}^+, v_{a_j}^+\right)\right), \left(\Delta^{-1}\left(s_{\varphi_j}, \rho_j\right), \left(u_{a_j}, v_{a_j}\right)\right)\right)}{d\left(\left(\left(\Delta^{-1}\left(s_{\varphi_j}, \rho_j\right)\right)^+, \left(u_{a_j}^+, v_{a_j}^+\right)\right), \left(\Delta^{-1}\left(s_{\varphi_j}, \rho_j\right)\right)^-, \left(u_{a_j}^-, v_{a_j}^-\right)\right)}, \tag{17}$$

$$P_i = \max\left\{\omega_j \frac{d\left(\left(\left(\Delta^{-1}\left(s_{\varphi_j}, \rho_j\right)\right)^+, \left(u_{a_j}^+, v_{a_j}^+\right)\right), \left(\Delta^{-1}\left(s_{\varphi_j}, \rho_j\right), \left(u_{a_j}, v_{a_j}\right)\right)\right)}{d\left(\left(\left(\Delta^{-1}\left(s_{\varphi_j}, \rho_j\right)\right)^+, \left(u_{a_j}^+, v_{a_j}^+\right)\right), \left(\Delta^{-1}\left(s_{\varphi_j}, \rho_j\right)\right)^-, \left(u_{a_j}^-, v_{a_j}^-\right)\right)}\right\}. \tag{18}$$

Here, d denotes the normalized Hamming distance and ω_j means the weight of attributes with these conditions, $0 \le \omega_j \le 1, \sum\limits_{j=1}^{n} \omega_j = 1$.

Step 5: Compute Q_i values as follows:

$$Q_i = v\frac{S_i - S_i^*}{S_i^- - S_i^*} + (1-v)\frac{P_i - P_i^*}{P_i^- - P_i^*}, \tag{19}$$

where

$$S_i^* = \min_i S_i, S_i^- = \max_i S_i, \tag{20}$$

$$P_i^* = \min_i P_i, P_i^- = \max_i P_i, \tag{21}$$

where v can be described as the decision-making mechanism coefficient. If $v > 0.5$, it denotes "the maximum group utility"; if $v < 0.5$, it denotes "the minimum regret", and if $v = 0.5$, it denotes "equality".

Step 6: According to the Q_i values, the optimal decision among the rank alternatives is the alternative with the minimum Q value.

4. Numerical Example and Comparative Analysis

4.1. Numerical Example

In this section, we shall provide a numerical example to evaluate the human factors in the process of construction project management by using the P2TL-VIKOR model. Assume that five possible construction projects $\Phi_i(i = 1,2,3,4,5)$ are to be selected and there are four evaluation criteria $\tau_j(j = 1,2,3,4)$ to evaluate these construction projects: ① τ_1 is the workers' proficiency; ② τ_2 is the workers' safety awareness; ③ τ_3 is the technical workers' quality; and ④ τ_4 is the workers' emergency capacity. The five possible construction projects $\Phi_i(i = 1,2,3,4,5)$ are to be evaluated through using P2TLNs with the four criteria by three experts, E^k (expert's weight $w = (0.29, 0.38, 0.33)$, which have an attributes weight of $\omega = (0.24, 0.17, 0.31, 0.28)^T$).

In order to carry out this evaluation, decision makers use language variables to express their evaluation, and the language variables of evaluation alternatives are shown in Table 1. The following steps are used to evaluate the human factors associated with the five construction projects using the proposed P2TL-VIKOR method:

Table 1. Linguistic variables and their fuzzy numbers.

Linguistic Variable	Pythagorean 2-Tuple Linguistic Numbers
Very low (VL)	$((s_0, 0), (0.1, 0.8))$
Low (L)	$((s_1, 0), (0.2, 0.7))$
Medium low (ML)	$((s_2, 0), (0.3, 0.6))$
Medium (M)	$((s_3, 0), (0.5, 0.5))$
Medium high (MH)	$((s_4, 0), (0.6, 0.3))$
High (H)	$((s_5, 0), (0.7, 0.2)$
Very high (VH)	$((s_6, 0), (0.8, 0.1))$

Step 1. Construct the evaluation matrix $R^{(3)} = \left(r_{ij}^3\right)_{5\times4} (i = 1,2,\ldots,5, j = 1,2,\ldots,4)$ of each decision maker as in Tables 2–4. Based on Tables 1–4 and Equation (9), the group Pythagorean 2-tuple linguistic fuzzy decision matrix is computed. Table 5 shows the results.

Table 2. Rating alternatives on each criterion by E_1.

	τ_1	τ_2	τ_3	τ_4
Φ_1	M	H	ML	L
Φ_2	VL	L	L	VH
Φ_3	L	VL	H	M
Φ_4	H	MH	ML	VH
Φ_5	VH	M	MH	VL

Table 3. Rating alternatives on each criterion by E_2.

	τ_1	τ_2	τ_3	τ_4
Φ_1	M	MH	H	ML
Φ_2	L	VL	L	H
Φ_3	H	M	M	VL
Φ_4	VH	ML	M	MH
Φ_5	ML	H	H	M

Table 4. Rating alternatives on each criterion by E_3.

	τ_1	τ_2	τ_3	τ_4
Φ_1	MH	VL	L	MH
Φ_2	M	VH	ML	L
Φ_3	ML	H	H	L
Φ_4	L	M	H	VH
Φ_5	VL	M	MH	VL

Table 5. The group Pythagorean 2-tuple linguistic decision matrix R.

	τ_1	τ_2
Φ_1	<(s_3, 0.33), (0.5369, 0.5165)>	<(s_3, −0.03), (0.555, 0.5879)>
Φ_2	<(s_1, 0.37), (0.3274, 0.6947)>	<(s_2, 0.27), (0.5452, 0.4297)>
Φ_3	<(s_3, −0.15), (0.5082, 0.4583)>	<(s_3, −0.21), (0.5332, 0.4518)>
Φ_4	<(s_4, 0.06), (0.6705, 0.3706)>	<(s_3, −0.09), (0.4786, 0.6552)>
Φ_5	<(s_3, −0.5), (0.5338, 0.7651)>	<(s_4, −0.24), (0.5935, 0.4316)>
	τ_3	τ_4
Φ_1	<(s3, −0.19), (0.5067, 0.4823)>	<(s2, 0.37), (0.4209, 0.5535)>
Φ_2	<(s1, 0.33), (0.2383, 0.7378)>	<(s4, −0.03), (0.6573, 0.4823)>
Φ_3	<(s4, 0.24), (0.6399, 0.4518)>	<(s1, 0.2), (0.3095, 0.8167)>
Φ_4	<(s3, 0.37), (0.5491, 0.4518)>	<(s5, 0.24), (0.743, 0.296)>
Φ_5	<(s4, 0.38), (0.6426, 0.3646)>	<(s1, 0.14), (0.3303, 0.7139)>

Step 2. Determine the R_j^+ and R_j^- by Equations (13) and (14).

$$R_j^+ = \left\{ \begin{array}{c} \langle (s_4, 0.06), (0.6705, 0.3706) \rangle \\ \langle (s_4, -0.24), (0.5935, 0.4297) \rangle \\ \langle (s_4, 0.38), (0.6426, 0.3646) \rangle \\ \langle (s_5, 0.24), (0.743, 0.296) \rangle \end{array} \right\},$$

$$R_j^- = \left\{ \begin{array}{c} \langle (s_1, 0.37), (0.3274, 0.7651) \rangle \\ \langle (s_2, 0.27), (0.4786, 0.6552) \rangle \\ \langle (s_1, 0.33), (0.2383, 0.7378) \rangle \\ \langle (s_1, 0.14), (0.3095, 0.8167) \rangle \end{array} \right\}.$$

Step 3. Compute S_i and P_i values by Equations (17) and (18).

$$S_1 = 0.5917,\ S_2 = 0.7793,\ S_3 = 0.5228,\ S_4 = 0.2563,\ S_5 = 0.4590,$$

$$P_1 = 0.2186,\ P_2 = 0.3100,\ P_3 = 0.2790,\ P_4 = 0.1362,\ P_5 = 0.2724.$$

Step 4. Calculate Q_i values as follows (Let $v = 0.4$):

$$Q_1 = 0.5411,\ Q_2 = 1.0000,\ Q_3 = 0.6968,\ Q_4 = 0.0000,\ Q_5 = 0.6253.$$

Step 5. According to the Q_i values, the optimal decision among the rank alternatives is the alternative with the minimum Q value: $\Phi_4 > \Phi_1 > \Phi_5 > \Phi_3 > \Phi_2$. Thus, the most optimal alternative is Φ_4.

4.2. Comparative Analyses

In this part, we will make some comparative analyses to compare in our proposed P2TL-VIKOR model the P2TLWA and P2TLWG operators defined by Wei, Lu, Alsaadi, Hayat and Alsaedi [50], and the P2TL-TODIM method proposed by Huang and Wei [22].

The comparison results of different methods are as follows.

It is clear from Table 6 that the results are slightly different in ranking of alternatives, but the best alternative is always Φ_4 by comparing the values of our proposed P2TL-VIKOR method with the P2TLWA/P2TLWG operators and the P2TL-TODIM model. Notably, in practical MADM problems, the P2TL-VIKOR method fully considers the conflict between attributes, which is more reasonable and scientific. All these methods have their good advantages: (1) P2TLWA operators emphasize the group influences; (2) the P2TLWG operator emphasizes individual influences; (3) the P2TL-TODIM method based on the prospect theory is exactly a kind of method that considers the influence of the experts' psychological behaviors factors on the decision results; and (4) the P2TL-VIKOR method takes into account the contradictory criteria such as the objectivity of decision makers and the complexity of the decision environment, so as to obtain more useful and scientific evaluation information.

Table 6. Rank of alternatives by using different methods.

Methods	Order
P2TLWA	$\Phi_4 > \Phi_5 > \Phi_3 > \Phi_1 > \Phi_2$
P2TLWG	$\Phi_4 > \Phi_1 > \Phi_5 > \Phi_3 > \Phi_2$
P2TL-TODIM	$\Phi_4 > \Phi_5 > \Phi_3 > \Phi_1 > \Phi_2$
P2TL-VIKOR	$\Phi_4 > \Phi_1 > \Phi_5 > \Phi_3 > \Phi_2$

5. Conclusions

Human factors are not only the leading factors affecting the quality of construction projects, but also the most basic and core factors in the quality assurance system; so, the evaluation of human factors in construction projects is particularly critical. Human factor evaluation in construction projects is a MADM problem, and the information available for decision making is vague or uncertain in nature. Therefore, we used language variables to express the preferences of experts. The Pythagorean 2-tuple linguistic sets (P2TLSs) can well reflect uncertain or fuzzy information and solve these kinds of problems, and the original VIKOR is characterized by handling conflicting attributes. Naturally, we combined the Pythagorean 2-tuple linguistic sets with VIKOR, and the recommended method was systematically applied to the human factor evaluation of five construction projects to find an optimal construction project. The comparative study shows that the proposed MADM algorithm is feasible. This method is very effective and useful for decision making.

The main contributions of this study is fourfold: (1) the Pythagorean 2-tuple linguistic VIKOR (P2TL-VIKOR) method is designed to tackle the Pythagorean 2-tuple linguistic MAGDM issues; (2) a case study for evaluating human factors in construction project management is designed to show the developed approach; (3) some comparative studies are provided with the existing methods to give effect to the rationality of P2TL-VIKOR; and (4) the proposed method can also successfully contribute to the selection of suitable alternatives in other selection issues.

In the future, the proposed method can be expanded to deal with other decision-making issues, such as the selection of green suppliers, the location of waste disposal station, and so on.

Author Contributions: T.H., G.W. (Guiwu Wei), J.L., C.W. (Cun Wei) and R.L. conceived and worked together to achieve this work, T.H. compiled the computing program by Excel and analyzed the data, T.H. and G.W. (Guiwu Wei) wrote the paper. Finally, all the authors have read and approved the final manuscript.

Funding: The work was supported by the National Natural Science Foundation of China under Grant No. 71571128 and the Humanities and Social Sciences Foundation of Ministry of Education of the People's Republic of China (14YJCZH091). The APC was funded by Humanities and Social Sciences Foundation of Ministry of Education of the People's Republic of China (14YJCZH091).

Conflicts of Interest: The authors declare no conflicts of interest.

References

1. Ning, X.; Wang, L.G. Construction Site Layout Evaluation by Intuitionistic Fuzzy TOPSIS Model. In *Frontiers of Green Building, Materials and Civil Engineering*; Sun, D., Sung, W.P., Chen, R., Eds.; Scientific.net: Baech, Switzerland, 2011; Volumes 71–78, pp. 583–588.
2. Lu, Y. Research of models and methods of construction project information management with 2-tuple linguistic information. *Inf. Int. Interdiscip. J.* **2012**, *15*, 3911–3916.
3. Gu, X.; Zhao, P.; Wang, Y. Models for multiple attribute decision making based on the Einstein correlated aggregation operators with interval-valued intuitionistic fuzzy information. *J. Intell. Fuzzy Syst.* **2014**, *26*, 2047–2055.
4. Wu, S.J.; Wang, J.; Wei, G.W.; Wei, Y. Research on construction engineering project risk assessment with some 2-tuple linguistic neutrosophic hamy mean operators. *Sustainability* **2018**, *10*, 1536. [CrossRef]
5. Zadeh, L.A. Fuzzy sets. *Inf. Control* **1965**, *8*, 338–356. [CrossRef]
6. Atanassov, K.T. Intuitionistic fuzzy sets. *Fuzzy Sets Syst.* **1986**, *20*, 87–96. [CrossRef]
7. Li, Z.X.; Gao, H.; Wei, G.W. Methods for multiple attribute group decision making based on intuitionistic fuzzy dombi hamy mean operators. *Symmetry* **2018**, *10*, 574. [CrossRef]
8. Wei, G.W.; Wang, J.; Lu, M.; Wu, J.; Wei, C. Similarity Measures of Spherical Fuzzy Sets Based on Cosine Function and Their Applications. *IEEE Access* **2019**, *7*, 159069–159080. [CrossRef]
9. Wei, G.W. 2-Tuple intuitionistic fuzzy linguistic aggregation operators in multiple attribute decision making. *Iran. J. Fuzzy Syst.* **2019**, *16*, 159–174.
10. Wu, L.P.; Wang, J.; Gao, H. Models for competiveness evaluation of tourist destination with some interval-valued intuitionistic fuzzy Hamy mean operators. *J. Intell. Fuzzy Syst.* **2019**, *36*, 5693–5709. [CrossRef]
11. Gao, H.; Wu, J.; Wei, C.; Wei, G.W. MADM method with Interval-valued Bipolar Uncertain Linguistic Information for Evaluating the Computer Network Security. *IEEE Access* **2019**, *7*, 151506–151524. [CrossRef]
12. Yager, R.R. Pythagorean Fuzzy Subsets. In Proceedings of the Joint IFSA World Congress and NAFIPS Annual Meeting (IFSA/NAFIPS), Edmonton, AB, Canada, 24–28 June 2013; pp. 57–61.
13. Lu, J.P.; Tang, X.Y.; Wei, G.W.; Wei, C.; Wei, Y. Bidirectional project method for dual hesitant Pythagorean fuzzy multiple attribute decision-making and their application to performance assessment of new rural construction. *Int. J. Intell. Syst.* **2019**, *34*, 1920–1934. [CrossRef]
14. Tang, M.; Wang, J.; Lu, J.P.; Wei, G.W.; Wei, C.; Wei, Y. Dual hesitant pythagorean fuzzy heronian mean operators in multiple attribute decision making. *Mathematics* **2019**, *7*, 344. [CrossRef]
15. Wang, J.; Gao, H.; Wei, G.W. The generalized dice similarity measures for Pythagorean fuzzy multiple attribute group decision making. *Int. J. Intell. Syst.* **2019**, *34*, 1158–1183. [CrossRef]
16. Wei, G.W.; Wang, J.; Wei, C.; Wei, Y.; Zhang, Y. Dual hesitant pythagorean fuzzy hamy mean operators in multiple attribute decision making. *IEEE Access* **2019**, *7*, 86697–86716. [CrossRef]
17. Zhang, X.L.; Xu, Z.S. Extension of TOPSIS to multiple criteria decision making with Pythagorean fuzzy sets. *Int. J. Intell. Syst.* **2014**, *29*, 1061–1078. [CrossRef]

18. Zhang, X.L. Multicriteria Pythagorean fuzzy decision analysis: A hierarchical QUALIFLEX approach with the closeness index-based ranking methods. *Inf. Sci.* **2016**, *330*, 104–124. [CrossRef]
19. Ren, P.J.; Xu, Z.S.; Gou, X.J. Pythagorean fuzzy TODIM approach to multi-criteria decision making. *Appl. Soft Comput.* **2016**, *42*, 246–259. [CrossRef]
20. Bolturk, E. Pythagorean fuzzy CODAS and its application to supplier selection in a manufacturing firm. *J. Enterp. Inf. Manag.* **2018**, *31*, 550–564. [CrossRef]
21. Chen, T.Y. Remoteness index-based Pythagorean fuzzy VIKOR methods with a generalized distance measure for multiple criteria decision analysis. *Inf. Fusion* **2018**, *41*, 129–150. [CrossRef]
22. Huang, Y.H.; Wei, G.W. TODIM method for Pythagorean 2-tuple linguistic multiple attribute decision making. *J. Intell. Fuzzy Syst.* **2018**, *35*, 901–915. [CrossRef]
23. Ilbahar, E.; Karasan, A.; Cebi, S.; Kahraman, C. A novel approach to risk assessment for occupational health and safety using Pythagorean fuzzy AHP & fuzzy inference system. *Saf. Sci.* **2018**, *103*, 124–136.
24. Khan, M.S.A.; Abdullah, S.; Ali, M.Y.; Hussain, I.; Farooq, M. Extension of TOPSIS method base on Choquet integral under interval-valued Pythagorean fuzzy environment. *J. Intell. Fuzzy Syst.* **2018**, *34*, 267–282. [CrossRef]
25. Perez-Dominguez, L.; Rodriguez-Picon, L.A.; Alvarado-Iniesta, A.; Cruz, D.L.; Xu, Z.S. MOORA under Pythagorean fuzzy set for multiple criteria decision making. *Complexity* **2018**, *2018*, 2602376. [CrossRef]
26. Xue, W.T.; Xu, Z.S.; Zhang, X.L.; Tian, X.L. Pythagorean fuzzy LINMAP method based on the entropy theory for railway project investment decision making. *Int. J. Intell. Syst.* **2018**, *33*, 93–125. [CrossRef]
27. Opricovic, S.; Tzeng, G.H. Compromise solution by MCDM methods: A comparative analysis of VIKOR and TOPSIS. *Eur. J. Oper. Res.* **2007**, *156*, 445–455. [CrossRef]
28. Benayoun, R.; Roy, B.; Sussman, B. ELECTRE: Une méthode pour guider le choix en présence de points de vue multiples. *Rev. Franaise Inf. Rech. Opérationnelle* **1966**, *3*, 31–56.
29. Mareschal, B.; Brans, J.P.; Vincke, P. Prométhée, a new family of outranking methods in multicriteria analysis. *ULB Inst. Repos.* **1984**.
30. Wei, G.W.; Wang, H.J.; Lin, R.; Zhao, X.F. Grey relational analysis method for intuitionistic fuzzy multiple attribute decision making with preference information on alternatives. *Int. J. Comput. Intell. Syst.* **2011**, *4*, 164–173. [CrossRef]
31. Hwang, C.L.; Yoon, K. Multiple Attribute Decision Making. In *Methods and Applications A State-of-the-Art Survey*; Springer: Berlin, Germany, 1981.
32. Wang, J.; Wei, G.W.; Lu, M. TODIM method for multiple attribute group decision making under 2-tuple linguistic neutrosophic environment. *Symmetry* **2018**, *10*, 486. [CrossRef]
33. Wei, G.W. TODIM method for picture fuzzy multiple attribute decision making. *Informatica* **2018**, *29*, 555–566. [CrossRef]
34. Gomes, L.; Lima, M. TODIM: Basics and application to multicriteria ranking of projects with environmental impacts. *Found. Comput. Decis. Sci.* **1979**, *16*, 113–127.
35. Devi, K. Extension of VIKOR method in intuitionistic fuzzy environment for robot selection. *Expert Syst. Appl.* **2011**, *38*, 14163–14168. [CrossRef]
36. Du, Y.; Liu, P.D. Extended fuzzy VIKOR method with intuitionistic trapezoidal fuzzy numbers. *Inf. Int. Interdiscip. J.* **2011**, *14*, 2575–2583.
37. Park, J.H.; Cho, H.J.; Kwun, Y.C. Extension of the VIKOR method for group decision making with interval-valued intuitionistic fuzzy information. *Fuzzy Optim. Decis. Mak.* **2011**, *10*, 233–253. [CrossRef]
38. Chatterjee, K.; Kar, M.B.; Kar, S. Strategic Decisions Using Intuitionistic Fuzzy Vikor Method for Information System (IS) Outsourcing. In Proceedings of the 2013 International Symposium on Computational and Business Intelligence, New Delhi, India, 24–26 August 2013; pp. 123–126.
39. Liao, H.C.; Xu, Z.S. A VIKOR-based method for hesitant fuzzy multi-criteria decision making. *Fuzzy Optim. Decis. Mak.* **2013**, *12*, 373–392. [CrossRef]
40. Liu, H.C.; Liu, L.; Wu, J. Material selection using an interval 2-tuple linguistic VIKOR method considering subjective and objective weights. *Mater. Des.* **2013**, *52*, 158–167. [CrossRef]
41. Wei, G.; Zhang, N.A. A multiple criteria hesitant fuzzy decision making with shapley value-based VIKOR method. *J. Intell. Fuzzy Syst.* **2014**, *26*, 1065–1075.

42. Bausys, R.; Zavadskas, E.K. Multicriteria decision making approach by VIKOR under interval neutrosophic set environment. *Econ. Comput. Econ. Cybern. Stud. Res.* **2015**, *49*, 33–48.
43. Dammak, F.; Baccour, L.; Alimi, A.M. A comparative analysis for multi-attribute decision making methods: TOPSIS, AHP, VIKOR using intuitionistic fuzzy sets. In Proceedings of the 2015 IEEE International Conference on Fuzzy Systems, Istanbul, Turkey, 2–5 August 2015.
44. Liao, H.C.; Xu, Z.S. Approaches to manage hesitant fuzzy linguistic information based on the cosine distance and similarity measures for HFLTSs and their application in qualitative decision making. *Expert Syst. Appl.* **2015**, *42*, 5328–5336. [CrossRef]
45. You, X.Y.; You, J.X.; Liu, H.C.; Zhen, L. Group multi-criteria supplier selection using an extended VIKOR method with interval 2-tuple linguistic information. *Expert Syst. Appl.* **2015**, *42*, 1906–1916. [CrossRef]
46. Buyukozkan, G.; Feyzioglu, O.; Gocer, F. Evaluation of Hospital Web Services Using Intuitionistic Fuzzy AHP and Intuitionistic Fuzzy VIKOR. In Proceedings of the 2016 IEEE International Conference on Industrial Engineering and Engineering Management (IEEM), Bali, Indonesia, 4–7 December 2016; pp. 607–611.
47. Zhang, F.Y.; Luo, L.; Liao, H.C.; Zhu, T.; Shi, Y.K.; Shen, W.W. Inpatient admission assessment in West China Hospital based on hesitant fuzzy linguistic VIKOR method. *J. Intell. Fuzzy Syst.* **2016**, *30*, 3143–3154. [CrossRef]
48. Hu, J.H.; Pan, L.; Chen, X.H. An interval neutrosophic projection-based VIKOR method for selecting doctors. *Cogn. Comput.* **2017**, *9*, 801–816. [CrossRef]
49. Wang, L.; Zhang, H.Y.; Wang, J.Q.; Li, L. Picture fuzzy normalized projection-based VIKOR method for the risk evaluation of construction project. *Appl. Soft Comput.* **2018**, *64*, 216–226. [CrossRef]
50. Wei, G.; Lu, M.; Alsaadi, F.E.; Hayat, T.; Alsaedi, A. Pythagorean 2-tuple linguistic aggregation operators in multiple attribute decision making. *J. Intell. Fuzzy Syst.* **2017**. [CrossRef]
51. Wei, G.W. Pythagorean fuzzy interaction aggregation operators and their application to multiple attribute decision making. *J. Intell. Fuzzy Syst.* **2017**, *33*, 2119–2132. [CrossRef]
52. Herrera, F.; Martinez, L. A 2-tuple fuzzy linguistic representation model for computing with words. *IEEE Trans. Fuzzy Syst.* **2000**, *8*, 746–752.

Article

A Compact Representation of Preferences in Multiple Criteria Optimization Problems

Francisco Salas-Molina [1], David Pla-Santamaria [2,*], Ana Garcia-Bernabeu [2] and Javier Reig-Mullor [3]

1. Department of Management "Juan José Renau Piqueras", Faculty of Economics, Universitat de València, Av. Tarongers s/n, 46022 Valencia, Spain; francisco.salas-molina@uv.es
2. Department of Economics and Social Sciences, Higher Polytechnic School of Alcoi, Universitat Politècnica de València, Ferrándiz y Carbonell, 03801 Alcoi, Spain; angarber@upv.es
3. Department of Economics and Financial Studies, Faculty of Social Sciences and Law, Universitas Miguel Hernández, 03202 Elche, Spain; javier.reig@umh.es

* Correspondence: dplasan@upv.es

Received: 24 September 2019; Accepted: 6 November 2019; Published: 11 November 2019

Abstract: A critical step in multiple criteria optimization is setting the preferences for all the criteria under consideration. Several methodologies have been proposed to compute the relative priority of criteria when preference relations can be expressed either by ordinal or by cardinal information. The analytic hierarchy process introduces relative priority levels and cardinal preferences. Lexicographical orders combine both ordinal and cardinal preferences and present the additional difficulty of establishing strict priority levels. To enhance the process of setting preferences, we propose a compact representation that subsumes the most common preference schemes in a single algebraic object. We use this representation to discuss the main properties of preferences within the context of multiple criteria optimization.

Keywords: subjective preferences; analytic hierarchy process; lexicographic orders; powerset

---

1. Introduction

Setting preferences in multiple criteria optimization problems may have an important influence on the final selection of the best solutions. However, it is usually assumed in the literature that preferences are somehow given by the decision-maker to the analyst. Then, the focus is placed on methodology disregarding some important features that may result critical in eliciting solutions. These features include (1) cardinality, when preferences are given in terms of weights attached to criteria; (2) ordinality, when preferences are expressed as a criteria order; and (3) clustering, when preferences are grouped in hierarchies or noncontinuous levels with non-finite preferences among criteria.

Eliciting weights that accurately represent the preferences of decision-makers is a key issue in multiobjective optimization [1]. To some extent, expressing weights implies the assumption of some knowledge about the impact that these weights have on the performance of alternative decisions. Indeed, decision-makers usually express their preferences for alternative solutions in terms of the set of criteria under consideration [2]. Furthermore, weight setting in the context of multiobjective optimization is also an interactive process in which decision-makers play a relevant role [3].

Several methodologies have been proposed to compute the relative priority of criteria when preference relations can be expressed either by ordinal or by cardinal information. On the one hand, the analytic hierarchy process (AHP) proposed in [4] introduces relative priority levels and cardinal preferences. A broad range of recent AHP applications can be found in many different disciplines dealing with environmental problems [5–7], health [8], finance [9], and product development [10]. On the other hand, lexicographical orders [11,12] combine both ordinal and cardinal preferences and present the additional difficulty of establishing strict priority levels where preferences among criteria are noncontinuous. Some recent examples of lexicographic optimization applications include the optimization of water resources planning [13], electricity market clearing [14], and combined heat and power scheduling [15].

Despite sharing a common goal, namely, establishing priorities between decision-making objectives, there is no unifying approach to represent and setting preferences derived from the analysis of AHP and lexicographic orders. Furthermore, decision-makers can express their opinions by means of ordinal or cardinal information, but they need to represent it in a suitable way to make computations in the optimization process [16]. In this paper, we provide a compact representation of preferences that generalizes cardinal, ordinal, and clustering approaches to manage the relative importance of goals in a single algebraic object. This compact representation is based on the concept of powerset, which includes all possible subsets that can be formed from a given set of goals. Our representation aims to improve the interactive process of setting preferences. It facilitates the understanding and it allows further analysis on the main properties of different preference setting options. In addition, we explore the main properties of powerset preference rules within the context of multiple criteria optimization problems. Summarizing, three important contributions must be highlighted:

- We introduce the concept of powerset preference rule that generalizes previous approaches such as AHP and lexicographical orders.
- We provide a unique and compact representation of preferences for objectives based on both ordinal and cardinal information.
- We propose two theoretical results on the properties of powerset preference rules, namely, alignment, and preference order.

As the influence of preference setting in the solution of an optimization problem may be remarkable, we argue that this analysis may result fruitful for decision-makers in several ways. First, we establish a link between AHP and lexicographical orders, as both methodologies are more related than it may seem at first glance. Indeed, both methods aim to express preferences for different objectives. Second, we solve the problem of combining both ordinal and cardinal information by relying on the concept of powerset preference rule. We also combine both clusters of objectives and complex hierarchical structures in a single algebraic object. Third, from the analysis of the main properties of powerset preference rules, decision-makers can better understand the implications and importance of weight setting through the common interactive process in multiobjective optimization. Detection of inconsistencies and a more effective way to develop this interactive process are the main benefits that decision-makers can derive from the proposal described in this paper. In addition, researchers have now the opportunity to propose new theoretical results, as this compact representation and the definition of powerset preference rules pave the way to formal reasoning.

In addition to this introduction, this paper includes Section 2, where useful background on setting preferences multicriteria optimization problems is given. Section 3 is the central part of this paper where the concept of powerset preference rule is introduced. Section 4 explores the main properties of powerset preference rules, and Section 5 concludes.

2. Setting Preferences in Multicriteria Optimization Problems

The concept of attribute refers to interesting values for a decision-maker that related to an objective reality that can be measured [12]. Let us assume that these measures can be expressed by means of a general mathematical function $g_i(\boldsymbol{x})$ that ultimately depends on vector of solutions $\boldsymbol{x}$ within feasible set $\mathcal{X}$. From the set of all possible attributes, decision-makers select those that result of interest for them. In addition, they establish the desired direction of improvement (maximization or minimization) and some aspiration levels (if any). As a result, attributes are transformed in given a set of criteria $\mathcal{G} = \{g_1(\boldsymbol{x}), g_2(\boldsymbol{x}), \ldots, g_q(\boldsymbol{x})\}$. From this set, a general multicriteria optimization problem can be formulated under a maximization point of view:

$$\max \quad [g_1(x), g_2(x), \ldots, g_q(x)] \tag{1}$$

subject to

$$x \in \mathcal{X}. \tag{2}$$

Each function $g_i(x)$ is a measure of the achievement of the i-th criterion. However, the presence of multiple criteria requires setting specific preferences, priorities, or weights to express the relative importance of the achievement of each criterion. For instance, if $g_1(x)$ measures return and $g_2(x)$ measures safety of an investment x, but return is assessed by a decision-maker to be twice more important than safety, we can reasonably use the following weighted objective function,

$$w_1 \cdot g_1(x) + w_2 \cdot g_2(x) = 0.67 \cdot g_1(x) + 0.33 \cdot g_2(x). \tag{3}$$

Note that Equation (3) is a scalarizing function [17] used as a surrogate to measure global achievement according to the particular preferences for a decision-maker. These preferences are expressed by means of cardinal information, i.e., by means of weights w_1 and w_2. However, preferences for criteria can also be expressed using ordinal information by establishing strict priority levels of criteria, as in the case of lexicographical optimization [11,12]. The fulfillment of some criteria are immeasurably preferred to the fulfillment of another set of criteria. Criteria are then classified in groups, priority levels, or clusters and the preference among clusters may be either finite (AHP) or infinite (lexicographical orders). In addition, if several criteria are at the same priority level, cardinal information is used to expressed preferences among criteria within the level [18].

To facilitate the analysis, criteria (and clusters of criteria) can be organized in complex structures such as hierarchies or networks. The analytic hierarchy process (AHP) proposed by [4], and its generalization—the analytic network process (ANP) [19]—is the most common approach to set preferences in a multicriteria context. The relative importance of criteria and the final priorities are computed considering the relationship between clusters of criteria and the specific priority established among criteria within a cluster.

Summarizing, we find in the literature three main approaches to set preferences in a multicriteria optimization context: (1) A basic approach when the number of criteria is small as described in Equation (3); (2) AHP: when complex structures are used to establish cardinal priorities; and (3) lexicographical orders: when ordinal information is required to set strict priority levels between clusters and, at the same time, cardinal information is used to establish numerical preferences among criteria within a cluster. The main features of these approaches are summarized in Table 1. Next, we consider these last two approaches in more detail.

Table 1. Features of main approaches to set preferences.

Approach	Cardinal	Ordinal	Clustering
Basic	✓		
AHP/ANP	✓		✓
Lexicographical	✓	✓	✓

2.1. Analytic Hierarchy Process

The AHP [4] has become one of the most widely used tools in the resolution of complex decision-making problems. All the pairwise comparisons generated by the relative weights of the criteria represent judgments made by decision-makers. The AHP provides a systematic process to incorporate factors such as logic, experience or knowledge, emotion, and a sense of optimization into a decision-making methodology [20]. A detailed description of the AHP method can be found in [21]. AHP has been applied in a broad range of application areas [22,23]. There has been a steady-state increase in its usage since its introduction, due to its ease of application. A large body of research concerning with AHP applications in different disciplines can be found in literature including, but not limited to, wastewater services [7], new product development [10], Internet finance [9], sustainable and renewable energy [24,25], water resources management [5,6], agriculture [26], health [8], nuclear power [27], climate change [28], and presidential elections [29].

A useful tool to set and work with preferences in AHP is a comparison matrix. For each cluster of criteria of size n, decision-makers establish specific preferences by setting element a_{ij} of matrix $A \in \mathbb{R}^{n \times n}$ to a numeric value. As an illustrative example, let us consider the concept of sustainability. The intersection of economic, social, and environmental criteria is usually known as sustainability [30]. While designing public policies, a government may be interested in sustainable policies, namely, those that consider not only economic but also social and environmental aspects. As a result, we are dealing with a multiple criteria decision-making problem, but we need to set preferences in the achievement of the economic, social, and environmental criteria. AHP relies on a numerical scale and a pairwise comparison matrix to set these priorities. The numerical scale ranges from 1 (both criteria are equally important) to 9 (one criterion is extremely more important than another one). An example of the pairwise comparison between economic, social and environmental criteria is summarized in Table 2.

Table 2. Establishing priorities between economic, social, and environmental criteria.

	Economic	Social	Environmental	Sum	Priorities
Economic	1	1/7	1/3	1.5	0.09
Social	7	1	3	11.0	0.65
Environmental	3	1/3	1	4.3	0.26
Total				16.8	1.00

According to the numerical scale proposed by Saaty, the social criterion is very strongly more important than the economic one and the environmental criterion is moderately more important than the economic one. Finally, the social criterion is moderately more important than the environmental one. To obtain the final priorities following the approximate method proposed by Saaty, we sum the values of each row and divide by the sum of all judgments. Alternatively, we can normalize the judgments in any column by dividing each entry by the sum of that column and by averaging normalized values for each row. However, rather than the method to establish priorities, in this paper, we focus on the representation of preferences and its main properties. The algebraic object equivalent to Table 2 is the following matrix,

$$A = \begin{bmatrix} 1 & 1/7 & 1/3 \\ 7 & 1 & 3 \\ 3 & 1/3 & 1 \end{bmatrix}. \tag{4}$$

Note that the AHP implies reciprocal judgments, meaning that $a_{ij} = 1/a_{ji}$. In addition, consistency is a desired property. Here, perfect consistency means that $a_{ik} = a_{ij} \cdot a_{jk}$. That is, if the social criterion is moderately more important than the environmental criterion and, at the same time, the environmental criterion is moderately more important than the economic criterion, the social criterion should be extremely more important than the economic criterion for consistency. Note that consistency does not perfectly holds in matrix A, as $a_{13} = 0.33$ is not exactly $a_{12} \cdot a_{23} = 0.43$. Although some level of inconsistency can be accepted, decision-makers must check this property when setting preferences. An additional requirement of the AHP is the homogeneity of elements to be compared, as ranges are restricted to the scale of 1 to 9. We cannot compare a grain of sand to a mountain.

However, the main feature of the AHP is that preferences are established hierarchically, meaning that more general criteria subsume specific criteria. A hierarchy can be represented graphically as a parent node and several child nodes linked to it as shown in Figure 1. This graph defines the relative importance that subcriteria (growth, risk, and profits) have on a parent node (economic). Priorities among the subcriteria are then established by clusters (growth, risk, and profits on the one hand, and social, economic, and environmental performance on the other hand). Then, the relative importance of lower-level criteria (growth, risk, and profits) on the overall goal (sustainability) is weighted by the priority of the economic criterion with respect to same-level criteria (social and Environmental). As a result, we need a comparison matrix for each of the clusters in the hierarchy. Finally, each criterion in the hierarchy has an influence in the overall criterion (Sustainability).

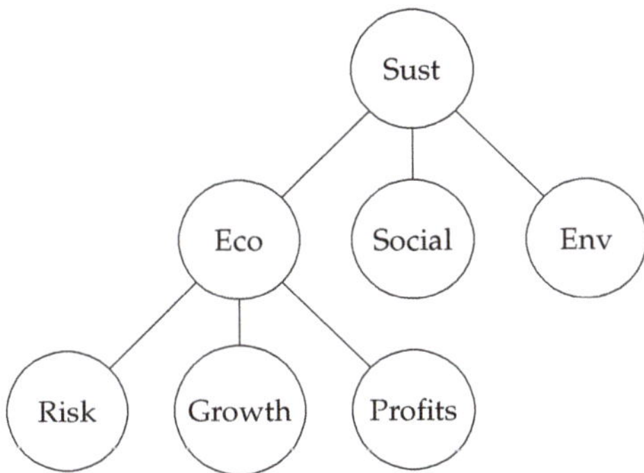

Figure 1. A graph representing a hierarchy of criteria: Sustainability (Sust), Social, Environmental (Env), Economic (Eco), and its three subcriteria: Risk, Growth, and Profits.

2.2. Lexicographic Orders

The lexicographic approach to multiobjective optimization was first introduced by [31] and later developed mainly by [18,32]. Lexicographic orders [1,11,12,33] are useful when a decision-maker has a predefined ordering of criteria. Because of that, lexicographic optimization is also known as preemptive optimization. In lexicographic orders, optimization is organized in strict priority levels, meaning that higher priority levels are infinitely more important than lower priority levels for optimization purposes. There is no trade-off between the achievement goals in different priority levels. The achievement of goals in a higher priority level are immeasurably preferred to the achievement of goals in a lower priority level. However, the achievement levels of other priority level goals can also be determined, therefore allowing an interesting sensitivity analysis [34]. As a result, optimization takes place in a sequential manner. Optimal solutions for higher priority levels are considered as invariant input data for lower level optimization.

Some examples of the use of lexicographic orders in multiple criteria optimization are the following. The authors of [32] proposed a lexicographic goal program to select projects or investment opportunities from a given finite set to maximize profits as a first priority and market share as a second priority. The authors of [34] tackled natural resource planning problems with economic, environmental, and social criteria in the development of forest energy plantations in Eastern Ontario by means of lexicographic goal programming. The authors of [13] described the optimization of water resources planning for Lago Maggiore in Italy to maximize flood protection, minimize supply shortage for irrigation, and maximize electricity generation where the order of objectives is strict and prescribed by law. More recently, the authors of [14] proposed a multi-objective lexicographic optimization framework for electricity market clearing, and the authors of [15] relied on lexicographic orders for combined heat and power scheduling.

As criteria placed in a higher priority level are strictly preferred to those placed in a lower priority level, cardinal, or numerical information is not appropriate to set preferences between priority levels. Instead, we require ordinal information to establish preferences, namely, a binary relation between priority levels. In the hierarchy represented in Figure 1, if we assume that the Economic criterion is strictly preferred to the Social criterion and, at the same time, the Social criterion is strictly preferred, we are dealing with the following lexicographic maximization problem,

$$\text{Lex max} \quad [g_1(x),\ g_2(x),\ g_3(x)] \tag{5}$$

subject to

$$x \in \mathcal{X} \tag{6}$$

where $g_1(x)$, $g_2(x)$, and $g_3(x)$ represent, respectively, the achievement of the Economic, Social, and Environmental criteria. Here, the order of goals in expression (5) is very important, as it is only when goal $g_1(x)$ is optimized that goal $g_2(x)$ is considered, and, in turn, it is only when $g_2(x)$ is optimized that $g_3(x)$ is considered. Goals are then ordered as the words in a dictionary. However, we can find numerical preferences among the criteria within the same priority level. In the example shown in Figure 1, a conservative manager may consider risk twice more important than growth and profits. The lexicographic maximization of Equation (5) should then be expanded to consider the relative importance of the subcriteria that form the Economic criterion.

$$\text{Lex max} \quad [0.5g_4(x) + 0.25g_5(x) + 0.25g_6(x),\ g_2(x),\ g_3(x)] \tag{7}$$

where $g_4(x)$, $g_5(x)$, and $g_6(x)$ represent, respectively, the achievement of the Risk, Growth, and Profits criteria. As a result, lexicographic multicriteria optimization requires representing preferences by means of both cardinal (numerical) and ordinal (binary) information. This information can be summarized in numerical matrices, as in the case of the AHP, and also in Boolean matrices to set priority levels. Next, we propose a compact representation of preferences that subsumes AHP and Lexicographic orders in a single algebraic object.

3. A Compact Preference Representation

In Section 1, we considered two different ways of representing preferences: (1) the measurable way, when preferences are expressed by cardinal information (AHP), and (2) the immeasurable way, when preferences are expressed by a ordinal information (lexicographical orders). In addition, both the AHP and lexicographical orders imply the definition of some degree of hierarchical dependence among criteria. These two approaches require the use of several matrices to represent the global set of preferences for multiple criteria. Next, we rely on the concept of powerset preference rule to provide a compact representation of preferences by means of a single algebraic object, namely, a matrix.

3.1. A Compact Representation of AHP Preferences

In multiple criteria optimization problems, setting the preferences for the criteria that are important to decision-makers is a key issue. Given a set of criteria $\mathcal{G} = (g_1(x), g_2(x), \ldots, g_q(x))$, the main goal is to establish a set of priorities or weights for each criterion to derive the solutions. These priorities state when and how much a criterion is preferred to another. To this end, AHP [4] is based on binary relations among criteria.

Definition 1. ***Binary relation***. *Given a set $\mathcal{G}$, a binary relation is a subset of $\mathcal{G} \times \mathcal{G}$.*

Let us consider a binary relation $\succeq$, meaning at-least-as-good-as, defined on set $\mathcal{G}$ with cardinality $q = |\mathcal{G}|$. In the case of AHP [4], a pairwise comparison is a map from a binary relation to a numerical value: $\varphi : \mathcal{G} \times \mathcal{G} \to \mathbb{R}$. The partial result is a positive real variable a_{ij}, which reveals some degree of preference within the interval $[1/9, 9]$ among criteria i and j, namely, $g_i \succeq g_j$. Additionally, the global result is a $q \times q$ matrix of preferences A as in matrix (4). In this example, goal g_2 is assessed by the decision-maker to be strongly more important than g_1, and g_3 moderately more important than g_1. According to the numerical equivalence described in [4], this assessment implies setting element a_{21} of matrix A to 7 and w_{31} to 3. Similarly, g_2 is assessed to be moderately more important than g_3. This assessment implies setting element w_{23} to 3. For reciprocity, element a_{ij} is set to the inverse of element a_{ji}, and elements a_{ii} are set to 1, meaning that goals are equally important. For coherence, element

$a_{ik} \approx a_{ij} \cdot a_{jk}$, meaning that if goal i is more important than goal j and j is more important than k, then i should be more important than k. Summarizing, preferences are expressed by means of a binary relation and a preference rule (see, e.g., in [4]).

Definition 2. ***Preference rule***. *Given a set of criteria $\mathcal{G}$, a preference rule is a map $\varphi : \mathcal{G} \times \mathcal{G} \to Y$, where Y is a set of numbers.*

This numerical assessment (preference rule) is usually summarized in a matrix. Due to its hierarchical structure, AHP requires the use of a different matrix to establish preferences for criteria within the same level. In Figure 1, a root node (g_1, Sustainability) subsumes the Social, Economic, and Environmental criteria within the same level. Risk, Growth, and Profits belong to a different level. However, we can integrate all matrices in a single block matrix. Let A_1 be the 3×3 pairwise comparison matrix for the first level, with Social, Economic, and Environmental criteria identified as g_2, g_3, and g_4, respectively. Let A_2 be the 3×3 pairwise comparison matrix for the second level with Risk, Growth, and Profits, denoted by g_5, g_6, and g_7, respectively. By considering an additional 3×3 matrix A_0 with all its elements set to zero, we can summarize these preferences by means of a 6×6 block matrix as follows,

$$A = \begin{bmatrix} A_1 & A_0 \\ A_0 & A_2 \end{bmatrix}. \tag{8}$$

Note that the size of A_0 is chosen to satisfy the requirement that A is a square matrix, and that in the case where the sizes of A_1 and A_2 are different, we would require two matrices, namely, A_0 and A_0', with different sizes. This preference representation does not provide information about the hierarchical structure described in Figure 1. As it is typical in graph theory, we can represent this hierarchical structure using an adjacency matrix Z with element z_{ij} set to 1 if there is a relation between criteria i and j, zero otherwise. By convention, we consider that nodes are not connected to themselves by setting $z_{ii} = 0$.

$$Z = \begin{bmatrix} 0 & 1 & 1 & 1 & 0 & 0 & 0 \\ 1 & 0 & 0 & 0 & 1 & 1 & 1 \\ 1 & 0 & 0 & 0 & 0 & 0 & 0 \\ 1 & 0 & 0 & 0 & 0 & 0 & 0 \\ 0 & 1 & 0 & 0 & 0 & 0 & 0 \\ 0 & 1 & 0 & 0 & 0 & 0 & 0 \\ 0 & 1 & 0 & 0 & 0 & 0 & 0 \end{bmatrix}. \tag{9}$$

Note that matrix Z is symmetric, as elements z_{12}, z_{13}, and z_{14} set to one require that their reciprocals (z_{21}, z_{31}, and z_{41}, respectively) are also set to one, showing that there is a relation between criteria g_1 and subcriteria g_4, g_5, and g_6. By merging matrices A and Z into a larger matrix, we could obtain a global representation of preferences and the relationship among criteria. However, to provide a more compact representation of AHP preferences, we use the concept of powerset [35].

Definition 3. ***Powerset***. *Given a set of criteria $\mathcal{G}$, the powerset of $\mathcal{G}$, denoted by $2^{\mathcal{G}}$, is the set of all subsets of $\mathcal{G}$, including the empty set and $\mathcal{G}$ itself.*

As an illustrative example, given $\mathcal{G} = (g_1, g_2)$, the powerset of $\mathcal{G}$ is

$$2^{\mathcal{G}} = (\emptyset, g_1, g_2, (g_1, g_2)). \tag{10}$$

Note that if $q = |\mathcal{G}|$ is the cardinality of $\mathcal{G}$, the cardinality of $2^{\mathcal{G}}$ is $|2^{\mathcal{G}}| = 2^q$. Then, to establish a hierarchical structure in which a subset of goals are clustered for comparison purposes, we next introduce the concept of powerset preference rule as follows.

Definition 4. ***Powerset preference rule.*** *Given a set of criteria $\mathcal{G}$, a powerset preference rule is a map $\varphi : 2^{\mathcal{G}} \times 2^{\mathcal{G}} \rightarrow Y$, where Y is a set of numbers.*

Let us consider the following powerset preference rule $\varphi : 2^{\mathcal{G}} \times 2^{\mathcal{G}} \rightarrow \mathbb{R}$,

$$\varphi(Q_i, Q_j) = \begin{cases} a_{ij} \in [1,9] & \text{if} \quad Q_i \succeq Q_j, \\ 1/a & \text{if} \quad Q_i \preceq Q_j, \\ 0 & \quad\quad otherwise. \end{cases} \tag{11}$$

where Q_i and Q_j are subsets of $\mathcal{G}$, and a_{ij} is an integer number restricted to the interval $[1,9]$, as it is customary in AHP to describe a cardinal binary relation $\preceq$ among pairs of goals. This powerset preference rule allows us to compactly represent AHP preferences including clusters (subsets) of goals that are related in a hierarchical mode. Then, by replacing the economic goal with subset (R,G,P), denoting Risk, Growth, and Profits, we can compactly represent AHP preferences as shown in Table 3. Subset (R,G,P) is at the same hierarchical level than the Social and Environmental goals. In turn, Risk, Growth, and Profits within subset (R,G,P) are equally important according to the information provided by the decision-maker. As a result, the hierarchical structure of AHP and its preferences are both completely and compactly described by a powerset preference rule as in Equation (11).

Table 3. A compact preference representation of analytic hierarchy process (AHP) where (R, G, P) = (Risk, Growth, Profits).

	(R, G, P)	Soc	Env	Risk	Growth	Profits
(R, G, P)	1	1/7	1/3	0	0	0
Soc	7	1	3	0	0	0
Env	3	1/3	1	0	0	0
Risk	0	0	0	1	1	1
Growth	0	0	0	1	1	1
Profits	0	0	0	1	1	1

Note that Table 3, as an example of the output derived form the preference rule in Equation (11), is a different object than the supermatrix described in [19]. A supermatrix summarizes the influences of a set of elements in a cluster of criteria on any other element. These influences are represented by weight vectors derived from pairwise comparisons between criteria. Here, we represent these pairwise comparisons (including individual criteria and sets of criteria) and the particular relationship between criteria in a more compact way by means of a single algebraic object. Nevertheless, the supermatrix can also be derived from the representation of preferences proposed in this work.

To better illustrate our proposal, we next use a real example from recent literature. Consider the credit evaluation for Internet finance companies in [9]. The authors use a survey within an AHP configuration described in Table 4 to propose a framework for evaluating credit indexes of Internet start-ups in China. Multiple pairwise comparisons are performed and stored in separate tables, as shown in Appendix C in [9], to keep the integrity of the hierarchical structure. By using the concept of powerset, we propose to summarize all the pairwise comparisons in single table regardless of its size. Even though this approach may present difficulties in user visualization, it represents a clear advantage when processing information automatically through the use of computers.

Criteria C1–C8 and also B4–B12 in Table 4 can be regarded as operational criteria, as indicators (or measures) for all of them are going to be used to evaluate the overall performance of a company. On the other hand, criteria such as A1 or B3 can be called cluster criteria, as there is no specific indicator linked to them. Their role in the hierarchy is to gather other operational criteria into a cluster. These operational criteria (C1–C8 and B4–B12) form set $\mathcal{G}$, and from this set we can select the elements of powerset $2^{\mathcal{G}}$ that are required to establish the pairwise comparisons in the usual way of AHP. This subset of $2^{\mathcal{G}}$ is summarized in Table 5. A matrix with these 24 elements both in rows and columns is the single algebraic object required to establish the pairwise comparison. Note that set (C1–C4) is equivalent to B1 but note also that a key point of our proposal is the strict use of elements of $2^{\mathcal{G}}$ to be able to identify the hierarchy of AHP, namely, set (C1–C4). Indeed, B1 is only a cluster criteria: an auxiliary element to gather other criteria.

Table 4. Hierarchy of criteria $\mathcal{G}$ described in [9].

Criterion Layer	Index Layer	Second Index Layer
A1 Financial status	B1 Debt Solvency	C1 Debt/assets ratio C1 Current ratio C3 Quick ratio C4 Cash ratio
	B2 Profitability	C5 Profit ratio C6 Return on assets
	B3 Operational capacity	C7 Inventory turnover C8 Receivable turnover
A2 Credit status	B4 Loan repayment B5 Payment capability B6 Taxes capability	
A3 Enterprise development	B7 Operators quality B8 Staff quality B9 Prospects	
A4 Internet financial status	B10 Customer evaluation B11 Logistics B12 Turnover index	

Table 5. Elements of powerset $2^{\mathcal{G}}$ derived from Table 4.

Id	Elements of $2^{\mathcal{G}}$
1	(C1)
2	(C2)
3	(C3)
4	(C4)
5	(C1, C2, C3, C4) = B1
6	(C5)
7	(C6)
8	(C5, C6) = B2
9	(C7)
10	(C8)
11	(C7, C8) = B3
12	(C1, C2, C3, C4, C5, C6, C7, C8) = A1
13	(B4)
14	(B5)
15	(B6)
16	(B4, B5, B6) = A2
17	(B7)
18	(B8)
19	(B9)
20	(B7, B8, B9) = A3
21	(B10)
22	(B11)
23	(B12)
24	(B10, B11, B12) = A4

3.2. A Compact Representation of Lexicographical Preferences

Recall from Section 2 that lexicographic orders define strict priority levels, where the achievement of goals are organized in subsets with immeasurable (ordinal) preferences for high priority levels with respect to low priority levels. However, a measurable (cardinal) preference is established among goals within the same priority level. To manage the ordinal preferences of subsets of goals, we can rely again on the concept of powerset preference rule. To illustrate the use of a powerset preference rule, so as to represent lexicographical orderings, we use an example of lexicographic goal programming described in [12] (p. 35), which we slightly modify for illustrative purposes.

Assume that a decision-maker is dealing with five goals. The decision-maker establishes the next priority levels: Q_1 with goals g_4 and g_5; Q_2 with goal g_1; Q_3 with goal g_2; and Q_4 with goal g_3. As we are dealing with lexicographic goal programming, the decision-maker aims to minimize either positive (ρ_j) or negative (η_j) deviations, or both, from specific targets. Then, the following program describes the optimization problem,

$$\text{Lex } \min\left[(0.67\rho_4+0.33\rho_5),(\rho_1),(\eta_2),(0.5\eta_3+0.5\rho_3)\right] \tag{12}$$

subject to

$$g1: x_1+x_2+\eta_1-\rho_1=300 \tag{13}$$

$$g2: 1000x_1+3000x_2+\eta_2-\rho_2=400{,}000 \tag{14}$$

$$g3: x_1+x_2+\eta_3-\rho_3=400 \tag{15}$$

$$g4: x_1+\eta_4-\rho_4=300 \tag{16}$$

$$g5: x_2+\eta_5-\rho_5=200 \tag{17}$$

with non-negative decision variables x_i for $i=1,2$ and also non-negative goal deviation variables ρ_j and η_j for $j=1,2,3,4,5$.

Let us first consider the following powerset preference rule $\varphi: 2^{\mathcal{G}}\times 2^{\mathcal{G}}\to\{0,1\}$,

$$\varphi(Q_j,Q_k)=\begin{cases}1 & \text{if } \; Q_j\succeq Q_k,\\ 0 & \quad\text{otherwise,}\end{cases} \tag{18}$$

where Q_j and Q_k are subsets of $\mathcal{G}$. We can build a matrix of powerset preferences that directly derives from rule in Equation (18), as shown in Table 6.

Table 6. A matrix representation of a lexicographical order.

	(ρ_4,ρ_5)	(ρ_1)	(η_2)	(η_3,ρ_3)
(ρ_4,ρ_5)	1	1	1	1
(ρ_1)	0	1	1	1
(η_2)	0	0	1	1
(η_3,ρ_3)	0	0	0	1

Table 6 expresses an order in the priority of the achievements. As $G_1\succeq G_2$, $G_1\succeq G_3$, and $G_1\succeq G_4$ hold, but $G_2\succeq G_1$, $G_3\succeq G_1$, and $G_4\succeq G_1$ do not hold, we conclude that G_1 has the top priority. A similar reasoning leads to the lexicographical ordering $G_1\succ G_2\succ G_3\succ G4$, expressing a priority in the achievement of G_1 with respect to G_2, G_3, and G_4. In addition, $G_j\sim G_k$ for $j=k$, as we assume that $G_j\succeq G_k$ and $G_k\succeq G_j$ simultaneously holds. Note also that within priority levels Q_1 and Q_4, the decision-maker establishes a cardinal preference among positive and negative deviations. In priority level Q_1, positive deviation ρ_4 is twice more important than ρ_5. In priority level Q_4, negative deviation η_3 is equally important than ρ_3. We can integrate these cardinal preferences with ordinal preferences described in Equation (12) by considering a more general powerset preference rule $\varphi: 2^{\mathcal{G}}\times 2^{\mathcal{G}}\to\mathbb{R}$:

$$\varphi(Q_j, Q_k) = \begin{cases} a_{jk} \geq 1 & \text{if} \quad Q_j \succeq Q_k, \\ 0 \leq a_{jk} < 1 & \text{otherwise.} \end{cases} \tag{19}$$

Table 7 is a particular realization of the powerset preference rule described in Equation (19). From its observation, we are able to build objective function in Equation (12) provided that we are told that an additive rule is followed within priority levels, as is customary in goal programming. More precisely, the ordering of priority levels G_1, G_2, G_3, and G_4 is expressed in the same way as in Table 6. Then, cardinal preferences between pairs of goals within priority levels are expressed using a numerical scale as in AHP. Let us consider four additional priority levels that we identify with single decision variables denoting goals: $G_5 = (\rho_4)$, $G_6 = (\rho_5)$, $G_7 = (\eta_3)$, and $G_8 = (\rho_3)$. As $\varphi(\rho_4, \rho_5) = 2$, we infer that ρ_4 is twice more important than ρ_5, as is described in Equation (12). Moreover, as $\varphi(\eta_3, \rho_3) = 1$, we infer that both deviations are equally important for the decision-maker. The rest of entries in Table 7, where $\varphi(G_j, G_k) = 0$ and $\varphi(G_k, G_j) = 0$, denote that no preference has been established among subsets of goals (including singletons). These subsets are then incomparable [36].

Table 7. A compact representation of a lexicographical order.

	(ρ_4, ρ_5)	(ρ_1)	(η_2)	(η_3, ρ_3)	(ρ_4)	(ρ_5)	(η_3)	(ρ_3)
(ρ_4, ρ_5)	1	1	1	1	0	0	0	0
(ρ_1)	0	1	1	1	0	0	0	0
(η_2)	0	0	1	1	0	0	0	0
(η_3, ρ_3)	0	0	0	1	0	0	0	0
(ρ_4)	0	0	0	0	1	2	0	0
(ρ_5)	0	0	0	0	1/2	1	0	0
(η_3)	0	0	0	0	0	0	1	1
(ρ_3)	0	0	0	0	0	0	1	1

Generalizing, we aim to minimize a vector of priorities levels $(G_1, G_2, \ldots, G_n)$, where each priority level $G_j = f(\mathcal{G}_j)$, with $\mathcal{G}_j \subseteq \mathcal{G}$, is usually a linear combination of goals that have to be minimized at priority j. Each pair of priority levels—G_j and G_k—are related by a powerset preference rule $\varphi(G_j, G_k)$ that can express ordinal (among priority levels) and cardinal relations (among particular goals or singletons) at the same time. As $\mathcal{G}$ is a subset of $2^{\mathcal{G}}$ and Boolean numbers are also real numbers, the powerset preference rule $\varphi : 2^{\mathcal{G}} \times 2^{\mathcal{G}} \rightarrow \mathbb{R}$ suffices to compactly express lexicographical orders by means of a single algebraic object, namely, a matrix. Note also that AHP is a particular case of lexicographical orders with one priority level $G_1 = f(\mathcal{G})$ that can be expressed as a linear combination of weighted goals according to a cardinal preference hierarchy.

4. Properties of Powerset Preference Representations

Binary relation $\succeq$ within a powerset preference rule, described in Equation (19), presents some basic properties such as completeness, transitivity, and antisymmetry. The transition from a set to a powerset does not affect these properties, as there is no difference in comparing singletons (subsets with cardinality one) and comparing subsets with either one or more than one element (subsets with cardinality greater or equal than one). As a result, the following definitions are natural extensions of well-known set properties to include the concept of powerset.

Definition 5. ***Completeness****. Given a powerset $2^{\mathcal{G}}$, we say that $\succeq$ defined on $2^{\mathcal{G}}$ is complete when for all $G_1, G_2 \subseteq 2^{\mathcal{G}}$, and we have that $G_1 \succeq G_2$, $G_2 \succeq G_1$, or both.*

Definition 6. ***Transitivity****. Given a powerset $2^{\mathcal{G}}$, we say that $\succeq$ defined on $2^{\mathcal{G}}$ is transitive when for all $G_1, G_2, G_3 \subseteq 2^{\mathcal{G}}$, if $G_1 \succeq G_2$ and $G_2 \succeq G_3$, then $G_1 \succeq G_3$.*

Definition 7. ***Antisymmetry***. *Given a powerset $2^{\mathcal{G}}$, we say that $\mathcal{G}$ is antisymmetric when for all $G_1, G_2 \in 2^{\mathcal{G}}$, if $G_1 \succeq G_2$ and $G_2 \succeq G_1$, then $G_1 \sim G_2$, where relation "$\sim$" means is-indifferent-to.*

Then, a powerset preference rule φ from Definition 4 allows characterization of all the subsets of $\mathcal{G}$ that are relevant for an agent. In addition, there are some interesting features that deserve to be highlighted. For consistency, AHP pairwise comparisons require a matrix where $a_{ij} = 1/a_{ji}$ and $a_{ik} \approx a_{ij} \cdot a_{jk}$. These conditions also hold for comparisons between single objectives, but not for priority levels in lexicographical orders where consistency is respected when if $a_{ij} = 1$, then $a_{ji} = 0$ for $i \neq j$, denoting strict preference for higher priority levels, i.e., $G_i \succ G_j$.

We next consider additional properties of powerset preference rules described in Section 3, allowing us to characterize the relationships between goals or subset of goals. These properties present the advantage that can be expressed algebraically by means of powerset preference rule φ described in Equation (19).

Definition 8. ***Weak preference***. *Given $G_i, G_j \subseteq 2^{\mathcal{G}}$, and a powerset preference rule $\varphi(G_i, G_j)$ defined on $2^{\mathcal{G}}$, as in Equation (19), we say that G_i is weakly preferred than G_j, denoted by $G_i \succeq G_j$, when $\varphi(G_i, G_j) \geq 1$.*

Weak preference is particularly important in AHP where relationships between either goals or subsets or goals are expressed numerically within a given scale. Note that here weak means finite or measurable even in the case of the existence of extremely more important goals than others. The reason is to set a clear differentiation with strict preference.

Definition 9. ***Strict preference***. *Given $G_i, G_j \subseteq 2^{\mathcal{G}}$, and a powerset preference rule $\varphi(G_i, G_j)$ defined on $2^{\mathcal{G}}$ as in Equation (19), we say that G_i is strictly preferred than G_j, denoted by $G_i \succ G_j$, when $\varphi(G_i, G_j) = 1$ and $\varphi(G_j, G_i) = 0$.*

Strict preference is the type of relation established between priority levels in lexicographical orders where higher priority levels are infinitely or immeasurably preferred to lower priority levels. Both in AHP and lexicographical orders, the achievement of some goals may be indifferent.

Definition 10. ***Indifference***. *Given $G_i, G_j \subseteq 2^{\mathcal{G}}$ and a powerset preference rule $\varphi(G_i, G_j)$ defined on $2^{\mathcal{G}}$, as in Equation (19), we say that G_i is indifferent to G_j, denoted by $G_i \sim G_j$, when $\varphi(G_i, G_j) = 1$ and $\varphi(G_j, G_i) = 1$.*

Indifference appears when comparing one goal to itself for obvious reasons. However, it is likely that we also find indifference in the achievement of two particular goals either within a priority level in lexicographical orders or within a cluster of goals in AHP. In these cases, the decision-maker is neutral with respect the achievement of indifferent goals. On the contrary, some goals may be incomparable.

Definition 11. ***Incomparability***. *Given $G_i, G_j \subseteq 2^{\mathcal{G}}$ and a powerset preference rule $\varphi(G_i, G_j)$ defined on $2^{\mathcal{G}}$, as in Equation (19), we say that G_i is incomparable to G_j, denoted by $G_i \; ? \; G_j$, when $\varphi(G_i, G_j) = 0$ and $\varphi(G_j, G_i) = 0$.*

The use of powerset preference rules as a compact representation in the form of a table usually implies comparing goals that are incomparable. Here, incomparability means that there is no need to define preferences between two subsets of goals. There is no need to define preferences between the Risk and the Social goal in AHP (Table 3), as risk is a subgoal of the Economic cluster of goals. Similarly, there is no need to define preferences between particular goals of different priority levels in lexicographical orders.

As an additional contribution of this paper, we next discuss two properties that derive from powerset preference rules of the type $\varphi(G_i, G_j)$ in Equation (19), namely, alignment and preference order. These properties allow us to characterize powersets of criteria, therefore improving coherence and the interactive process of setting preferences by decision-makers.

Proposition 1. ***Alignment***. *Given a powerset preference rule $\varphi(G_i, G_j)$ defined on $2^{\mathcal{G}}$, $\varphi(G_i, G_j)$ is aligned if and only if $\varphi(G_i, G_j) \neq \varphi(G_j, G_i)$ for some pair $G_i, G_j \subseteq 2^{\mathcal{G}}$.*

Proof. From Definition 2, $\varphi(G_i, G_j)$ is a numerical assessment of the preference of G_i over G_j. If $\varphi(G_i, G_j) = \varphi(G_j, G_i)$, then the numerical assessment is inconclusive. Therefore, alignment necessarily requires that $\varphi(G_i, G_j) \neq \varphi(G_j, G_i)$. □

Note that alignment characterizes lexicographical orders, as the presence of at least two priority levels implies alignment for one of them. This feature requires that the reciprocals are different as in $\varphi(\rho_1, \eta_2) \neq \varphi(\eta_2, \rho_1)$ in Table 7. The absence of alignment implies that there is no preference among subsets of goals. Consider the special case of AHP matrix representation with all elements of the main diagonal set to one, zero otherwise. In this case, the condition $\varphi(G_i, G_j) = \varphi(G_j, G_i)$ holds because no preference has been established. No alignment or positioning of decision-makers for the achievement of a subset of goals is expressed. Note also that alignment is a different concept than weak and strict preference from Definitions 8 and 9. Indeed, strict preference $G_i \succ G_j$ implies alignment, but alignment does not imply strict preference because the expression of alignment $\varphi(G_i, G_j)$ is not limited to zero-one as in the case of the AHP numerical scale. Furthermore, weak preference $G_i \succeq G_j$ is inconclusive with respect to alignment because it is unknown if $G_j \succeq G_i$ holds.

Proposition 2. ***Preference order***. *Given a powerset preference rule φ defined on set $2^{\mathcal{G}}$ with cardinality q and a sequence of integers $1 \leq i \leq j \leq k \leq q$ such that $G_i, G_j, G_k \subseteq 2^{\mathcal{G}}$, φ defines a total order of preferences $\{G_i, G_j, G_k\}$ if and only if φ is positive and monotonically increasing in the interval of integers (i, k), i.e., if and only if $0 < \varphi(G_i, G_j) \leq \varphi(G_j, G_k)$.*

Proof. A total order is a binary relation over a set satisfying completeness, antisymmetry, and transitivity. The first condition, $\varphi(G_i, G_j) > 0$, ensures completeness of subsets; the second condition, $\varphi(G_i, G_j) = \varphi(G_i, G_j)$, permits antisymmetry when $i < j = k$; and the third condition, $\varphi(G_i, G_j) \leq \varphi(G_j, G_k)$, allows transitivity when $i < j < k$. Therefore, $\{G_i, G_j, G_k\}$ is a total order of preferences. □

Monotonicity in Proposition 2 depends on the actual arrangement of the elements of powerset $2^{\mathcal{G}}$. As a result, some permutations of $2^{\mathcal{G}}$ may be monotonically increasing in (i, k) and some others may not be. Consider again the AHP example described in Table 3 within the interval $(1, 3)$. The ordered sequence $\{(R, G, P), Soc, Env\}$ is not monotonically increasing, but the permutation $\{(R, G, P), Env, Soc\}$ is monotonically increasing. As a result, a monotonically increasing sequence of goals in AHP is an ordered sequence such that goals are at-least-as-important-as preceding goals. In lexicographical orders, a monotonically decreasing sequence of priority levels is an ordered sequence of levels, such that priority levels are infinitely more important than subsequent priority levels. In Table 6, the sequence $\{(\rho_4, \rho_5), (\rho_1), (\eta_2), (\eta_3, \rho_3)\}$ is monotonically decreasing in $(1, 4)$, showing a preemptive ordering of priority levels established by a decision-maker.

Alignment and preference order are two examples of interesting theoretical results that may help practitioners to achieve a better understanding of complex schemes of preferences. These properties combined with a unique representation of preferences enable both decision-makers and analysts to enhance the interactive process of multiple criteria optimization [3]. Interactive processes rely on iterative algorithms in which preference elicitation and optimization stages are repeated until a satisfactory solution is obtained. As a result, a global overview of preferences for objectives and a sound preference rule based of set theory support decision-makers to obtain better solutions.

5. Conclusions

The relative importance of goals in multiple criteria optimization problems may have an impact on the final solutions proposed to decision-makers. Either cardinal or ordinal information is required to establish preference among goals that, in addition, may be structured in clusters (hierarchically or by means of strict priority levels). To improve the interactive process of setting preferences in multiple criteria optimization problems, in this paper, we show that a single compact representation summarizing both AHP and lexicographical orders is possible. To this end, we introduce the concept of powerset preference rule. As the powerset derived from a set of goals includes all possible subsets of goals, a powerset preference rule allows us to compactly represent preferences of both AHP and lexicographic orders. In addition, we study the link between AHP and lexicographic orders, showing that the former is a special case of the latter.

This compact representation also facilitates the understanding of the main properties of different preference setting options. In this sense, we discuss the main relations that can be established among the goals of multiple criteria optimization problems. More precisely, we study the case of alignment of objectives that can be identified through a given powerset preference rule. We also characterize preference orders, describing a ranking of priorities that are easier to understand by decision-makers, especially when a large number of objectives are considered.

Although the compact representation described in this paper may help decision-makers to better understand the impact of preference setting on the results derived from optimization, our proposal is limited to providing a preference scheme. This scheme can be used as an input to optimization methods, and it provides a more effective way to develop the interactive process of preference eliciting and optimization. In addition, it allows to deploy mechanisms of formal reasoning that may lead to detect inconsistencies in preferences or to establish interesting theoretical properties. The search for further theoretical results derived from this compact representation is a natural extension of this work.

Author Contributions: Conceptualization, F.S.-M., D.P.-S., A.G.-B. and J.R.-M.; Data curation, F.S.-M., D.P.-S., A.G.-B. and J.R.-M.; Formal analysis, F.S.-M., A.G.-B. and J.R.-M.; Investigation, F.S.-M., D.P.-S., A.G.-B. and J.R.-M.; Methodology, F.S.-M., D.P.-S., A.G.-B. and J.R.-M.; Project administration, F.S.-M. and D.P.-S.; Resources, D.P.-S.; Supervision, F.S.-M., A.G.-B. and J.R.-M.; Validation, F.S.-M.; Visualization, J.R.-M.; Writing—original draft, F.S.-M. and D.P.-S.; Writing— review and editing, F.S.-M., D.P.-S., A.G.-B. and J.R.-M.

Funding: This research received no external funding.

Conflicts of Interest: The authors declare no conflict of interest.

References

1. Jones, D.; Tamiz, M.; *Practical Goal Programming*; Springer: Berlin, Germany, 2010; Volume 141.
2. Ballestero, E.; Pla-Santamaria, D. Selecting portfolios for mutual funds. *Omega* **2004**, *32*, 385–394. [CrossRef]
3. Miettinen, K.; Ruiz, F.; Wierzbicki, A.P. Introduction to multiobjective optimization: Interactive approaches. In *Multiobjective Optimization*; Springer: Berlin, Germany, 2008; pp. 27–57.
4. Saaty, T.L. *The Analytic Hierarchy Process*; Mc Graw-Hill: New York, NY, USA, 1980.
5. Gdoura, K.; Anane, M.; Jellali, S. Geospatial and AHP-multicriteria analyses to locate and rank suitable sites for groundwater recharge with reclaimed water. *Resour. Conserv. Recycl.* **2015**, *104*, 19–30. [CrossRef]
6. Kavurmaci, M.; Üstün, A.K. Assessment of groundwater quality using DEA and AHP: A case study in the Sereflikochisar region in Turkey. *Environ. Monit. Assess.* **2016**, *188*, 258. [CrossRef] [PubMed]
7. Nam, S.N.; Nguyen, T.T.; Oh, J. Performance Indicators Framework for Assessment of a Sanitary Sewer System Using the Analytic Hierarchy Process (AHP). *Sustainability* **2019**, *11*, 2746. [CrossRef]
8. Nguyen, T.; Nahavandi, S. Modified AHP for gene selection and cancer classification using type-2 fuzzy logic. *IEEE Trans. Fuzzy Syst.* **2015**, *24*, 273–287. [CrossRef]
9. Gu, W.; Basu, M.; Chao, Z.; Wei, L. A unified framework for credit evaluation for internet finance companies: Multi-criteria analysis through AHP and DEA. *Int. J. Inf. Technol. Decis. Mak.* **2017**, *16*, 597–624. [CrossRef]

10. Oliveira, G.A.; Tan, K.H.; Guedes, B.T. Lean and green approach: An evaluation tool for new product development focused on small and medium enterprises. *Int. J. Prod. Econ.* **2018**, *205*, 62–73. [CrossRef]
11. Romero, C. *Handbook of Critical Issues in Goal Programming*; Pergamon Press: Oxford, UK, 1991.
12. Ballestero, E.; Romero, C. *Multiple Criteria Decision Making and Its Applications to Economic Problems*; Springer Science & Business Media: New York, NY, USA, 1998.
13. Weber, E.; Rizzoli, A.E.; Soncini-Sessa, R.; Castelletti, A. Lexicographic optimisation for water resources planning: The case of Lake Verbano, Italy. In *Integrated Assessment and Decision Support, Proceedings of the First Biennial Meeting of the International Environmental Modelling and Software Society, Lugano, Switzerland, 24–27 June 2002*; IEMSS: Lugano, Switzerland, 2002; pp. 235–240.
14. Aghaei, J.; Amjady, N.; Shayanfar, H.A. Multi-objective electricity market clearing considering dynamic security by lexicographic optimization and augmented epsilon constraint method. *Appl. Soft Comput.* **2011**, *11*, 3846–3858. [CrossRef]
15. Ahmadi, A.; Ahmadi, M.R.; Nezhad, A.E. A lexicographic optimization and augmented ϵ-constraint technique for short-term environmental/economic combined heat and power scheduling. *Electr. Power Compon. Syst.* **2014**, *42*, 945–958. [CrossRef]
16. González-Arteaga, T.; Alcantud, J.C.R.; de Andrés Calle, R. A new consensus ranking approach for correlated ordinal information based on Mahalanobis distance. *Inf. Sci.* **2016**, *372*, 546–564. [CrossRef]
17. Miettinen, K.; Mäkelä, M.M. On scalarizing functions in multiobjective optimization. *OR Spectr.* **2002**, *24*, 193–213. [CrossRef]
18. Ignizio, J.P. Generalized goal programming an overview. *Comput. Oper. Res.* **1983**, *10*, 277–289. [CrossRef]
19. Saaty, T.L.; Vargas, L.G. *Decision Making with the Analytic Network Process*; Springer: Berlin, Germany, 2006; Volume 282.
20. Sitorus, F.; Cilliers, J.J.; Brito-Parada, P.R. Multi-criteria decision making for the choice problem in mining and mineral processing: Applications and trends. *Expert Syst. Appl.* **2019**, *121*, 393–417. [CrossRef]
21. Saaty, T.L.; Vargas, L.G. *Models, Methods, Concepts & Applications of the aNalytic Hierarchy Process*; Kluwer Academic Publishers: Norwell, MA, USA, 2001.
22. Zyoud, S.H.; Fuchs-Hanusch, D. A bibliometric-based survey on AHP and TOPSIS techniques. *Expert Syst. Appl.* **2017**, *78*, 158–181. [CrossRef]
23. Subramanian, N.; Ramanathan, R. A review of applications of Analytic Hierarchy Process in operations management. *Int. J. Prod. Econ.* **2012**, *138*, 215–241. [CrossRef]
24. Singh, R.P.; Nachtnebel, H.P. Analytical hierarchy process (AHP) application for reinforcement of hydropower strategy in Nepal. *Renew. Sustain. Energy Rev.* **2016**, *55*, 43–58. [CrossRef]
25. Štreimikienė, D.; Šliogerienė, J.; Turskis, Z. Multi-criteria analysis of electricity generation technologies in Lithuania. *Renew. Energy* **2016**, *85*, 148–156. [CrossRef]
26. Abdollahzadeh, G.; Damalas, C.A.; Sharifzadeh, M.S.; Ahmadi-Gorgi, H. Selecting strategies for rice stem borer management using the Analytic Hierarchy Process (AHP). *Crop Prot.* **2016**, *84*, 27–36. [CrossRef]
27. Erdoğan, M.; Kaya, İ. A combined fuzzy approach to determine the best region for a nuclear power plant in Turkey. *Appl. Soft Comput.* **2016**, *39*, 84–93. [CrossRef]
28. Chen, Y.; Liu, R.; Barrett, D.; Gao, L.; Zhou, M.; Renzullo, L.; Emelyanova, I. A spatial assessment framework for evaluating flood risk under extreme climates. *Sci. Total Environ.* **2015**, *538*, 512–523. [CrossRef] [PubMed]
29. Zammori, F. The analytic hierarchy and network processes: Applications to the US presidential election and to the market share of ski equipment in Italy. *Appl. Soft Comput.* **2010**, *10*, 1001–1012. [CrossRef]
30. Carter, C.R.; Rogers, D.S. A framework of sustainable supply chain management: Moving toward new theory. *Int. J. Phys. Distrib. Logist. Manag.* **2008**, *38*, 360–387. [CrossRef]
31. Charnes, A.; Cooper, W.W. *Management Models and Industrial Applications of Linear Programming*; John Wiley and Sons: New York, NY, USA, 1961.
32. Ignizio, J.P. An approach to the capital budgeting problem with multiple objectives. *Eng. Econ.* **1976**, *21*, 259–272. [CrossRef]
33. Ehrgott, M. *Multicriteria Optimization*; Springer Science & Business Media: New York, NY, USA, 2005; Volume 491.
34. Lonergan, S.; Cocklin, C. The use of lexicographic goal programming in economic/ecolocical conflict analysis. *Socio-Econ. Plan. Sci.* **1988**, *22*, 83–92. [CrossRef]

35. Pinter, C.C. *A Book of Set Theory*; Dover Publications: New York, NY, USA, 2014.
36. González-Pachón, J.; Romero, C. Properties underlying a preference aggregator based on satisficing logic. *Int. Trans. Oper. Res.* **2015**, *22*, 205–215. [CrossRef]

MDPI
St. Alban-Anlage 66
4052 Basel
Switzerland
Tel. +41 61 683 77 34
Fax +41 61 302 89 18
www.mdpi.com

Mathematics Editorial Office
E-mail: mathematics@mdpi.com
www.mdpi.com/journal/mathematics

www.ingramcontent.com/pod-product-compliance
Lightning Source LLC
LaVergne TN
LVHW071523170726
843515LV00008B/2047
* 9 7 8 3 0 3 9 4 3 6 0 7 1 *